T0192605

INTERCONNECTION AND THE INTERNET

Selected Papers from the
1996 Telecommunications
Policy Research Conference

TELECOMMUNICATIONS

A Series of Volumes Edited
by Christopher H. Sterling

INTERCONNECTION AND THE INTERNET

Selected Papers from the 1996 Telecommunications Policy Research Conference

Edited by

Gregory L. Rosston
Stanford University

David Waterman
Indiana University

Routledge
Taylor & Francis Group
New York   London

First published by
Lawrence Erlbaum Associates,
10 Industrial Avenue
Mahwah, NJ 07430

Transferred to Digital Printing 2010

Routledge
Taylor & Francis Group
270 Madison Avenue
New York, NY 10016

Routledge
Taylor & Francis Group
2 Park Square
Milton Park, Abingdon
Oxon OX14 4RN

Cover design by Jennifer Sterling

Library of Congress Cataloging-in-Publication Data

　　　Telecommunications Policy Research Conference (24th : 1996)
Interconnection and the internet : selected papers from the 1996
Telecommunications policy Research Conference / edited by Gregory L.
Rosston, David Waterman.
　　　　p.　　cm. (Telecommunications)
　　Includes bibliographical references and index
　　ISBN 0-8058-2847-8 (cloth : alk. paper). — ISBN 0-8058-2848-6 (pbk. :
alk. paper)
　　　1. Telecommunications policy – Research – United States – Congresses.
2. Information superhighway – United States – Congresses.　　3. Information
networks –Government policy – United States – Congresses.　　4. Internet
(Computer network) – United States – Congresses.　　5. Computers – United
States – Congresses.　　I. Rosston, Gregory L.　　II. Waterman, David.　　III.
Title.　　IV. Series : Telecommunications (Mahwah, N.J.)
HE7781.T44　　1997
384.3'0973—dc21
　　　　　　　　　　　　　　　　　　　　　　　　　　97-22934
　　　　　　　　　　　　　　　　　　　　　　　　　　CIP

Contents

Authors

Philip Aspden
*Center for Research
on the Information Society*

Gerald W. Brock
The George Washington University

L. Jean Camp
Harvard University

David D. Clark
MIT Laboratory for Computer Science

Gerald R. Faulhaber
University of Pennsylvania

Robert M. Frieden
Pennsylvania State University

John Haring
Strategic Policy Research, Inc.

Heather E. Hudson
University of San Francisco

James Katz
Bellcore

Michael L. Katz
University of California at Berkeley

Jeffrey K. MacKie-Mason
University of Michigan

Jon M. Peha
Carnegie Mellon University

Donna M. Riley
Carnegie Mellon University

Jeffrey H. Rohlfs
Strategic Policy Research, Inc.

Susan Dente Ross
Washington State University

Gregory L. Rosston
Stanford University

Durga P. Satapathy
Carnegie Mellon University

Supriya Singh
*Centre for International Research on
Communication and Information
Technologies*

Jonathan Weinberg
Wayne State University Law School

Kimberly White
Price Waterhouse

David Waterman
Indiana University

Frank A. Wolak
Stanford University

Acknowledgments

Throughout its 24-year history, a high proportion of the papers presented at the Telecommunications Policy Research Conference (TPRC) have resulted in journal articles and other publications. Although conference proceedings were also published in some earlier years, there had never been an ongoing publication outlet for research presented at the TPRC. In 1994, the efforts of Christopher H. Sterling, the series editor of the Lawrence Erlbaum Associates Telecommunications Series, Hollis Heimbouch of LEA, and TPRC Board members John Haring, Bridger Mitchell and Jerry Brock, resulted in a new arrangement for LEA to publish a series of research paper collections from the TPRC. This is the third volume of the series and it consists of selected papers from the 24th annual conference, held in October 1996.

We are grateful to the members of the TPRC organizing committee, session chairs and moderators, discussants, and other Friends of the TPRC who served as referees to help us select among the many excellent papers presented at the conference.

We especially thank Jerry Brock, editor of the first volume in the series and co-editor of the second volume, for his generous advice to us.

The remarkable pace of technology and policy events in telecommunications has made timely production of this volume especially important. We thank Cyndi Connelley, Roxie Glaze, and James Quigley for their expert copyedit entry and typesetting work. We are especially grateful to Linda Bathgate, Kathy Scornavaca, Kathleen O'Malley, and their colleagues at LEA for guiding the production of the book to its conclusion.

Finally, we owe a special thanks to the John R. and Mary Markle Foundation. Their support made the TPRC possible, and without that, there would be no book.

Gregory Rosston
David Waterman

Preface

David Waterman
Indiana University

On behalf of the organizing committee, I welcome you to this volume of selected papers from the 24th annual Telecommunications Policy Research Conference (TPRC). It was my honor and pleasure to serve as chair of the
conference organizing committee.

The TPRC is an annual forum for dialogue among scholars and the policymaking
community on a wide range of telecommunications issues. The conference was begun in 1972 as a small gathering of researchers from the government and academic
communities. Since then, it has grown and matured to reflect the rapid development—and the growing influence—of policy-relevant telecommunications research
from a variety of disciplinary perspectives. The TPRC is widely regarded to be the
premier telecommunications policy research event in the United States.

It would be hard to imagine a more interesting and exciting time to hold the
TPRC than October 1996, when the meeting from which the papers in this book
are selected took place. The February passage of the Telecommunications Act of
1996, and the Federal Communications Commission's subsequent implementation orders, inspired two plenary sessions and a preconference tutorial. These policy events were also the subject or the backdrop of a large number of papers
presented in the 16 regular conference sessions. Another plenary session involved
Internet censorship issues in the wake of the District Court's overturning of Title
V (the Communications Decency Act) of the 1996 Act. Internet-related research
paper presentations—concerning online speech, Internet pricing and architecture
issues, electronic commerce, and other subjects—have grown enormously as a
proportion of the full conference in the past 2 or 3 years. Other regular session papers at the 1996 TPRC reflected recent policy, business, and research develop-

ments in a variety of other areas, including spectrum management, telephony and video delivery outside the United States, cable regulation, the political economy of regulation, and infrastructure investment.

Without the voluntary efforts of many individuals, the 1996 TPRC could never have been possible. I extend my appreciation and congratulations to the members of the 1996 TPRC organizing committee for their first-class, energetic work. Those members are Thomas Aust of Citicorp., Tim Brennan of the University of Maryland, Baltimore County, Wendy Gordon of Boston University, Evan Kwerel of the Federal Communications Commission, Jeff MacKie-Mason of the University of Michigan, Paul Resnick of AT&T Labs, Lisa Rosenblum of Cablevision Systems Corporation, and Jean Paul Simon of France Telecom.

The organizing committee is in turn grateful to everyone who submitted paper or session proposals to us. We received over 140 proposals for the 1996 conference, approximately a one-third increase over the previous year and a conference record. An unusually large number of very promising proposals could therefore not be included in the program.

The committee also thanks the session chairs for their important contributions in selecting participants and moderating the presentations at the conference. We further benefited from the help of many other friends of the TPRC. This year, as in the past, these individuals gave advice and worked behind the scenes to insure the continuing high quality of this conference.

The administrative work that actually made the 1996 TPRC happen—negotiating with service providers, processing abstracts and papers, cajoling the committee, session chairs, and presenters to deliver the goods on time, and managing the event itself—was handled by Dawn Higgins and Lori Rodriguez. As in years past, their highly professional work was invaluable, and we sincerely thank them for it.

Finally, but by no means least, we express our thanks to the John and Mary R. Markle Foundation. The quality and diversity of the TPRC is expensive to provide. The Markle Foundation, which has supported the TPRC generously for a number of years, again supplied a major proportion of the income used to pay for the 1996 conference.

Introduction

Gregory L. Rosston
Stanford University

David Waterman
Indiana University

The past several years—and especially the year immediately preceding the 24th Annual Telecommunications Policy Research Conference (TPRC) held in October 1996—have provided a remarkably fertile research environment in telecommunications. Historic policy events marked the preceding year, and the transformation of electronic communications via the Internet continued to accelerate.

In February, President Clinton signed the Telecommunications Act of 1996 into law. Among other things, this extensive rewrite of the 1934 Communications Act opened up previously regulated telephony and video markets to competition from all players. Pursuant to the act, the Federal Communications Commission (FCC) followed with several implementation orders. Prominent among these was the Commission's August 1996 Local Competition Order, which set out specific standards for the pricing of telephone network interconnection and transport. Very soon after, the Eighth Circuit Appeals Court stayed the FCC order, leaving interconnection policy hanging in the balance.

Also confronting a judicial barrier was Title V of the 1996 Act, known as the Communications Decency Act (CDA), which prescribed conditions for the control of pornography on the Internet. A three-judge federal district court panel held Title V unconstitutional in June 1996, and the Supreme Court upheld the ruling in June 1997.

Independently of government action, development of the Internet continues to provide exceptional research opportunities. The system for pricing of Internet access and services to support the network infrastructure remains in flux, and there are many policy options. Electronic commerce on the Internet is growing rapidly, but numerous alternative methods of payment have left both service providers and consumers in a state of uncertainty. "Filtering" software and other systems for controlling Internet content are developing rapidly and remain the subject of First Amendment controversy. More generally, debates about the adequacy of existing intellectual property law to serve the evolving digital communication environment are intensifying.

Several of the 15 selected papers of this volume specifically respond to policy developments preceding the 1996 TPRC. Several others wade into the controversies surrounding development of the Internet. All 15 of the selected papers reflect the spirit and the energy that have accompanied the recent policy upheavals and the rapid economic, technological, and social developments in telecommunications.

We group the selected papers into four main sections.

I. INTERCONNECTION AND COMPETITION

The first three of four chapters in this section focus on telephone network interconnection and FCC implementation of the 1996 Act's local competition provisions. The commission's August 1996 Local Competition Order established federal guidelines governing the behavior of incumbent local exchange carriers (ILECs) with respect to their provisioning and pricing of interconnection and other services to competing local exchange carriers (CLECs).

Gerald W. Brock begins his chapter, entitled "Local Competition Policy Maneuvers," by tracing the legal and political developments leading up to the 1996 Act's provisions on local competition and the FCC Local Competition Order. He then characterizes the guiding principle of the FCC order as an attempt to "level the playing field" by removing the competitive advantages that large, established networks have because of network externalities and economies of scale, scope, and density. Brock maintains that to achieve this goal, the FCC imposed three different requirements in its order: network interconnection at "forward-looking" economic cost, the unbundling of network elements (e.g., the local loop) by ILECs, and the right for CLECs to obtain the rights to resell local telecommunications services (e.g., local dialtone service) at a fixed discount from the ILEC's retail prices. Brock predicts that if the appeals court decision staying the FCC order holds, state-by-state interconnection negotiations will proceed more slowly, with more emphasis on the interests of the facilities-based carriers, than if the FCC rules are upheld.

The next chapter, by Michael L. Katz, is entitled "Economic Efficiency, Public Policy, and the Pricing of Network Interconnection under the Telecommunications Act of 1996." Katz argues that although both the supporters and the opponents of the FCC interconnection pricing guidelines have overstated the guidelines' effects, the government has a useful role to play. Katz

draws on the economics of bargaining to argue that without the FCC intervention, ILECs have insufficient incentives to reach socially efficient outcomes; they might, for example, find it advantageous to delay or weaken
competition by refusing to interconnect with CLECs. Katz proceeds to discuss what economic theory has to say about what telephone network interconnection standards might look like. In doing so, he considers several
issues, including marginal cost pricing, Ramsey pricing, and universal service objectives. In concluding his chapter, Katz asserts that competitive markets will not develop quickly enough to abandon interconnection price
regulation. He then examines options that regulators have to move prices to
efficient levels without sacrificing the firms' incentives to reduce costs.

John Haring and Jeffrey H. Rohlfs offer an alternative perspective on the
interconnection debate. In "Telecommunications Pricing for Efficient Local
Competition," these authors focus on the appropriate price for interconnection that should be paid by a new entrant. In their theoretical framework,
Haring and Rohlfs consider the cost savings that an ILEC realizes, given alternative assumptions about how a call is made and switched. They show that
setting the interconnection price equal to these cost savings will generally
lead to efficient entry decisions. Haring and Rohlfs then examine several alternative interconnection pricing approaches that have been proposed. They
argue that certain of these proposed mechanisms, such as symmetrical (or reciprocal) pricing and pricing slightly above incremental cost, will lead to uneconomic entry or other inefficient behavior—unless local telephone rates
are rebalanced and deaveraged to eliminate historical cross-subsidies. They
argue that until rates are rebalanced and deaveraged, use of the Efficient
Component Pricing Rule may be a reasonable interim measure.

The final chapter in this section considers a quite different problem: the sharing
of spectrum by potentially competing firms or users. Current spectrum management policy typically gives exclusive rights of access to domains of the spectrum
by license holders, with very limited opportunities for nonlicense holders. In
"Spectrum Sharing Without Licenses: Opportunities and Dangers," Durga P.
Satapathy and Jon M. Peha consider the incentives of users to share spectrum with
others, and they propose rules to achieve efficient use of the spectrum through
sharing. They postulate that there are some services (such as wireless local-area
networks, LANs) for which unlicensed spectrum, if shared, would be more efficient. Satapathy and Peha then develop an analytical framework to evaluate the effects of various rules on the incentives of spectrum users to create efficient
hardware and to use spectrum in a socially optimal way. A central element of
Satapathy and Peha's framework is that with unlicensed spectrum, a "tragedy of
the commons" may result. Etiquette rules may mitigate the tragedy of the commons problems, although such rules also impose a cost by restricting usage.

II. INTERNET GROWTH AND COMMERCE

The five chapters in this group are an eclectic group that bear in different ways on policy issues. These chapters consider the characteristics of Internet users and nonusers, how usage-sensitive pricing should be implemented to support the Internet infrastructure, the viability of electronic money and alternative electronic payment mechanisms, and the effects of Internet growth on the use of postal services.

The first chapter, by James Katz and Philip Aspden, "Motivations for and Barriers to Internet Usage: Results of a National Opinion Survey," presents findings covering users, nonusers, and former users of the internet. Although other recent surveys have tended to focus only on users and on issues of commercial potential, Katz and Aspden's work centers on demographic and other characteristics of both users and nonusers.

Katz and Aspden find evidence of a "digital divide," in that Internet users tend to be male, younger, wealthier, and more educated. Also, even when age, income, and education differences are accounted for, Blacks and Hispanics have a greater tendency to be nonusers. Most significantly, those groups are disproportionately unaware of the Internet. In terms of motivations, Katz and Aspden report that sociopersonal development, along with social and work networks, appears to be important for stimulating interest in the Internet and for supporting Internet users. With respect to usage barriers, the authors' results support the call for an easier-to-use Internet. They report that in addition to cost, difficulty in using the Internet is a prominent barrier, even among experienced users.

For many of the same reasons that the interconnection provisions of the 1996 Telecommunications Act pose difficult policy questions, the issue of how user payments should support the interconnected networks of the Internet is also difficult and actively debated. "Combining Sender and Receiver Payments in the Internet,' by David D. Clark, joins a growing literature on Internet pricing. Clark's beginning premise is that the current system, in which payments to access providers support the Internet infrastructure, is limited because it allows no explicit way to measure the willingness of different users to pay for its use. One commercial Web site, for example, might be willing to pay substantial fees to reach receiving users, whereas another might believe its information is of high enough value that receivers will pay to reach it. Similarly, different users attach different values to a site's content. To remedy this inflexibility, Clark proposes a "zone-based cost-sharing" plan, in which sending and receiving users identify certain geographic zones within which they are willing to pay. In common with other schemes, usage-sensitive payments are applied to enhanced services, such as assured capacity in times of congestion. The distinguishing feature of Clark's plan is that both senders and receivers of information contribute to the cost of the transmission.

For many years, automatic teller machines have given bank customers the opportunity to make commercial transactions electronically. The exponential development of commercial activity on the Internet, however, has given new life to a very fertile area of research: How should electronic transactions be conducted, and how do the alternative methods affect users? To date, payment for services over the Internet has not advanced to the point that individuals or businesses feel confident about the accuracy or security of transactions.

Jeffrey K. MacKie-Mason and Kimberly White, in "Evaluating and Selecting Digital Payment Mechanisms," consider the service provider's choice among the numerous electronic payment mechanisms that are becoming available. The authors categorize 10 of the most highly visible mechanisms (e.g., NetBill, Ecash, CyberCash) according to 30 criteria (e.g., "easily exchangeable," "low transaction delay," and "privacy"), and show how a decision maker can evaluate and select among alternative mechanisms using a systematic rational choice approach. MacKie-Mason and White show that even with such a large number of criteria, their selection process typically requires only a few steps and requires relatively little information about each mechanism. The authors then apply their analysis to the University of Michigan Digital Library. Through that example, they show how their approach could be used to select different mechanisms to serve different needs of a single digital service provider.

Supriya Singh approaches the electronic money issue from a user's perspective in "The Social Impact of Electronic Money." Drawing on qualitative survey research conducted with Anglo-Celtic married couples in Australia, Singh reports that trust, or a sense of control, is an especially important determinant of willingness to use online money. Factors leading to trust in electronic money include, for example, the ability to track and substantiate a transaction and knowledge of the service provider. Singh also argues that the type of information yielded by a transaction (e.g., a record, no record, a running balance) is important to an understanding of how people mix the use of different types of money—electronic and physical—for different types of transactions. Singh argues that to be successful, banks and other providers of online services must understand how the use of electronic money is shaped by social relations and cultural values. In conclusion, Singh claims that the absence of that understanding from the users' perspective has contributed to banks' declining market share in the payments business in the United States. The banks have been asking questions, she observes, from the providers' rather than the users' perspective.

The final chapter in this group, "Electronic Substitution in the Household-Level Demand for Postal Delivery Services," by Frank A. Wolak, is concerned with the effects of Internet usage on postal service. It has been widely predicted that the Internet, through its provision of rapid, low-cost communication, will have substantial demand-reducing effects on postal services. Focusing on this hypothesis, Wolak analyzes the changes in demand for postal services over the 1986-1995 period. He finds that as penetration of personal computers has risen over this period, the demand for postal service has decreased, and that higher income households

are increasingly moving away from postal services. Wolak draws two main conclusions from his research: first, that the price increases of postal services enacted
in January 1995 (from 29 to 32 cents for a first-class letter) should lead to a significant fall in household demand, and second, that annual increases in the pervasiveness of PCs at historical rates should lead to reductions in household
expenditures on postal services at least as large as those that would result from a
postal price increase of about 10%.

III. INTERNET REGULATION AND CONTROL

The three chapters in this section turn to specific issues of government or private
regulation of Internet content and business activity. Following the pattern of
countless other new technologies, many existing regulations do not seem to fit the
Internet, and new or proposed regulations specifically designed for it have immediately generated controversy or court challenges.

One promising Internet technology creating policy debate is Internet telephony.
Although Internet telephony requires a substantial user investment for a personal
computer, modem, and other equipment, the per-minute cost of the service is far
less than that for conventional telephone service. To date, Internet telephony is essentially unregulated.

In "Can and Should the FCC Regulate Internet Telephony?" Robert M. Frieden analyzes the policy implications of an important recent petition to the FCC submitted
by America's Carriers Telecommunications Association (ACTA), a group of smaller
long distance carriers. The ACTA petition argues that the commission should begin
regulating Internet telephony suppliers as conventional common carriers.

Frieden analyzes "level playing field" arguments put forth by the ACTA petition
and considers effects of differential regulation between private providers and common carriers. He argues that the FCC can and should refuse to regulate Internet
telephony providers as common carriers. Frieden concludes that regulatory asymmetry is beneficial in this case because it promotes competition and disciplines
prices. He further argues that as long as the FCC classifies Internet telephony providers as "telecommunications carriers", they would have to contribute to universal service funding, thus removing the threat to universal service without
subjecting Internet telephony to common carrier regulation.

As experience with the Communications Decency Act has demonstrated, how
the courts should decide First Amendment-related issues involving the Internet remains quite unresolved. In "Bedrooms, Barrooms, and Boardrooms on the Internet," L. Jean Camp and Donna M. Riley consider how Internet services should be
classified for the purpose of determining First Amendment speech rights. Historically, the legal system in the United States has evolved four categories of "media
types" to guide First Amendment jurisprudence: broadcaster, publisher, distributor, and common carrier. Camp and Riley's central claim is that media type clas-

sifications are ill-suited to classification of electronic network services. Using a series of illustrations, they argue, for example, that electronic newsgroups have functions involving all four media types. Instead, Camp and Riley argue that electronic network services are more usefully classified as "virtual spaces," in which standards of acceptable speech are analogous to prevailing standards in physical space counterparts. For example, different Usenet news groups might be classified as "bedrooms," "workspaces," "classrooms," or "public forums," depending on the appropriateness of certain types of speech in those environments. They consider a variety of case studies in which their classifications could be applied to particular network services.

Especially since the portions of the 1996 Act restricting the transmission of pornography over the Internet have been challenged in court (and many think they are unenforceable even if legal), the ability of voluntary ratings and parental discretion to control Internet content has become an important subject of debate.

Jonathan Weinberg, in "Rating the Net," analyzes the potential of "filtering" or "blocking" software designed to allow Internet users to voluntarily erect their own barriers to pornographic, violent, or other unwanted content. As Weinberg notes, for example, rating services such as "SafeSurf," "RSACi," and "SurfWatch" are becoming increasingly popular and sophisticated. Weinberg acknowledges that rating systems provide "an impressive solution to the problem of sexually explicit speech on the Net." He argues, however, that those benefits come at considerable costs. Intermediaries, such as employers and libraries, will tend to increase their control over what is seen and read. Some sites may be excluded because of political choices made by rating services administrators.

Weinberg also contends that because such a large number of sites need to be categorized, inaccuracies are inevitable and it will not be possible to achieve both consistency and a sensitivity to subtle differences in the sites. In any case, it will be extremely difficult to maintain up-to-date ratings for all sites. That leads Weinberg to observe that the practical necessity of excluding unrated sites may "flatten speech," making it relatively difficult for newer, smaller, noncommercial, non-mass-audience sites to remain accessible.

IV. TELECOMMUNICATIONS AND POLITICS

The undeniable discrepancies between theoretical ideals and the actual results of government regulation have led to an extensive literature that analyzes the political environment within which regulations are established and carried out. As the three chapters in this section demonstrate, telecommunications provides no shortage of fertile subject matter in that endeavor.

Susan Dente Ross's chapter, "Bell Had a Hammer: Using the First Amendment to Beat Down Entry Barriers," was a prizewinner in the 1996 TPRC student paper competition. Ross studies the legal strategy by which telephone companies in the

United States managed to expedite their entry into both video transmission and content creation—in spite of the Congressional mandate in the 1992 Cable Act that telcos should not exercise editorial control by also creating the content. She reviews the history of telephone company success in overturning the Cable Act's video creation ban on constitutional, First Amendment grounds; the ban, the telephone companies argued, violated their corporate rights of free speech. Ross argues, however, that the telephone companies' obvious objective was simply economic, and that the First Amendment arguments were no more than a convenient strategic device to achieve that objective. Ross concludes that although opening the video market to full competition may be good economics, neither legal precedent nor logic supports the elimination of structural economic regulation on First Amendment grounds. The result has been "muddied First Amendment jurisprudence" and unresponsiveness to public desires as they were expressed through the Congress.

Gerald R. Faulhaber, in "Voting on Prices: The Political Economy of Regulation," follows by developing and testing a "median voter" model of regulation. In doing so, he completely departs from the dominant "asymmetric information" paradigm of political economy models. In those models, regulation tends to breaks down or works poorly because the regulated firms can exploit the strategic advantage of having more or better information than the regulators. Instead, Faulhaber makes the simple assumption that the actions of regulators do in fact reflect the desires of the median (or typical) voter, who in turn is assumed to maximize his or her economic surplus from the consumption of regulated services. Using this model, he shows theoretically that under many circumstances, regulation is even less efficient than unregulated monopoly because majority voting induces the regulator to excessively subsidize "mass" services with the revenues of more narrowly consumed "specialized" services. Faulhaber tests his model using a 1960–1993 telecommunications database. He shows that the historical shift in the toll-to-local telephone rate subsidy in the United States can be explained by the change over time in the median voter's relative demands for toll services versus local services.

Heather E. Hudson's chapter, "Restructuring the Telecommunications Sector in Developing Regions: Lessons from Southeast Asia," concludes this section of the book with an international perspective. Hudson examines recent telecommunications policy developments in five Southeast Asian countries and draws relevant parallels with more industrialized nations. Governments in all five subject Asian countries—although very diverse in size, wealth, and other characteristics—are pursuing policies to stimulate investment in their information infrastructures. Pursuit of some other objectives, however, such as tariff reform, has fallen short because it has been difficult to establish regulatory bodies that are autonomous from influence of the dominant carrier or other private interests. Economic incentives for dominant carriers to extend networks to rural and other less profitable areas have also been inadequate in some countries. Of particular interest, several Asian countries are facing a fundamental policy conflict: Their governments are attempting to control access to information by banning satellite antennas, employing content controls, or other policies. An inevitable result of information infrastructure investment, however, is increased access to information. Hudson observes in conclusion that telecommunications planners and policymakers must recognize that sharing and using information should be the ultimate purpose of telecommunications policy reform.

THE EDITORS

The editing of this book was conducted while Greg Rosston was Deputy Chief Economist of the Federal Communications Commission, Washington, D.C. David Waterman was Associate Professor in the Department of Telecommunications at Indiana University, Bloomington.

INTERCONNECTION
AND COMPETITION

Local Competition Policy Maneuvers

Gerald W. Brock
The George Washington University

Local telephone service has long been assumed to be a natural monopoly, even after other parts of the telecommunication industry were opened to competition. The fundamental goal of the antitrust settlement agreement of 1982 that resulted in the 1984 divestiture of AT&T was to separate the natural monopoly local competition services from the potentially competitive remainder of the industry in order to prevent AT&T from using its monopoly power in local telephone service to impede competition in other services. Many states granted a legal monopoly to the incumbent local exchange carriers (LECs) and refused permission to companies seeking to compete with the incumbent LECs.

In recent years, the assumption that local exchange telephone service is a natural monopoly that should be provided without competition has been changed into an assumption that all portions of the telecommunication market can be served competitively with an appropriate policy framework. Although competitive rhetoric is now widely accepted and seldom challenged in the United States, there is less agreement on the appropriate policy framework for fully establishing competition. This chapter examines the 1996 controversies surrounding the development of local competition policy through the Telecommunications Act of 1996, the Federal Communications Commission (FCC) effort to develop detailed regulations for implementing that law, and the successful (at least temporarily) opposition to the FCC rules.

DEVELOPMENT OF LOCAL COMPETITION POLICY

The Communications Act of 1934[1] (1934 Act) formalized a jurisdictional division of authority in the United States in which the federal government (through the

---

1. *The Communications Act of 1934*, codified at Title 47 of *United States Code*, Section 151, and following.

Federal Communications Commission) was responsible for regulating interstate telephone service and the state governments (through the various state public utility commissions) were responsible for regulating intrastate telephone service. Although many disputes over jurisdiction occurred, there was no question that under the 1934 Act the states had exclusive authority for regulating intrastate service.[2] Accordingly, in recent years through 1995 several states developed frameworks for initiating competition into the previously monopolized local exchange telephone markets. Similar issues were examined independently in each of the state proceedings, and various outcomes were reached.

The February 1996 signing of the Telecommunications Act of 1996[3] (1996 Act) imposed federal policy on local competition. The 1996 Act was structured as an amendment to the 1934 Act rather than as a complete replacement for it. It established federal control over the development of local exchange competition but did not repeal provisions from the 1934 Act, reserving oversight of intrastate communications to the states. That left a confused jurisdictional situation that has not been fully resolved as of the end of 1996.

The essential policy structure of the 1996 Act is to open the local exchange market to competition in exchange for eliminating the restrictions (including the prohibition on providing long distance service) imposed on the Bell Operating Companies by the antitrust settlement and divestiture agreement of 1982. Sections 251 and 252 of the 1996 Act provide the legal structure for local competition. Section 251 requires incumbent local exchange carriers to provide interconnection "at any technically feasible point...equal in quality to that provided by the local exchange carrier to itself or to any subsidiary...on rates, terms and conditions that are just, reasonable, and nondiscriminatory"; to provide unbundled access "to network elements...at any technically feasible point"; and "to offer for resale at wholesale rates any telecommunications service that the carrier provides at retail to subscribers who are not telecommunications carriers," among other requirements. That section explicitly instructs the Federal Communications Commission (FCC) "to establish regulations to implement the requirements of this section."

Section 252 provides that the starting point for agreements between potential entrants and incumbent local exchange carriers for interconnection, unbundled elements, or services for resale is negotiation between the carriers. In the case of difficulty in reaching agreement, carriers are authorized to seek nonbinding mediation or binding arbitration from the appropriate state commission. Section 252 in-

---

2. The origination and termination of interstate calls used physically intrastate local exchange carrier facilities but was classified as interstate communications subject to the FCC jurisdiction. In 1993, jurisdiction over commercial mobile radio services (including cellular) was granted to the federal government without regard to the physical routing of the calls.

3. The *Telecommunications Act of 1996*, Public Law no. 104-104, 110 Statutes 56. The local competition provisions of the Act discussed in this chapter are codified in Title 47 of *United States Code*, Sections 251 and 252.

cludes a "most favored nation" (MFN) clause applicable to both voluntary and arbitrated agreements that requires any agreement for "interconnection, service, or network element" to be made available "to any other requesting telecommunications carrier upon the same terms and conditions as those provided in the agreement." That section also provides generalized pricing guidelines binding on state arbitrators that require prices for the transport and termination of traffic to "provide for the mutual and reciprocal recovery by each carrier of costs associated with the transport and termination on each carrier's network facilities of calls that originate on the network facilities of the other carrier...on the basis of a reasonable approximation of the additional costs of termination such calls"; that requires prices for unbundled network elements to be "based on the cost...of providing the...network element"; and that requires prices for wholesale services for resale to be set "on the basis of retail rates charged to subscribers for the telecommunications service requested, excluding the portion thereof attributable to any marketing, billing, collection, and other costs that will be avoided by the local exchange carrier."

Section 252 is federal law that establishes national requirements for intrastate local competition with instructions to state commissions for arbitration and approval of agreements. The section is silent as to what responsibilities the FCC has to promulgate regulations to clarify its provisions. It is the normal right and responsibility of the FCC to issue regulations for the implementation of any amendment to the 1934 Act, but this case is confused because the section deals with intrastate matters reserved generally to the states under the original provisions of the 1934 law; and although the section clearly reduces the freedom of states to develop their own local competition policies, it does not explicitly instruct the FCC to issue additional regulations implementing the section.

The narrow legal issue of whether or not the FCC should issue regulations interpreting Section 252 has considerable policy and competitive significance. Potential entrants generally believe that they have better competitive opportunities under uniform national rules interpreted by the FCC, whereas incumbent LECs generally believe that they have better competitive opportunities under separate proceedings in each state controlled by state commissions that are generally less enthusiastic about competition than the FCC. Delegation of authority to the states to interpret and implement the 1996 Act tends to reduce the significance of that act for promoting competition because the states had authority to develop local competition policies on their own prior to the 1996 Act.

In proceedings before the FCC regarding interpretation of the local competition provisions of the 1996 Act, the incumbent LECs generally argued that the FCC should not prescribe detailed regulations. The incumbent LECs contended that the statute provided the necessary general guidance and left implementation to private negotiations with state arbitration as necessary, leaving no place for detailed federal regulations. The potential entrants generally argued that private negotiations would fail to reach good results because incumbent LEC dominance gave them a much stronger bargaining position than potential entrants, and that consequently national rules were necessary in order to fully implement the statutory local competition provisions.

In August 1996, the FCC adopted a detailed 700-page order to implement the local competition provisions of the 1996 Act.[4] The order largely adopted the positions of the potential entrants and generally rejected the proposals of the incumbents. The FCC made the following key decisions:

1. Jurisdiction: The FCC asserted wide jurisdiction under the new law and established a federal policy on local competition by adopting rules intended to be binding on state arbitrators.

2. Symmetry: The law's "mutual and reciprocal recovery" of costs for transport and termination of traffic was interpreted as requiring symmetric payments for the mutual termination of traffic between two interconnected carriers when the termination service provided by both companies covers a similar geographic area. Consequently, no payment for termination services will be made if traffic is balanced, whereas payments will be made from the carrier with excess originating traffic to the carrier with excess terminating traffic when traffic is unbalanced.

3. Pick and choose in most favored nation clause: The law's requirement that any agreement between two carriers be made available on the same terms to any other carrier was interpreted to include portions of an agreement. A potential competitor not only has the right to adopt an agreement reached with another carrier in its entirety but may choose individual provisions of the agreement without accepting other provisions of the agreement.

4. Unbundled element pricing: The law's requirement that incumbent LECs make unbundled elements available to competitors at "cost" was interpreted to mean that the price of unbundled elements should be set at forward-looking long-run incremental cost. The FCC created a new term, *total element long run incremental cost* (TELRIC), to describe its pricing methodology. The TELRIC approach is an adaptation of the total service long run incremental cost (TSLIRC) methodology already used in state regulatory proceedings with elements substituted for services. The FCC stated that it expected common costs to be minimal under its methodology, but that forward-looking common costs could be allocated to individual elements: "Under a TELRIC methodology, incumbent LECs' prices for interconnection and unbundled network elements shall recover the forward-looking costs directly attributable to the specified element, as well as a reasonable allocation of forward-looking common costs...Directly attributable forward-looking costs include...the investment costs and expenses

---

4. Implementation of the Local Competition Provisions in the Telecommunications Act of 1996, CC Docket no. 96-98, FCC 96-325, released August 8, 1996.

related to primary plant used to provide that element...[and] also include the incremental costs of shared facilities and operations" (paragraph 682).

5. Call completion pricing: The symmetric rates for the transport and termination of traffic received from another carrier must be based on the same forward-looking incremental cost TELRIC methodology as the rates for unbundled elements. For calls delivered to a LEC end office, the relevant costs are the traffic-sensitive local switching cost: "Only that portion of the forward-looking, economic cost of end-office switching that is recovered on a usage-sensitive basis constitutes an 'additional cost' to be recovered through termination charges" (paragraph 1057). Rates for termination may include a limited allocation of common cost but may not include a charge for lost contribution: "Rates for termination established pursuant to a TELRIC-based methodology may recover a reasonable allocation of common costs...that is no greater proportionally than that allocated to unbundled local loops, which,...should be relatively low. Additionally, we conclude that rates for the transport and termination of traffic shall not include an element that allows incumbent LECs to recover any lost contribution to basic, local service rates represented by the interconnecting carriers' service."

6. Proxies: The FCC specified proxy rates to be used until properly documented studies are available, including flat-rated loop cost proxies that vary by state (ranging from $9.83 for Massachusetts to $25.36 for North Dakota), a local switching rate between $0.002/minute and $0.004/minute, and a wholesale rate for resellers that is between 17 and 25% below the corresponding retail rate level. The proxies are default rates to be used in negotiation and arbitration until adequate cost studies are available.

The incumbent LECs vigorously opposed the FCC August 1996 interpretation of the 1996 Act's local competition provisions, first by seeking relief from the FCC and then by seeking judicial relief from the Appeals Court. A sample of the LEC concerns is contained in the GTE and Southern New England Telephone petition to the FCC:

The Commission's rules rest on a series of errors that ensure the rules will be overturned in whole or in part upon review. In the first place, the Commission has no authority under the Act to promulgate rules governing pricing, since the Act assigns that responsibility specifically to the states. Even if it had authority to regulate pricing, moreover, the standard the Commission has chosen, Total Long Run Incremental Cost plus a so-called "reasonable" allocation for joint and common costs—or "TELRIC plus"—would accomplish an uncompensated taking of property in violation of the Fifth Amendment. The Commission's pricing rules thus violate the statutory command that rates be "just" and "reasonable" and based on "cost." In addition, the Commission has acted arbitrarily in setting default prices that are not themselves based on the methods the Commission has prescribed for determining rates. Even that series of errors, however, does not exhaust the list of legal infirmities underpinning the rules. In a series of substantive prescriptions concerning unbundling, interconnection and resale the Commission has imposed requirements that plainly exceed the mandates of the Act.[5]

The FCC refused to stay its rules, and many opponents (including all of the Regional Bell Operating Companies, GTE, and numerous state public utility commissions) appealed the order to several different circuits of the Appeals Court. The opponents sought an immediate stay of the rules pending judicial review of the alleged serious substantive errors that could cause irreparable harm to those appealing the order. The appeals were consolidated before the Eighth Circuit (St. Louis) Appeals Court by random chance among the circuits in which appeals were filed, in accordance with established procedure for cases filed in multiple circuits. The court granted a temporary stay of the order pending oral argument, and then in October 1996 granted a stay of the pricing rules and the "pick and choose" interpretation of the most favored nation requirements until a full substantive proceeding before the Appeals Court is completed. An important part of the stay order was a finding that the petitioners were likely to prevail on the merits after full judicial consideration. In reaching the conclusion that the FCC pricing rules should be stayed, the court indicated its belief that the states and not the FCC have the authority to interpret the pricing provisions in Section 252 of the Act:

> The petitioners allege primarily that the FCC exceeded its jurisdiction by imposing national pricing rules for what is essentially local service. They argue that the text and the structure of the Act give the States, not the FCC, authority over the pricing of intrastate telephone service...The sections of the Act that directly authorize the state commissions to establish prices are devoid of any command requiring the state commissions to comply with FCC pricing rules (or, for that matter, authorizing the FCC to issue any pricing rules). This absence indicates a likelihood that Congress intended to grant the state commissions the authority over pricing of local telephone service, either by approving or disapproving the agreements negotiated by the parties, or, when the parties cannot agree, through compulsory arbitration, thereby preserving what historically has been the States' role.[6]

Formally, the Appeals Court stay simply preserves the status quo while the issues are being litigated. However, the court's finding that the opponents are likely to prevail on the merits is a strong indication that the court is unlikely to uphold the FCC pricing and "pick and choose" rule after the appeals proceedings are completed

---

5. Joint Motion of GTE Corporation and the Southern New England Telephone Company for Stay Pending Judicial Review, FCC, CC Docket no. 96-98, August 28, 1996, p. 6.

6. Order Granting Stay Pending Judicial Review, no. 96-3321, October 15, 1996, pp. 12, 13 (http://www.wulaw.wustl.edu/8th.cir/Opinions/FCC/963321.008).

(probably in the second half of 1997). With the stay in effect at the end of 1996, the FCC pricing "rules" have no legal effect but remain "suggestions" to the states. Informal reports from some ongoing state arbitration proceedings suggest that some states that opposed the mandatory application of the FCC rules are taking them seriously as useful guidelines and suggestions subject to state interpretation.

IMPLICATIONS OF FCC APPROACH TO LOCAL COMPETITION

The guiding principle behind the FCC decision appears to be that the size of a network should have no effect on competitive advantages because all networks should be able to share the benefits of network externalities, economies of scale, and economies of scope. The potential competitive advantages of a large network include the following:

1. The network externality that makes a large network more valuable to customers than a small network.
2. Reduced costs from the utilization of ordinary economies of scale in transmission and other facilities.
3. Economies of scope among various services provided over the same network.
4. Economies of density from complete geographic coverage of the service area.

The interconnection portion of the order attempts to socialize the network externality. By requiring carriers to provide transport and termination services to each other based on forward-looking incremental cost, and by requiring symmetric payments between carriers for terminating services provided to the other, the interconnection rules attempt to prevent established carriers from using the network externality for their competitive advantage. In an unconstrained bargaining framework, the interconnection agreements are unlikely to be symmetric because the entrants value interconnection with the incumbents more highly than the incumbents value interconnection with the entrants. With unconstrained bargaining, the incumbent could either prevent entry altogether or extract substantial payments from the entrant for the privilege of interconnection. On the other hand, if multiple networks of reasonably equal size already existed, we would expect them to interconnect voluntarily and make symmetric payments for services provided to each other because each places an equal value on interconnection. The FCC has attempted to promote the development of a competitive local exchange market by imposing interconnection rules that resemble what a free bargaining situation would reach under a competitive market structure rather than under the current market structure.

With interconnection (including transport and termination of traffic) mandated at symmetric rates based on incremental cost, there would still be economies of scale and scope to provide advantages to the incumbent networks relative to entrants. Unbundling at incremental cost provides an additional constraint on the incumbents and an additional opportunity for the entrants because it deprives the incumbents of the advantages of scale and scope. Insofar as a network can be deployed with economies of scale or scope, the largest firm would gain from doing so in the absence of an unbundling requirement. The ability of entrants to purchase individual network elements at forward-looking incremental cost allows them to share in the economies of scale and scope and puts a severe constraint on the strategies available to the incumbent. Past policy has required unbundling of elements that were assumed to be part of a monopoly structure, but not of elements assumed to be part of a competitive structure. Unbundling is one way to limit the pricing freedom of a monopolist and to allow greater customer or competitor freedom in utilizing the facilities of a monopolist. Insofar as the unbundling requirements are permanent rather than a transitional measure to move toward competition, they represent a substantial increase in regulatory oversight over what are expected to become competitive markets.

Resale can be considered as unbundling on a customer service level rather than on a physical element level. With resale, one is allowed to gain access to a particular service and to put the entrant's brand name on it. It deprives the incumbent of exclusive reputation effects and of the advantages of complete geographical coverage. The reseller can build its own reputation for quality by reselling high-quality incumbent facilities. An entrant with limited physical facilities can market as if it had full coverage and fill in the gaps with resold service.

In summary, the FCC order on local competition placed three different limitations on the incumbent's freedom to utilize its established position for competitive advantage:

1. The requirement for interconnection (including transport and termination) at incremental cost limits the use of the network externality for competitive advantage and tends to equalize the competitive position of small and large networks.
2. The requirement that incumbent LECs provide unbundled elements at incremental cost limits the exploitation of physical economies of scale and scope as a competitive advantage.
3. The requirement that incumbent services be made available for resale at a default rate of 17–25% discount from the retail price limits the exploitation of the marketing advantages of scale and scope by allowing a brand name to be widely distributed across a wide range of customers and products even without the development of extensive physical facilities.

If the FCC framework for local competition were fully implemented, it would have significant implications for the way services are priced in the future. The interconnection conditions themselves tend to force prices toward cost. However, they do not necessarily force individual component prices toward cost. Just as the highly competitive airline industry still practices extensive price discrimination, it

seems likely that with interconnection conditions alone, there would still be substantially varying price-cost margins among various customers and services.
There are two different reasons for this. The first is transactions costs. Companies
often find it worthwhile to advertise relatively simple price plans, or prices that are
averaged across a wide variety of customers, in order to simplify their marketing,
billing, and other management functions. Even without regulatory restrictions, it
is unlikely that incumbent LECs will undertake extensive deaveraging to fit fine
differences in their cost structure for serving different customers. The second reason is that packages of services can be delivered to a customer without necessarily
having each component price adjusted exactly to the cost of that component. Vertical services can still be charged at higher than the cost-based rate if each party
has to attract the customer first before it can sell vertical services.

Requiring resale discounts and favorable unbundled element rates along with
interconnection severely limits the pricing freedom of the LECs. Unbundled elements can be combined to create the equivalent of vertical services. If the unbundled elements are priced at TELRIC plus a limited allocation of forward-looking
common costs, and vertical services created from those elements can be sold to
customers who are receiving basic service through resale, then the prices of vertical services will be brought down toward cost. The combination of resale, unbundled elements, and interconnection, all at favorable prices, will tend to force prices
into alignment with costs in a quite strict way. Under the FCC rules, the freedom
of state public utility commissions to set local pricing policy (including geographical averaging, business versus residential prices, and vertical service pricing) will
be severely constrained. That freedom will be eroded in any case by increasing
competition, but full implementation of the FCC rules would probably lead to significant changes in price policy even before the entrants gain a significant share of
the market.

If it is possible to put together a complete service from unbundled elements,
then the significant price differential between business and residence services
seems likely to disappear. If business services are priced above cost as the LECs
contend, then it should be possible to put together a cheaper equivalent in unbundled elements. That would tend to bring the price of business services down
to a price determined by the rates for the unbundled elements and associated
transactions costs. It would remain possible to have a subsidized service, as long
as the source of the subsidy is explicit and unavoidable. However, it seems that
implementation of the FCC rules would be likely to raise residential rates significantly and to cut most other rates, including the rates for vertical services and
business services.

The FCC framework implies that the TELRIC prices will largely determine the
pricing freedom in the output market. Insofar as TELRIC is interpreted strictly as
forward-looking incremental costs plus a limited contribution to forward-looking
common costs, the prices for a combination of unbundled elements will in many

cases be below the price of the corresponding retail service. Prices based on the forward-looking incremental cost of elements comprising the service may be below the current retail price, either because the service is being used to subsidize other services or because the forward-looking costs are lower than the historical costs that form the basis for state rate regulation. The limit on retail service prices will be the cost of putting together a combination of unbundled elements or the cost of pure facilities based competition. The pricing of unbundled elements may affect the overall price structure of retail services as well as the competitive interaction of incumbent LECs and entrants.

IMPLICATIONS OF THE APPEALS COURT DECISION

If the initial approach of the Appeals Court is upheld, the FCC pricing rules will not take effect and pricing will be determined by individual negotiations and by the vari　s decisions of state arbitrators. We have some evidence of how that process is likely to proceed through the decisions reached in the initial agreements between incumbent LECs and competitors under the 1996 Act without the constraint of the FCC rules.[7] Of the first eight agreements reached under the 1996 Act, three were completed by voluntary negotiations and five were completed by action of the state regulatory commissions of Florida, Alabama, Massachusetts, Ohio, and Texas. Some of the initial agreements covered only a portion of the issues and left unresolved controversies for later determination. In general, the early negotiators found significant agreement on the nonprice issues such as points of interconnection, directory listings and directory assistance, and access to poles, ducts, and rights-of-way, but found substantial and continuing disagreements on pricing issues.

For transport and termination of traffic originated on another network, the early agreements provided for reciprocal compensation, with payment for traffic delivered to an end-office ranging from zero (bill and keep) to 1.9 cents per minute (MCI–BellSouth agreement for Tennessee). Several of the agreements provided for payments of 1.0 cent per minute for traffic delivered to an end office, well above the FCC proxy rate of 0.2–0.4 cents per minute. Because higher transport and termination rates are generally considered an advantage for incumbents and a disadvantage for entrants, the early negotiated or state-arbitrated agreements are less favorable to entrants than the FCC rules. However, the disadvantage for entrants of the higher rates is mitigated to some extent by the reciprocity provision and the MFN provisions contained in the agreements. Early entrants have generally had higher traffic outbound from their customers to the incumbent networks than inbound from the incumbent network to their customers. Consequently, even

7. The initial agreements are summarized and evaluated in V. W. Davis and M. E. Clements, (1996), Convergence and Controversy in Early Interconnection Agreements, National Regulatory Research Institute Report 96-27, October.

with reciprocity, the net payment from the entrant to the incumbent rises with an increase in the price of transport and termination, but with reciprocity and MFN provisions in the agreements, entrants have an incentive to seek customers with heavy inbound traffic when rates are high, thereby causing the LEC to seek to reduce the symmetric interconnection rate. Although the incentives for facilities-based entry are lower in the early negotiated agreements than in the FCC rules, there is no reason to believe that competition will be stifled if interconnection occurs on terms similar to the initial negotiated agreements rather than on terms specified by the FCC rules.

The initial agreements for unbundled elements were substantially less favorable to the entrants than the structure adopted in the FCC rules. Several of the agreements specifically note that the parties could not agree on unbundled elements. Incumbents objected both to the detail of unbundling requested by potential entrants and to the pricing requested. In general, the initial requests of potential entrants largely paralleled the proposal in the FCC rules (a detailed list of unbundled elements available at prices based on forward-looking incremental cost), but those requests were opposed by the incumbents and not granted by the state arbitrators. Some agreements were quite vague on unbundled elements, with both the specific elements and the prices to be determined later.

Difficulties in negotiations for resale of service have been similar to the difficulties in negotiations for unbundled elements. Potential resale entrants have requested the right to resell all services, including promotions and special pricing plans, and have requested substantial discounts from the incumbent's retail price for wholesale services to be resold. Incumbents have sought to restrict the range of services available for resale and to maintain a narrow price margin between wholesale and retail prices. Several of the initial agreements left either the prices or the set of services to be resold for further negotiations. Those that did include a specific wholesale discount from the retail rate ranged from 6 to 10%, far lower than the FCC proxy discount of 17 to 25% off the retail rate.

Compared to the FCC framework, the initial negotiated and arbitrated agreements are slightly less favorable to full facilities-based entrants and substantially less favorable to those who planned to enter by procuring unbundled elements or wholesale services from the incumbents. The most critical requirement for facilities-based carriers is interconnection (including transport and termination of traffic) on reasonable terms and conditions. The law itself limits the freedom to utilize the network externality as a strategic weapon by requiring the transport and termination of traffic at "additional cost." Thus although the early state-supervised agreements utilize higher prices than the FCC proxies, they are not fundamentally different because both are interpreting the relatively specific requirements of the law. In contrast, the law simply requires the unbundled elements to be provided at "cost." The FCC explicitly rejected arguments that the law favored facilities-based carriers over resale carriers and that it ought to provide incumbents with greater

pricing freedom for unbundled elements and wholesale services than for the transport and termination of traffic. Instead, the FCC declared that unbundled elements
should be priced according to the same forward-looking incremental cost methodology as the transport and termination of traffic. The states have disagreed with
that part of the FCC decision and have given the incumbents much greater freedom in their provision of unbundled elements and wholesale services for resale
than for the transport and termination of traffic.

Under the FCC framework, there was a possibility that the rates for wholesale
services and unbundled elements were so attractive that they would prevent the entry of facilities-based carriers. Insofar as entrants found it less expensive to pro-
cure their network from the incumbent LECs than to build their own, they would
have less incentive to enter with their own facilities. The less favorable terms in
the early state-supervised agreements for wholesale services and unbundled elements make non-facilities-based entry difficult if not impossible. Entrants could
still use unbundled elements and resale to expand their range of services and begin
marketing while facilities are being developed, but would be unlikely to make a
long-term business success without their own facilities. Thus competition is likely
to occur under either framework, but under the Appeals Court view of state-supervised pricing it is likely to occur more slowly and with more emphasis on facilities-based carriers, whereas under the FCC framework it is likely to occur more
rapidly and with more emphasis on resale carriers.

CONCLUSION

The development of local competition rules in the United States has been chaotic,
with initial state efforts superseded by a Congressionally developed federal law,
that law interpreted (and some say, misinterpreted) and developed into detailed
rules by the FCC, the Appeals Court issuing a tentative decision that the FCC rules
were outside its authority, and the resumption of private negotiations supervised
by state commissions as the route to local competition policy. More changes are
likely. The multiple government authorities and continuing changes even after de-
finitive regulations have been issued is a characteristic of U.S. policy that is often
confusing to international observers. James Q. Wilson has provided a colorful description of the U.S. policy process in general that is relevant to this case:

> Policy making in Europe is like a prizefight: Two contenders, having
> earned the right to enter the ring, square off against each other for a pre-
> scribed number of rounds; when one fighter knocks the other one out, he
> is declared the winner and the fight is over. Policy making in the United
> States is more like a barroom brawl: Anybody can join in, the combat-
> ants fight all comers and sometimes change sides, no referee is in charge,
> and the fight lasts not for a fixed number of rounds but indefinitely or
> until everybody drops from exhaustion.[8]

The author has elsewhere described the actions of multiple conflicting government agencies that created U.S. telecommunication policy prior to the passage of the 1996 Act.[9] Although that policy process lacked the rational structure that some might desire, it was effective at representing the interests of many different parties and providing a way to gradually change policy from one that assumed monopoly to one that assumed competition. It appeared possible that some of the past policy controversies arose from the failure of Congress to update the basic communications law passed in 1934, and that with the passage of the 1996 Act, the process would become much more straightforward. The experience to the end of 1996 with local competition policy suggests that the past practice of policymaking through the uncoordinated inputs of a variety of different policy institutions is continuing.

The formal legal problem that created confusion over the authority of the FCC to specify detailed rules for local competition was the failure of Congress to either explicitly grant or deny the FCC the authority to interpret Section 252 of the 1996 Act. If that failure were simply a drafting error, it would be a straightforward matter to pass a minor amendment clarifying the intent of the law. However, it appears that the ambiguity in the law was intentional in order to gain agreement from those who disagreed on the substantive question. After the FCC issued its August decision, two groups of legislators who had been closely involved in the passage of the bill presented briefs to the Appeals Court that was reviewing the FCC decision in order to clarify "the intent of Congress" in the law. One group, including Congressmen John Dingell, W. J. (Billy) Tauzin, Rick Boucher, and Dennis Hastert, supported the LECs and state utility commissions in their assertions that Congress had not intended for the FCC to issue pricing rules. The second group, including Congressmen Thomas Bliley and Edward Markey and Senators Ernest Hollings, Ted Stevens, Daniel Inouye and Trent Lott, supported the FCC in its assertion that Congress clearly intended for the FCC to issue pricing rules. The second group stated:

> Petitioners' [LECs and state public utility commissions] arguments to this Court reflect several critical misconceptions about the Act and the purposes underlying it...Perhaps most fundamentally, petitioners seek to remove the Federal Communications Commission (FCC) from any role whatsoever in implementing the Act's provisions requiring that incumbent local exchange carriers (LECs) charge their competitors rates that are "just, reasonable, and nondiscriminatory," and to leave that critical duty solely to the States. That result would be an extraordinary reversal of a clear Congressional decision. Indeed, it would directly negate the balanced partnership between the FCC and

---

8. James Q. Wilson, (1989), *Bureaucracy*, (pp. 299, 300), New York: Basic Books/ Harper Collins.

9. Gerald W. Brock, (1994), *Telecommunication policy for the information age*, Cambridge, MA: Harvard University Press.

the States that Congress sought to establish. While we take no position here on the substance of the FCC's rules, its overall authority to implement the local competition sections of the Act is plain . . . The Act is thus firmly predicated on the understanding that the FCC will continue to play its customary role as the nation's expert agency on telecommunications in interpreting provisions of the nation's telecommunications laws, and in applying its understanding of the economic and technical realities of the industry to fashion and enforce federal rules to implement Congressional intent.[10]

Along with ambiguity in the 1996 Act, a second reason for policy confusion was the fact that the substantive policy decision of the FCC strongly favored potential entrants over incumbents. Many major FCC decisions represent some form of compromise among the interested parties so that each identifiable group of significant power has an interest in seeing the decision implemented. It is hard to find benefits for either the state public utility commissions or the incumbent LECs from the FCC August 1996 decision. The LECs had the political influence and financial incentive to find a way to slow or stop the implementation of the FCC rules. The U.S. telecommunication policy process has so many different centers of power that a party that loses in one forum can seek to have the decision modified in another forum. In this case, the LECs were able to get a stay from the Appeals Court, but had that failed they probably could have found another forum to gain some relief from the FCC rules. Local competition policy is likely to stabilize only after a compromise is reached (through the combination of legislation, FCC rules, state decisions, and court decisions) that provides enough benefits to each powerful group so that the groups no longer have a strong incentive to attempt to overturn the policy.

ACKNOWLEDGEMENTS

Greg Rosston and David Waterman provided useful comments on an earlier version of this chapter. The author has served as a consultant to Comcast Corporation and Cox Enterprises in matters related to local competition. The views expressed in this chapter are those of the author alone and should not be attributed to any other individual or organization.

---

10. Iowa Utilities Board, et al., v. Federal Communications Commission, et al., U.S. Court of Appeals for the Eighth Circuit, No. 96-3321, Brief of Amici Curiae Bliley, Hollings, Stevens, Inouye, Lott, and Markey, December 23, 1996, pp. 5, 6.

Economic Efficiency, Public Policy, and the Pricing of Network Interconnection Under the Telecommunications Act of 1996

Michael L. Katz[1]
University of California at Berkeley

In February 1996, the U.S. Congress passed the Telecommunications Act of 1996 (the 1996 Act).[2] The 1996 Act is a significant milestone along the lengthy path toward telecommunications competition. For much of the last 60 years, public policy has been hostile to competition on the theory that carriers needed protection from competition in order to ensure their ability to invest in new facilities and their willingness to serve all segments of the public. Gradually, however, competition has been introduced to many telecommunications markets.[3] The 1996 Act takes the view that competition in all telephony markets is beneficial and should be promoted by public policy.[4] The legislation sets out

---

1. At the time this chapter was drafted, I served as a consultant to AirTouch Communications and the National Cable Television Association on related matters. The views expressed in this chapter are my own and should in no way be construed as representing those of either organization.

2. Telecommunications Act of 1996, Public Law No. 104-104, 110 Stat. 56. The 1996 Act amends the Communications Act of 1934, 47 U.S.C. §§ 151 *et seq.*

3. See Brock (1994) for an interesting history of the development of competition in U.S. telecommunications markets.

4. To the best of my knowledge neither the U.S. Congress nor the Commission has ever undertaken a serious study of whether oligopoly is indeed more efficient than a regulated monopoly in these markets.

a structure for regulatory reform that should accelerate the movement toward greater competition that was already underway.

The pricing of incumbent network facilities and services for use by entrants will play a critical role in opening markets to competition. Because of the costs of building overlapping networks, it generally is efficient for competing carriers to rely on one another's facilities to complete calls made by subscribers on one network to subscribers on another (what the 1996 Act labels transport and termination). Similarly, there may be services and network elements that it is efficient for the incumbent local carrier to provide to other carriers at wholesale in order: (a) to facilitate entry by reducing its sunk costs and risk, and (b) to allow entrants to specialize on those segments where entry is efficient, rather than being forced to come in as integrated providers who also operate in segments where having additional providers is inefficient.[5]

In order to promote competition, the 1996 Act mandates that incumbent local exchange carriers (incumbent LECs) provide elements of their networks (e.g., local loops) on an unbundled basis and that they enter into mutual compensation arrangements for the provision of transport and termination services. Acting under its interpretation of the mandate created by the 1996 Act, the Federal Communications Commission issued pricing guidelines for unbundled network elements and transport and termination in August 1996.[6] At the time that this chapter was being written, the Commission's pricing rules were undergoing legal challenge by some states and incumbent LECs.

The Commission, Congress, and the federal courts all face a number of economic issues as they grapple with the appropriate public policy toward the pricing of unbundled elements and transport and termination services. This chapter examines some of these issues from the perspective of economic efficiency. The next section first considers the role of government pricing guidelines. I argue that, although the effects of guidelines have been overstated by both their opponents and supporters, there is a useful role for government guidelines in promoting socially efficient outcomes from the bargaining between incumbent and entrant LECs.

The next two sections consider what pricing standards might promote efficiency. First is an examination of the use of cost-causative prices to generate proper signals to guide consumption and investment decisions. The following section then examines how to account for costs that cannot be recovered on a cost-causative basis: common costs (because there is no way to make a cost-causative as-

---

5. For a summary discussion of the benefits of network interconnection and public policy toward network interconnection, see Katz, Rosston, and Anspacher (1995).

6. *Implementation of the Local Competition Provisions in the Telecommunications Act of 1996*, CC Docket No. 96-98, *Interconnection between Local Exchange Carriers and Commercial Mobile Radio Service Providers*, CC Docket No. 95-185, First Report and Order, FCC 96-235 (released Aug. 28, 1996). Hereafter, *First Report and Order*.

signment) and universal service costs (because policymakers have chosen not to recover these expenses on a cost-causative basis). As discussed later, there are many areas in which economic theory provides less clear guidance than some have claimed. At the same time, there are areas in which economics has quite powerful prescriptions. Unfortunately, policymakers have ignored these prescriptions to date and give every indication that they will continue to ignore them in the future.[7] In particular, there has been an almost wholesale rejection of the notion that consumption efficiency should be taken into account in designing a system of common cost or universal service cost recovery. This is particularly ironic, and disappointing, in the case of universal service policy given that the stated intent of this policy is to increase consumer welfare.

REGULATORY GUIDELINES AND BARGAINING THEORY

One of the key features of the 1996 Act is that it sets up an arbitration process for situations in which the incumbent LEC and its would-be competitors cannot reach agreement over the pricing of unbundled elements or transport and termination.[8] Subject to a statutory timetable, either party may request arbitration by the relevant state commission if the private negotiations are not proceeding successfully. In its August 1996 decision, the Commission promulgated guidelines for the state commissions to follow in arbitrating disputes. The public debate leading up to, and following, the Commission's decision has been quite contentious, and it is worth examining the economics of bargaining to see what it says about these issues.

At the most fundamental level, one can begin with the question: What useful purpose, if any, might government intervention in private negotiations play? Many incumbent LECs argue that government intervention is unnecessary and, indeed, will constrain private bargaining in ways that lead to inefficient outcomes.[9] Other parties argue that guidelines are essential to opening markets to competitive entry.[10]

In order to address these issues, it is important to keep in mind two fundamental points from the economics of bargaining. First, a party's bargaining position depends on how that party would fare in the event that the current round of bargaining fails to reach an agreement. The better a party fares in the event that an agreement is not reached, the greater that party's bargaining power.

Consider the relative threat points of an incumbent LEC and a competitive LEC negotiating over unbundled elements or transport and termination in the absence

---

7. See, for example, Federal-State Joint Board on Universal Service, CC Docket No. 96-45, Recommended Decision (released November 8, 1996), which advocates lowering the subscriber line charge despite the fact that it is a non-traffic-sensitive charge for recovering non-traffic-sensitive costs.

8. See 47 U.S.C. §252.

9. These comments are summarized in the *First Report and Order*, paragraphs 50–51.

10. These comments are summarized in the *First Report and Order*, paragraphs 46–49.

of government oversight. If the parties fail to reach agreement, the competitive LEC may be forced out of business or deterred from entering the market in the first place. The incumbent LEC, however, can largely continue with business as usual. Moreover, because they weaken its competitors, bargaining breakdowns or delays may actually be benefits from the incumbent LEC's perspective, not costs. Thus, in the absence of government intervention, incumbent LECs will generally have much stronger bargaining positions than competitive LECs.

A second critical insight from the economics of bargaining is that, in the absence of impediments to efficient bargaining, parties have incentives to reach agreements that maximize their joint benefits, even when each party is concerned only with its own welfare. Under efficient bargaining, the parties will first act to maximize the total benefits to be divided between them and then bargain over the division of these gains. Thus, if industry profits would be increased by entry (say because competitive LECs can offer innovative, differentiated services that incumbent LECs cannot), then incumbent LECs and competitive LECs could be expected to reach agreements that facilitated competitive LEC entry. This logic suggests that government intervention is unnecessary. The argument is, however, seriously incomplete.

There are two broad reasons that incumbent LECs may seek inefficient terms for the provision of unbundled network elements and transport and termination. The first is that, in its role as a supplier with market power, an incumbent LEC may set inefficiently high prices in order to extract greater economic rents for itself. This is a variant of the standard monopoly distortion. But why not extract monopoly rents efficiently? One reason is that the bargaining between incumbent LECs and competitive LECs takes place under conditions of asymmetric information. This fact implies that an incumbent LEC cannot rely on lump-sum charges to transfer economic rents efficiently to itself from interconnecting providers. Instead, the incumbent LEC needs to rely on metering in the form of inefficiently high traffic-sensitive charges for services and flat rates for dedicated facilities.[11] While too high from a social perspective, these above-marginal cost prices allow the incumbent to extract the greatest profits from those competitive LECs who make the greatest purchases of the incumbent's facilities and services, and presumably generate the greatest value from them.

Public policy limitations on price discrimination also may have the (unintended) effect of inducing incumbent LECs to set charges that inefficiently distort consumption levels. Asymmetric information problems notwithstanding, public policy prohibitions may prevent an incumbent LEC from relying on individualized lump-sum charges to competitive LECs to transfer economic rents efficiently to itself from the competitive LECs. Instead, the incumbent LEC may be forced to

---

11. Even when facilities charges are flat rates, their levels may affect the decision whether to purchase a facility or the services that it provides.

offer the same pricing scheme to all competitive LECs. Again, the incumbent LEC implements a metering scheme that uses volume-sensitive charges to extract economic rents and thus suppresses traffic.

The second reason why an incumbent LEC may seek inefficient terms for the provision of unbundled network elements and transport and termination falls under the general heading of raising rivals' costs. By setting high charges for services sold to competitors, an incumbent LEC may be able to delay or weaken competition. Although the private parties have incentives to reach an agreement facilitating entry when it raises industry profits, entry may increase competition and thus reduce industry profits while raising consumer and total surplus.[12] In such cases, entry will not be in the joint interest of the negotiating parties, but will be in the public interest. There is thus a role for pricing guidelines in which the government represents consumers who are not otherwise party to the negotiations.

There is a complication. The 1996 Act's pricing guidelines are enforced only if one of the carriers requests enforcement. If bargaining is privately efficient, then one must ask why the incumbent LEC wouldn't simply "pay" the potential entrant not to compete, government guidelines notwithstanding. There are two parts to the answer. One is that any such payments would have to be disguised or would otherwise risk running afoul of the antitrust laws. Two, there may be multiple firms that could approach the incumbent LEC and demand to be compensated for staying out of the market. If policy guidelines give each competitive LEC a high degree of bargaining power, it can become expensive for the incumbent LEC to block competition by buying off all of the potential entrants.[13]

The fact that the pricing guidelines kick in only if one of the private parties seeks to have them applied raises several other policy issues. For instance, it suggests that the argument that government pricing guidelines will prevent private parties from reaching efficient agreements is flawed. Indeed, this line of reasoning suggests that regulatory guidelines are of little importance except in determining the levels of surplus that each party can demand from the other in bargaining.[14] One powerful implication of this line of reasoning is that the opponents of bill and keep were mis-

---

12. In an oligopolistic market, a firm generally benefits from an increase in the marginal costs faced by its rivals because such cost increases raise the rivals' profit-maximizing prices. Indeed, the oligopolists may collectively benefit if they can find a device to reduce competition among themselves. As discussed later, this effect arises even when the compensation scheme is symmetric if the carriers can separately influence the traffic flows in the two directions.

13. This point is examined formally in Appendix A. A remaining concern, however, is that carriers will use high transport and termination rates to induce both the incumbent and competitive LEC to raise their retail prices. The arbitration process set out in the 1996 Act does nothing to limit such behavior. It would, however, remain subject to the antitrust laws.

14. As noted earlier, the distribution of rents may have an effect on whether entry occurs and thus can matter for efficiency.

taken to be concerned that bill and keep arrangements for transport and termination would distort carriers' facilities investments.[15] Under efficient bargaining, two local carriers that exchange traffic would agree to construct facilities that minimize the sum of the costs incurred they incur to carry one another's traffic, regardless of the traffic-sensitive compensation rates for transport and termination.

Of course, it is not a given that bargaining will be privately efficient. As noted earlier, asymmetric information and public policy constraints may block the attainment of joint-profit-maximizing outcomes. Logically, there may be situations in which both (a) significant rents are created for a competitive LEC through some sort of inefficient behavior and (b) the incumbent LEC has no way to compensate the competitive LEC for behaving efficiently. Although in theory this is possible, I am unaware of any credible cases' having been put forth.

In summary, although the details of the guidelines may not matter, there is a useful role for regulatory pricing guidelines to play in reducing the chance that private negotiations will block efficient entry. Moreover, by framing the environment in which the negotiations take place, public policy can have this effect even in those instances where the states are not called upon to arbitrate a dispute. Further, to the extent that it stimulates public empirical studies as part of the policymaking process, the act of setting guidelines may itself reduce some of the informational asymmetries faced by the private parties and thus improve bargaining efficiency.

MISGUIDED CLAIMS ABOUT THE PRICING OF INCUMBENT LOCAL EXCHANGE FACILITIES AND INTERMEDIATE SERVICES

There was broad agreement among the parties to the Commission's proceeding that the notion of *total service long-run incremental cost*[16] (TSLRIC) plays a central role in the efficient pricing of unbundled network elements and transport and termination.[17] In its recommended guidelines, the Commission based unbundled network elements and transport and termination rates on forward-looking incremental costs under the new name of *total element long-run incremental costs* (TELRIC).[18] The big issues of contention concerned what, if any, additional loadings should be included to cover common costs, universal service costs, and embedded costs (i.e., past expenditures on network infrastructure that have not yet been recovered through service charges). Before turning to those issues, however,

---

15. For a summary, see *First Report and Order*, paragraph 1101.

16. Total service incremental cost is the incremental cost when the increment is defined to be an entire service (e.g., the full set of additional costs incurred to offer a set of customers exchange access service).

17. For a summary, see *First Report and Order*, paragraphs 635–642. Several parties objected to the possibility that this would be the sole basis for setting prices.

18. *First Report and Order*, paragraph 678. Appendix B of the present chapter addresses whether TSLRIC and TELRIC differ in more than name.

it is worth looking more closely at the arguments for pricing based on TSLRIC. I do this by putting forth several false, but apparently widely held, views.

"From the Perspective of Sending the Correct Signals to Consumers, Prices Should Be Set at Marginal Cost"

The prices of telephone services generate economic incentives that guide consumers' choices of calling levels, when to call, and which service providers to patronize (in those cases where there are competing ones). Economists' standard prescription for inducing consumption efficiency is to set price equal to marginal cost. However, marginal cost pricing of telephone services may not be optimal, due to both (a) the existence of externalities and (b) the fact that many of these services are inputs to other telecommunications services that are supplied under conditions of imperfect competition.

It has long been recognized that marginal cost pricing may be inefficient in the presence of consumption externalities. Telephony is subject to both *call externalities* (parties on both ends of a call generally benefit from the communication) and *network externalities* (an individual's decision to connect to the network confers benefits on other subscribers who can then call him or her). Each externality provides a theoretical justification for pricing services below marginal cost. Indeed, network externalities often are put forth as the ostensible reason for universal service policies that claim to promote network subscribership.

A second set of considerations—a set less widely recognized— arises because unbundled elements and transport and termination are intermediate goods, which serve as inputs to the production of services by other carriers. It appears likely that the retail services will continue to be provided under conditions of oligopoly, and we can thus expect to see significant margins between costs and retail prices. Consequently, even if it were efficient to price final goods at marginal cost, if might well be efficient to set the intermediate service prices below marginal cost in order to induce efficient pricing at the retail level.[19]

"From the Perspective of Sending the Right Signals to Providers, Prices Should Be Set at TSLRIC"

In addition to considering consumption incentives, it also is important to examine the effects of prices on provider investment incentives. In the presence of economies of scale and scope, setting prices at marginal costs will not allow a provider to recover its full costs. Hence, such a firm would have no incentive to make the investments needed to provide service. The Commission concluded that product-

---

19. See Laffont and Tirole (1994b). If retail prices below marginal cost are desirable due to externalities, then intermediate prices should be set even lower.

specific setup costs should be recovered from the rates charged for the specific service triggering those costs (i.e., pricing at average TSLRIC or TELRIC).[20]

Superficially, there are at least three arguments in favor of pricing at long-run average incremental cost. One, it is a relatively straightforward rule. Two, it seems to be "fair" because consumers of each service are responsible for the setup costs that their service triggers. Three, in a world with no long-lived, sunk investments and no product differentiation, it sends the proper economic signals to entrants: In such a world, if an entrant can profitably compete against a price set at long-run average incremental cost, then it is efficient for it to do.

There are, however, several reasons why this approach does not conform to first-best or second-best (i.e., prices constrained to allow firms to recover all costs of service provision) pricing principles. First, the assumptions that there are no long-lived, sunk investments and no product differentiation clearly are stringent and unrealistic. To see why the optimality of pricing at average TSLRIC breaks down without the assumption of no sunk investments, consider the following stylized example. There is an incumbent firm with sunk costs of K_I and incremental costs of m_I per unit of service. The sunk investment lasts forever. There is a potential entrant who can produce an identical service by making a sunk investment of K_E and bearing an incremental cost of m_E per unit. For simplicity, suppose that X units of service are inelastically demanded each period. Then entry is efficient if and only if $rK_E/X + m_E \leq m_I$, where r is the rate of interest. Entry is privately profitable if $rK_E/X + m_E \leq p_I$, where p_I is the incumbent's price. In this example, pricing at the incumbent's average TSLRIC, or $p_I = rK_I/X + m_I$, creates excessive entry incentives. The effect of such pricing is to induce the entrant to ignore the fact that the incumbent's fixed costs, K_I, are sunk.

Although this example is special, its implications are general. In the presence of sunk costs, pricing at long-run average incremental cost may distort the pattern or timing of entry. How can regulators resolve the apparent conflict between creating efficient entry incentives and allowing the incumbent to recover its costs? One way is to recognize the mistake in attempting to find a single, once-and-for-all price. Instead, the incumbent might be allowed to price at $rK_I/X + m_I$ prior to entry, with the understanding that it will be allowed to lower price to m_I if entry occurs. Notice that it is the postentry price, not the preentry price, that matters for entrant's investment incentives.

A second way to resolve the conflict is to require the entrant to pay rK_I/X per unit of service to the incumbent after entry. This is what the efficient component pricing rule tries to do. There are, however, many well-known problems with this rule.[21]

---

20. *First Report and Order*, Section VII.

21. For discussions and additional references, see Economides and White (1995) and Laffont, Rey, and Tirole (1996a).

A third approach is to recognize that it may be efficient to allow the firm to re-cover the setup cost of one service through the rates charged for other services.[22] Under second-best efficient pricing, the revenues collected for some services might well fail to cover their setup costs. This point raises a more general issue to which I return in the next section: Economic efficiency can be achieved only if policymakers take demand conditions into account.

A final defect with the principle of pricing at average TSRLIC arises when one again takes into account the fact that these are intermediate services. Unbundled elements may be complements, as well as substitutes for competitive LECs' out-puts. When considering the pricing of unbundled elements as complements to competitive LEC facilities, policymakers' concern is with the effects on entrant's investment incentives. High element prices may reduce the profitability of entry. Because potential entrants (who purchase these services) may have inefficiently low entry incentives (their entry may generate increases in consumer surplus that the entrants are unable to appropriate), it may be efficient to subsidize intermedi-ate service prices to correct inefficiently low entry incentives.[23]

"The Level of Transport and Termination Charges Doesn't Matter as Long as Traffic is Balanced."

The 1996 Act requires LECs to establish systems by which they mutually compen-sate one another for transport and termination.[24] Some commenters have argued that the level of transport and termination charges will have no effect as long as the traffic is balanced.[25] Even with balanced traffic, however, the level of transport and termination charges matters in those situations in which a carrier can differ-entially affect inbound and outbound traffic flows at the margin. For example, dif-ferent types of customers have different mixes of inbound and outbound calls. When transport and termination charges deviate from marginal costs, a carrier might seek customer types who have favorable mixes of calling. For example, if transport and termination charges are set above incremental cost (as proposed by the Commission), there will be stronger competition to attract those subscribers (e.g., pizza parlors) who have a greater propensity to take inbound calls than other customers, *ceteris paribus*.

Similarly, service prices may affect the calling patterns of carriers' existing sub-scribers. When transport and termination charges are not set at marginal cost, prof-

---

22. Of course, this approach has problems of its own. An incumbent LEC might be tempted to raise the prices of regulated services for which it faces relatively little competi-tion in order to subsidize services for which it does face competition.

23. See Laffont and Tirole (1994a).

24. 47 U.S.C. § 251(b)(5).

25. See *First Report and Order*, paragraph 1098.

it-maximizing carriers should take these changes in calling patterns into account. Consider a decrease in the price of local calling. Lowering the price of outbound calling has two effects. One, it attracts additional subscribers in competition with the other carrier. Two, it induces subscribers to make a greater number of outbound calls. Hence, the balance of traffic is affected at the margin. Because lowering the price triggers an increased payment to the other carrier, the incentive to lower the retail price is reduced and competition is weakened. Thus, two local exchange providers might well set high per-minute charges for terminating traffic on one another's networks as a means of giving each incentives to charge customers relatively high rates for local exchange services.[26]

COST RECOVERY NOT BASED ON COST CAUSATION

In its August 1996 decision, the Commission took the position that pricing at TEL-RIC will not fully cover costs and thus incumbent local exchange providers should be allowed to recover a "reasonable" allocation of common costs.[27] In addition, the Commission disallowed universal service costs and embedded costs. Instead, the recovery of these costs will be taken up in the upcoming (at the time this chapter was written) universal service proceeding. If they cannot be recovered on a cost-causative basis, how should these costs be recovered?

Common Costs: "A Minute Is a Minute"

By definition, common costs are costs that cannot be assigned to any one service on the basis of cost causation. Common costs do, however, have their own misunderstood principle. "A minute is a minute" is shorthand for the view that local calls, intrastate access, interstate access, LEC-CMRS interconnection, and enhanced service access all trigger the same costs and thus should all have same unit prices.[28] From the perspective of efficiently recovering common costs, this view is incorrect. Although the underlying incremental costs of different services may be quite similar (one needs to take into account traffic patterns), there are several reasons why the efficient recovery of common costs is unlikely to be proportional to incremental costs.

First, the inclusion of common-cost loadings has different competitive implications for different services, and the optimal degree of markup to cover common

---

26. See Armstrong (1996), Katz et al. (1995), and Laffont, Rey, and Tirole (1996a, 1996b) for additional discussions of these effects.

27. *First Report and Order* at paragraph 694.

28. Interestingly, to date the Commission has resisted applying this principle to enhanced service providers or even suggesting that it should apply to them. It is difficult to think of a plausible economic rationale for why Internet service providers and others should not cover at least the incremental costs of providing them telephone services.

costs should reflect the potential for harm to local exchange competition. Access charges are paid by long distance carriers to local exchange carriers. As long as the same charges are levied on all long distance carriers, the charges do not affect the competitive balance. In contrast, transport and termination charges are paid by one local exchange competitor to another. Thus, the charge level may well affect competition. Similarly, unbundled network elements may be purchased by carriers using those elements to compete with the incumbent LEC's services, again giving rise to competitive concerns from prices above incremental costs.

In addition to differences in the competitive concerns that their rates raise, services differ in the ways in which the burdens of their rates are distributed. Under the federal policy of geographic rate averaging, for example, the long-distance rates paid by end users do not reflect the particular access charges levied by their local exchange carriers.[29] Rather, rates are based on national averages. Hence, the burden of interstate access charges for one area of the country falls on consumers in all areas. The charges for local service, however, fall solely on subscribers in that local area. Policymakers may wish to take these effects into account.

The final—and probably the most important—reason why the efficient recovery of common costs is unlikely to be proportional to incremental costs is that different services have different demand elasticities and the efficient recovery of common costs takes these differences into account. This is the principle underlying so-called Ramsey pricing.[30]

The slogan that a minute is a minute is consistent with regulators' general reluctance to embrace Ramsey pricing. As noted earlier, in its August 1996 local competition order the Commission decreed that an incumbent LEC could include a "reasonable" allocation of its common costs in its rates for unbundled elements and transport and termination. The Commission went on to state that Ramsey pricing is not reasonable.[31] The Commission based its conclusion on the claim that the inverse elasticity rule would tend to stifle competition because those services that competitors really needed would tend to have inelastic demands. Although I have serious reservations about this argument (if the entrants' demands are so inelastic, it suggests that raising the price is not thwarting entry), the Commission is correct to be wary of the application of Ramsey pricing in this context.

As a theory of second-best monopoly pricing, Ramsey pricing is sound. Applied to actual markets, it suffers from two shortcomings. First, economists still have a poor understanding of how second-best pricing principles apply to multiproduct

---

29. 47 U.S.C. § 254(g).

30. See Ramsey (1927).

31. *First Report and Order* at paragraph 696. In an apparent contradiction of this statement, the rules that legally codify the Commission's decision state that the allocation of the costs of shared facilities is reasonable only if it is efficient (*First Report and Order*, Appendix B, amending C.F.R. Title 47, § 51.507). To an economist, this would appear to mandate the use of Ramsey pricing.

oligopolies, such as those that characterize telecommunications markets. With a monopoly, the market and firm-specific elasticities are the same thing. Moreover, there is no issue of distorting the allocation of production among carriers. With multiproduct oligopolies, market and firm-specific elasticities differ, raising both theoretical and practical issues. For example, second-best prices take market elasticities into account, but firms will respond to firm-specific elasticities. Thus, one can construct examples in which it is necessary for regulators to place price floors on competitive services in order to force firms to recover common costs from those services.[32] One needs a much greater understanding of these issues before advocating a policy of forcing competitive prices upward.

A second problem with the application of Ramsey principles arises from the difficulties in measuring demands, particularly for new services. In light of these difficulties, which are even greater for regulators than for firms, a policy that allows demand considerations to influence regulated prices may create scope for a regulated multiservice carrier to manipulate the margins on its services to discourage competition.

The problem of setting second-best prices in practice is made all the more difficult by the fact that the pursuit of efficiency may entail discouraging competition in some situations where entrants are undercutting prices that are being held above marginal costs to generate contribution toward the recovery of common costs. Unfortunately, it is impossible to set efficient prices without reference to demand conditions, yet regulators are poorly equipped to do so.

Universal Service Costs: "The Policy Serves the Public Interest"

Turn now to the question of how to recover universal service costs. There is an old joke that if you ask four economists a question, you will get five different answers. Some have observed a second, closely related feature of the world: If you do find a point on which all economists agree, no one else will agree with them.[33] Universal service reform appears to be one of those areas.

There is widespread agreement among economists that the current system—which relies on implicit cross-subsidies supported by the exercise of market power, and explicit subsidies raised from only a subset of providers and services and paid to only a subset of providers—both distorts competition and is threatened by it.[34] Universal service policy distorts competition when those firms who bear relatively low subsidy burdens or collect relatively high subsidy payments are able

---

32. This type of example arises when there is a service that has a very inelastic market demand (so it is efficient to raise its price to cover common costs), but any one carrier faces a highly elastic firm-specific demand curve for the service (so that competition will drive price toward incremental cost).

33. Prior to the rise of strategic trade theory, the optimality of free trade was one such point.

to attract traffic despite not offering the best combination of service cost and quality. Competition threatens cross-subsidies that rely on the exercise of market power as a source of funds.

Contributions toward universal service support constitute a tax levied on telecommunications users and providers. There is an extensive literature on efficient taxation from which one can derive several broad principles:[35]

- *Have as broad a tax base as possible.* Additional policy instruments provide a greater range of options. Moreover, the efficiency loss of a tax on any one good or service generally increases more than proportionately with the tax rate. Hence, from an efficiency perspective, it is preferable to have a low tax rate on a broad base than a high tax rate on a narrow base.[36] This principle, plus the general desirability of coordination, leads to the conclusion that universal service funds should be collected as an integrated part of the overall federal tax system. Political considerations, rather than a concern with the public welfare, appear to have ruled this out.

- *Rely on lump-sum taxation to the extent feasible.* A pure lump-sum tax (or one that depends on taxpayer characteristics that are beyond his or her control) is efficient—the person on whom it is levied can do nothing to affect the amount, and thus there is no incentive for the taxpayer to distort his or her actions. A near-lump-sum tax would be a desirable way to raise revenues to support universal service subsidies. Standard consumer theory demonstrates that a lump-sum tax is optimal if all consumers have similar tastes.[37] Of course, subscribers are heterogeneous, and it necessary to go

---

34. There also is widespread agreement that the failure to target subsidy funds toward qualifying individuals is inefficient and undermines the effectiveness of the program. From the perspective of economic theory, the provision of vouchers to qualifying individuals is an obvious solution. Policymakers do not agree, perhaps because it would make the costs and benefits of the policy too explicit.

35. For a summary of the optimal taxation literature, see Atkinson and Stiglitz (1980) and Auerbach (1985).

36. The mathematical intuition is as follows. Under the assumption that the commodity is competitively supplied at constant marginal cost, the deadweight loss associated with tax rate t on a base of B is DWL $= \frac{1}{2} t^2 \eta B$, where η is the price elasticity of demand of the service being taxed. (Here, I am being a little loose with respect to the aggregation of multiple services. For a more careful treatment, see Auerbach, 1985.) Fix the amount collected by the tax at R_0, so that $tB = R_0$. Then, $t = R_0/B$, and the deadweight loss of the tax is $\frac{1}{2} R_0^2 \eta/B$, which is decreasing in B.

37. If all consumers are identical, then a two-part tariff with a marginal price equal to marginal cost and a lump-sum fee set to cover fixed costs maximizes net consumer benefits and, thus, subscription. Hence, in this setting, such marginal cost pricing is optimal even in the presence of network externalities (although not call externalities). With call externalities, it would be efficient to increase the fixed charge further and lower the marginal price to a level below marginal cost in order to stimulate additional calls.

beyond the simple theory. Empirical studies, however, also support the view that it is not sound policy to raise universal service subsidies by taxing toll services.[38]

- *Where prices are distorted by the need to raise contribution, the responsiveness' of supply and demand to price must be taken into account.* The potential benefits and costs of accounting for demand considerations are similar to those for the recovery of common costs and are not discussed further here.
- *Don't tax distort production without a good reason.* Taxes can distort both consumption and production decisions. Because consumers eventually will bear the burden of the tax, it is in some sense inevitable that consumption will be distorted. But it may still be possible—and it is desirable—to maintain production efficiency. The implication of this fact is that policymakers should not tax intermediate services (like interconnection and access) unless there is a specific objective that could not be realized by taxing retail services.
- *Don't distort competition.* If two firms can provide substitute services for one another, the means of raising contribution toward universal service should not create additional efficiency losses by distorting competition between those providers.

Similar considerations apply to embedded costs. In terms of sending the right economic signals to consumers and providers, forward-looking costs are the most appropriate basis for setting prices. Hence, embedded costs should not be included in the calculation of cost-causative prices. But there also is fairly broad agreement that it is important to honor regulatory contracts—even implicit ones—in order to preserve government credibility and maintain an economic environment that is favorable toward investment.[39]

To the extent that such costs are recovered through the prices charged for various interconnection services, they will, in the end, affect the prices paid by subscribers. If policymakers believe that these legacy costs need to be recovered, then as with universal service costs this recovery should be done through direct charges to end users, rather than distorting production decisions. This approach is explicit and, done properly, second-best efficient. There are no sound economic grounds for hiding the recovery of these costs from consumers by embedding them in the rates end users pay for other services.

---

38. See, for example, Hausman, Tardiff, and Belinfante (1993). Crandall and Waverman (1995) estimate the annual consumption efficiency gains from rate rebalancing that lowers toll and raises local charges to be $8 billion.

39. Where there is little or no agreement is whether there was any such commitment in the present situation. I do not address this issue here. Instead, solely for purposes of discussion, I assume that there are some embedded costs that it would be appropriate for policymakers to allow ILECs to recover.

WHERE DO WE GO FROM HERE?

There are those who believe that competitive markets will develop sufficiently quickly that there will be no need to update regulatory ceilings on the prices for unbundled elements and transport and termination. Indeed, at least in their public statements, many members of Congress indicated that this was their expectation. I believe this view is mistaken. I do not expect to see widespread, facilities-based local exchange competition within the next few years, particularly in rural areas and in the provision of local loops. Thus, the way in which regulatory price ceilings are updated will be an important issue.

If prices are set on a one-time basis and the firm is free to keep any costs savings, then it has efficient incentives to pursue those savings. But under this regime, consumers share none of the benefits and the resulting prices are too high from the perspective of consumption efficiency. On the other hand, if cost reductions trigger reductions in regulated prices, then the incentives to innovate may be attenuated.

There are several options available to regulators that could push prices down to efficient levels without weakening the incentives to reduce costs. The key to all of them is to avoid tying the adjustments in a regulated carrier's rates too closely to the performance of that firm. One approach is to set an industrywide rate at which regulated rates fall, as is done with the so-called X-factor, which lowers the large local exchange carriers' interstate price caps at a set annual rate that is adjusted by the Commission only periodically based on overall industry performance. A further step in this approach would be to develop a purified X-factor for each carrier based on the performance of other local exchange providers, perhaps subject to adjustments for certain differences in topography and other factors thought to affect costs systematically. Alternatively, to the extent that some local exchange markets become competitive, it may be possible to use information generated in those markets to guide the regulation of prices in other markets. Of course, one must be alert to the possibility that this use of prices will distort their levels. But again, these misincentives can be eliminated by basing regulatory price ceilings on prices generated in competitive markets in which the regulated firm is not a major player.

APPENDIX A: BARGAINING POWER AND ENTRY

The following example illustrates the effects that the division of bargaining power can have on entry. Let $\pi(k)$ denote per-firm profits when there are k active LECs. Assume that $\pi(1) > 2\pi(2) > 0 > \pi(3)$. Suppose there are one incumbent and two potential entrants. Moreover, suppose that, absent regulation, the incumbent can refuse to offer interconnection to the potential entrants. Thus, assuming that interconnection is vital to an entrant's survival, the incumbent can insure itself profits of $\pi(1)$ by refusing to negotiate. The incumbent does not need to pay the entrant to keep it out, and given the assumption that entry lowers industry profits, the entrant will not be willing to buy its way into the market.

Now, suppose that, as the result of regulation, the potential entrant has sufficiently strong bargaining power that it can force an interconnection agreement yielding it profits of $\phi 2\pi(2)$, where ϕ is some constant between zero and one. Under this regime, the incumbent earns $(1 - \phi)2\pi(2)$ if entry occurs. There are, however, private gains to trade of $\pi(1) - 2\pi(2) > 0$ from stopping entry. Suppose that the incumbent offers payment D to deter the entrant from coming in. If this offer is accepted and entry is deterred, the resulting payoffs are $\pi(1) - D$ and D for the incumbent and potential entrant, respectively. The Nash bargaining outcome is the value of D that maximizes the product of the gains from trade, $[\pi(1) - D - (1 - \phi)2\pi(2)][D - \phi 2\pi(2)]$. The Nash bargaining solution thus is $D = \pi(1)/2 + (2\phi - 1)\pi(2)$. Note that D is increasing in ϕ and $\pi(1) - D > (1 - \phi)2\pi(2) > 0$ for all $\phi < 1$.

It appears that the incumbent will pay off the potential entrant, antitrust concerns aside. Recall, however, that there are multiple potential entrants. Each can ask for this amount after the other has agreed not to enter. If the incumbent has to pay D to each, it must compare $\pi(1) - 2D$ with $(1 - \phi)2\pi(2)$; $\pi(1) - 2D - (1 - \phi)2\pi(2) = -\phi 2\pi(2) < 0$. Thus, the incumbent will not find it profitable to deter entry.

Although there are many ways to model the bargaining between the incumbent and potential entrants, the insights of this simple model should hold more generally. If the incumbent does not have to offer interconnection, then it will not allow industry-profit-reducing entry no matter what the number of entrants. When the incumbent is required to offer interconnection on "reasonable" terms, it becomes much more unlikely that the incumbent will be able to block entry.

APPENDIX B: TSLRIC V. TELRIC

In its August 1996 decision, the Commission makes much of the distinction between TELRIC and TSLRIC, apparently on the belief that the problem of common cost allocation will be much less under a TELRIC regime. Is TELRIC an improvement over TSLRIC? In large part, the answer depends on what a TELRIC regime is, and at this point that is far from clear. The following simple model explores these issues under one interpretation of the Commission's decision.

Suppose that a local exchange carrier's total costs take the following form:

$$C = N + \Sigma_j S_j + \Sigma_i F_i + \Sigma_j \Sigma_i m_{ij} x_j$$

where N is corporate overhead, S_j is the setup cost associated with service j, F_i is the setup cost associated with element i, x_j is the number of minutes of traffic for service j, and m_{ij} is the traffic-sensitive cost triggered by a minute of traffic of service j over element i.

For such a firm, the TSLRIC of service j is

$$\mathrm{TSLRIC}_j = S_j + x_j \Sigma_i\, m_{ij}$$

Pricing each service to cover its TSLRIC (i.e., setting the price of service j equal to $\Sigma_i m_{ij} + S_j/x_j$) would lead to a shortfall of unrecovered common costs equal to $N + \Sigma_i F_i$.

The TELRIC of element i is

$$\mathrm{TELRIC}_i = F_i + \Sigma_j\, m_{ij} x_j$$

Pricing each element at TELRIC would lead to an unrecovered common cost shortfall equal to $N + \Sigma_j S_j$.

Either approach leaves a pool of common costs that must be allocated in some other way. However, one might argue that the so-called service setup costs should properly be interpreted as the costs of single-service elements. In this case, the S_j terms in the preceding notation all would be equal to zero and the setup costs entirely captured in the F_i terms. Hence, TELRIC pricing would recover all costs except corporate overhead.

So far, TELRIC looks pretty good. Two points are worth making. First, notice how the common cost problem is "solved." To the extent that end users purchase services, not elements, the retail provider has to determine how to recover the element setup costs, the F_i terms. In effect, the common cost problem has been pushed to the retail service providers. In some ways, this is a good thing because firms can be expected to take demand-side conditions into account in ways that regulators are unwilling to do.

The second point to note is that the analysis to this point assumes that a single firm buys each element in its entirety. This may be a sensible interpretation for local loop, but it is not a good model of local switching, where a number of different carriers will be purchasing services generated by a common facility (element). My reading of the Commission's decision is that TELRIC pricing for such elements means something like

$$p_{ij} = m_{ij} + F_i/\Sigma_j\, x_j$$

where the different services may be provided by different carriers.

Although such prices will recover costs, there is no theorem stating that it generally is efficient for each element to cover its own common cost and to do so through a uniform per-minute allocation of common costs among users. To the extent that the element setup costs are significant for important shared elements, TELRIC will mask the problem of cost allocation, but will not solve it. Ultimately, the extent of the problem is an empirical question. The jury is still out on whether

TELRIC will greatly reduce the problem of allocating common costs in pricing unbundled elements and transport and termination.

REFERENCES

Armstrong, M. (1996). *Network interconnection.* Unpublished manuscript.
Atkinson, A. B., & Stiglitz, J. E. (1980). *Lectures on public economics.* New York: McGraw-Hill.
Auerbach, A. (1985). The theory of excess burden and optimal taxation. In A. J. Auerbach & M. Feldstein (Eds.), *Handbook of public economics,* Vol. 1, pp. 61–128), Amsterdam: North Holland.
Brock, G. (1994). *Telecommunications policy for the information age.* Cambridge, MA: Harvard University Press.
Crandall, R. W., & Waverman, L. (1995). *Talk is cheap: The promise of regulatory reform in North American telecommunications.* Washington, DC: Brookings Institution.
Economides, N., & White, L. (1995). Access and interconnection pricing: How efficient is the "Efficient Component Pricing Rule"? *Antitrust Bulletin, 40,* 557–579.
Hausman, J., Tardiff, T., & Belinfante, A. (1993). The effects of the breakup of AT&T on telephone penetration in the United States. *American Economic Review, 83,* 178–184.
Katz, M., Rosston, G., & Anspacher, J. (1995). Interconnecting interoperable systems: The regulator's perspective. *Information, Infrastructure and Policy, 4,* 327–342.
Laffont, J. J., & Tirole, J. (1994a). *Creating competition through interconnection: Theory and practice,* Unpublished manuscript.
Laffont, J. J., & Tirole, J. (1994b). Access pricing and competition. *European Economic Review, 38,* 1673–1710.
Laffont, J. J., Rey, P., & Tirole, J. (1996a). Network competition: I. Overview and nondiscriminatory pricing. Unpublished manuscript.
Laffont, J. J., Rey, P., & Tirole, J. (1996b). Network competition: II. Price discrimination. Unpublished manuscript.
Ramsey, F. P. (1927). A contribution to the theory of taxation. *Economic Journal, 37,* 47–61.

Telecommunications Pricing and Competition

John Haring
Strategic Policy Research, Inc.

Jeffrey H. Rohlfs[1]
Strategic Policy Research, Inc.

There is a broad policy consensus that self-policing competition is preferable to regulated monopoly for economic governance of most markets.[2] Recent legislative and regulatory actions indicate that this consensus now extends to the local telecommunications industry. This consensus diverges sharply from historical practice in this economic sector, where thoroughgoing regulation and monopoly supply have been the rule rather than the exception.

Under competition, fear of losing sales motivates suppliers to discover and meet consumer demands effectively. Actual (and credible) potential supply alternatives make competition manifest and the fear of losing sales palpable. Legislative and

1. John Haring and Jeff Rohlfs are principals in Strategic Policy Research, Inc., an economics and telecommunications policy consulting firm located in Bethesda, MD. Dr. Haring formerly served as Chief Economist and Chief, Office of Plans and Policy, at the Federal Communications Commission. Dr. Rohlfs was formerly Department Head of Economic Modeling Research at Bell Labs. Strategic Policy Research served as a consultant to Bell Atlantic on related matters. The views expressed in this chapter are those of the authors alone and should not be attributed to any other individual or organization.

2. The preference for competition presumes that self-policing competition is feasible. For some goods, economies of scale or scope in production make effective competition infeasible. There is also a class of goods, so-called "public goods," whose efficient supply is not guaranteed by an effective competitive process. The technical characteristics of public goods—nonexcludability and nonrivalry in consumption—imply a potential tendency towards under supply under competitive market organization.

regulatory efforts have thus naturally and properly focused on making competition a reality (and competitive fear as real as possible). They seek to do so by lowering barriers to entry and enabling the efficient interconnection of new networks.

Issues involving the technical integration of new telecommunications supply capabilities are important, but they are not the whole story. The economic integration of new supply capabilities also presents a formidable challenge to policymakers. While issues of technical integration have received considerable attention in recent years,[3] issues of economic integration have been approached more gingerly. They have been resolved usually only on a rough-and-ready, interim basis (to the extent they have been resolved at all).[4]

The need for technical integration is self-evident: Interoperability of different systems requires effective technical integration of disparate supply capabilities. In the absence of technical integration, the phones (facsimile machines, modems, etc.) literally will not work. The need for economic integration is perhaps less obvious, but no less important. Without efficient economic integration, achieved through efficient pricing, the fruits of competition will be limited. Moreover, they will likely be distributed in ways that policymakers find objectionable and unjust. In this chapter we focus on the issues of economic integration posed by the shift to a competitive paradigm in the supply of local telecommunications services. We believe effective economic integration depends importantly on economically efficient pricing to both end users and competitors (including incumbents).

The challenges of economic integration arise for two sets of reasons:

1. Government regulation has historically pursued a variety of objectives apart from (and to some extent inconsistent with) economically efficient pricing of telecommunications services. As a result, regulation has created powerful economic incentives for competition to react to economically inefficient pricing practices (e.g., prices based on geographically averaged or fully allocated costs);

2. Effective delivery of service within a "network of networks" requires coordinated behavior among a large number of suppliers. In this context, suppliers do not exist (solely) in a competitive relation to one another. In economic terms, their relationship with one another is "vertical" in addition to "horizontal." Reconciling these vertical relationships with competitive rivalry poses a formidable problem for the responsible governmental authorities to address.[5]

---

3. Consider, for example, the FCC's focus on matters involving various conditions of network interconnection and collocation.

4. Issues of economic integration are addressed from a different perspective by M. Katz, G. Rosston, and J. Anspacher (1995). Interconnecting interoperable systems: The regulators' perspective, *Information Infrastructure and Policy 4*, 327–342.

5. Indeed, government intervention could itself create additional inefficiencies and produce opportunities for profitmaking arbitrage.

Problems of economic integration arising from the historical legacy of uneconomic pricing of telecommunications services under regulation and incentives for opportunistic behavior under conditions of multilateral monopoly do not have any easy solutions. Although telecommunications regulation was historically premised on control of monopoly, actual regulatory intervention sought to achieve a variety of objectives either unrelated or only indirectly related to control of monopoly power. The political pressures and concerns that motivated this regulation remain largely unmitigated by the onset of competition. Indeed, competition is likely to expose the tensions among the various inconsistent positions. Moreover, the governance changes demanded by the shift to a competitive paradigm may themselves present opportunities for efforts to extract additional private advantages from regulation.

There is already some experience (i.e., in international and domestic interexchange telecommunications) with economic integration of different carriers' supply capabilities for call completion. That experience provides little basis for optimism that more pervasive demands for economic integration can be effectively met. Current economic governance of international telecommunications is far from optimal. It allows countries to pursue policies that extract payments significantly above costs for the completion of international calls.[6] How can policymakers prevent this suboptimal equilibrium from occurring in the brave new world of competitive local telecommunications? This concern is more than theoretical. Long-distance carriers have already begun to complain about the supracompetitive levels of access prices charged by new entrants.[7] And long-distance calling itself has historically been burdened with various non-cost-based charges.[8]

CHAPTER OUTLINE

In this chapter we develop a framework for evaluating local interconnection pricing. We apply economic analysis to the particular circumstances of local interconnection and examine several alternatives for governance of interconnection pricing.

In all cases, we assume that the prices charged by entrants are not regulated. It is fair to say that no regulator is especially keen to undertake cost-based

---

6. See, J. Haring, H. M. Shooshan III and J. H. Rohlfs (1993). *The U.S. stake in competitive global telecommunications services: The economic case for tough bargaining.* Prepared for AT&T, December 16.

7. See, for example, oral testimony of Joe Gillan before the Public Service Commission of South Carolina in Docket No. 95-720-C, September 1995.

8. Even though these loadings and charges were often motivated by worthy objectives, they are nonetheless uneconomic. A relevant question is whether they are the least-cost means of achieving the objectives sought. In the competitive future, the answer to that question is almost surely, "No."

regulation of all (new) competitors in local telephony. There are, moreover, widely accepted general arguments against cost-based regulation (viz., it promotes cost inflation).

At the same time, we assume that regulation applies to the incumbent's retail prices, its call-termination charges, and what it must pay entrants for call termination. Alternatives for the incumbent's interconnection pricing include "bill and keep" (where a competitor's traffic is terminated without charge), symmetrical (or reciprocal) interconnection (under which the entrant and the incumbent local exchange carrier [LEC] have equal charges), and various cost-based regimes (in which charges are based on costs—either those incurred or those avoided). The regime adopted for interconnection charging is important because it will affect the prices different customers pay for calling and the economic efficiency with which scarce productive resources are deployed.

Our conclusion is that first best public policy would be to allow incumbent carriers to rebalance rates toward incremental costs and to deaverage rates to reflect cost differences. Such rebalancing and deaveraging would improve economic efficiency, in addition to making competition work better. Rebalancing and deaveraging should be implemented in both state and interstate jurisdictions. After they have been implemented—but only then—pricing interconnection slightly above incremental cost is reasonable. For competing local exchange carriers (CLECs), a symmetrical set of charges would then be an administratively simple scheme for reciprocal compensation of activities the LEC no longer performs when a competitive service is utilized. When charges are set to reflect costs, CLEC charges would thus be set to reflect LEC cost savings.

On the other hand, state or federal regulators may insist on maintaining an inefficient rate structure. Such pricing will become increasingly problematic as competition intensifies. Under these circumstances, pricing interconnection slightly above incremental cost invites inefficient arbitrage and (further) economic waste. In this world of second best, the Efficient Component Pricing Rule (ECPR) may be reasonable. It does not represent a the first best solution, but it does help avoid inefficient arbitrage. Although the ECPR has been subjected to several analytically valid criticisms, we do not regard these criticisms as necessarily debilitating. In particular, when like services are being supplied, the ECPR can be defended as less imperfect than other proposed approaches, including the Federal Communications Commission's (FCC) proposed pricing regime for competitive interconnection.[9]

---

9. Federal Communications Commission (1996). In the matter of implementation of the local competition provisions in the telecommunications act of 1996; interconnection between local exchange carriers and commercial mobile radio service providers. CC Docket Nos. 96-98 and 95-185, *First Report and Order*, adopted August 1, released August 8.

REGULATION FOR EFFICIENT COMPETITION

In a competitive regime, the principal focus of regulation should be to develop and enforce a set of competitive rules that allows the forces of competition to work effectively to maximize economic welfare. Such rules can be characterized as "promoting efficient competition."

"Efficient competition" is a shorthand term for competitive activity that improves the overall efficiency with which an industry or the wider economy operates. We believe that regulators should implement competitive rules that provide maximum scope for efficient competition. Rules that do not go far enough may forfeit some or all of the potential gains from competition. Rules that go too far and promote entry by inefficient competitors or increase costs of production will reduce economic welfare.

LOCAL INTERCONNECTION

Competitive entry will displace some current output of incumbent LECs. It will thereby reduce the need for inputs (e.g., facilities and labor) used to produce the displaced output. LECs will, however, be called on to provide entrants with interconnection services that were not previously provided. In this context, efficiency depends on which configuration of supply is the most efficient. Is it more efficient for the LEC to provide end-to-end service? Or is it more efficient for a CLEC to provide service to the customer, with the incumbent providing interconnection services to the entrant?

Efficiency naturally depends on the quality and quantity of services produced. It also depends on relative costs. In the latter context, the relevant comparison is the CLEC costs versus the LEC cost savings (reduction in overall operating costs) from providing interconnection to CLECs instead of end-to-end service. If entry does not affect the quality or quantity of services provided to end users, the LEC cost savings are a measure of the added value provided by the CLEC. Under these circumstances, efficiency depends on whether this added value exceeds the CLEC costs. In particular:

- If the CLEC costs are less than the LEC cost savings, then entry improves efficiency.
- If the CLEC costs exceed the LEC cost savings, then entry worsens efficiency.

For purposes of this analysis, we assume that quality and quantity do not change. This seems a reasonable first approximation for wireline CLECs. Such competitors generally offer services that are usually close (although not perfect) substitutes for LEC services. Furthermore, quantity of output is unlikely to change much as a result of wireline competition, because demand for local calling is typically estimated as fairly inelastic.[10]

Wireless competitors offer differentiated services that embody portability. We are also aware of instances where new entrants are deploying advanced wireline service capabilities (primarily to large business customers) more rapidly than incumbent LECs. Obviously, to the extent that there are significant quality and quantity changes, their effects need to be factored in evaluating consequences for economic efficiency. In general, effects of cost (and perhaps quality) variations are of a larger order of magnitude than quantity effects. The effects of entry on costs of production may vary and need to be evaluated on a net basis. Our focus is on the productive efficiency of call processing and completion in a competitive regime.[11]

LEC Cost Savings

In this section we examine LEC cost savings resulting from facilities-based CLEC entry. We consider cost savings associated with provision of loops and central-central office connections, as well as handling of calls.

Loops. The LEC does not need to provide loops to CLEC customers. Hence, the entire cost of previously supplied loops is saved. The costs savings may be moderate in the short run, if the saved loops are stranded.[12] In the long run, however, cost savings are likely to be significant. The savings include capital costs and operating costs such as maintenance and testing. Virtually all the cost savings are non-traffic sensitive (NTS). That is, they vary with the number of loops but do not depend on usage.

Central Office Connections. The LEC is also likely to save costs on connections to its switches. The CLEC is likely to connect broadband digital facilities to the trunk side of LEC switches. A single T-3 connection by a CLEC may thereby replace hundreds of analog-loop connections by end users. The LEC switch can therefore have fewer ports. Further costs are saved with respect to the main distributing frame, which physically connects switch ports to outside-plant facilities.

---

10. See, L. D. Taylor (1993). *Telecommunications demand in theory and practice.* Kluwer. Dordrecht, the Netherlands.

11. Issues of substitutability assume special importance and significance in a world in which regulators have chosen to depart from economically efficient charges.

12. Loop plant represents a sunk investment, that is, one with little salvage value. The cost of loop plant is incurred when the plant is deployed and cannot be avoided if the plant is subsequently not utilized. In the long run, additional cost savings occur as investments to replace loop plant are no longer needed.

These cost savings include both capital and operating costs. They are primarily NTS, even though they apply to central offices.[13]

Handling Calls. The LEC cost savings associated with handling calls depend on the nature of the call. One possibility is that the caller and the called party are both served by the same LEC central office. A substantial fraction of all calls (perhaps one-third) are of this type. This possibility is illustrated in Fig. 3.1.

Figure 3.1
Caller and Called Party Served by Same LEC Central Office

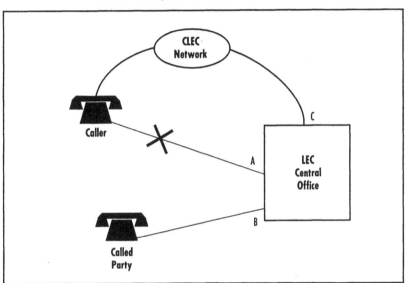

Prior to competitive entry, the LEC switched the call from line-side Port A to lineside Port B. After competitive entry, the caller no longer has a loop to Port A. Instead, the caller connects to the CLEC network, which in turn connects to the LEC network at trunkside Port C. The LEC saves loop costs. It saves additional costs to the extent that broadband digital CLEC connection is more efficient than analog loop connections. However, the LEC enjoys little or no cost savings in terms of call handling. The LEC must still switch the call. The only difference is that the call comes through a different port on the same switch.

CLEC entry, as illustrated in Fig. 3.1, does not reduce the LEC resources utilized to handle calls. That fact has important implications for interconnection pricing. To pro-

---

13. It is well understood that much of the cost of central offices is NTS. Indeed, prior to the AT&T divestiture, the Joint Board determined that 75% of the costs of digital switches are NTS. See, also B. M. Mitchell (1990). *Incremental costs of telephone access and local use*. Santa Monica, CA: RAND Corporation.

mote efficient competition, compensation arrangements for reciprocal interconnection should recover the cost of switching this call. They should not provide artificial incentives for CLECs to handle the type of call depicted in Fig. 3.1. Such incentives would induce the CLEC to incur costs to handle calls, whereas the LEC does not save costs—an increase in total costs (LEC plus CLEC) and likely loss of economic efficiency. On the other hand, compensation arrangements should provide incentives for CLECs to engage in welfare-enhancing activities. These activities may include the provision of loops, the utilization of low-cost central-office connections, and the handling of other types of calls (than depicted in Fig. 3.1), as discussed later.

Figure 3.2
CLEC Serves a Limited Geographic Area

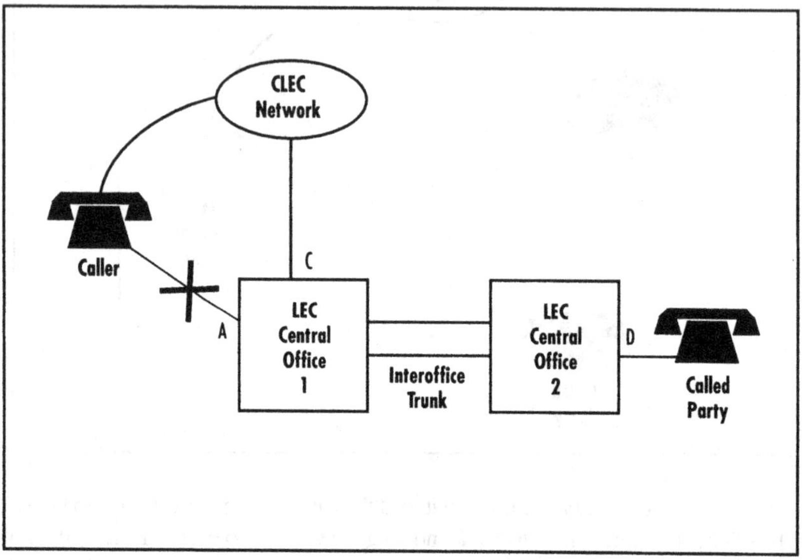

Figure 3.2 illustrates another type of local calling: The caller and the called party are served by different LEC central offices. However, the CLEC serves only a limited geographic area. Prior to competitive entry, the call came through line-side Port A of Central Office 1. The LEC switches the call to an interoffice trunk and transports the call to Central Office 2. There, the call is switched from the interoffice trunk to line-side Port D. After competitive entry, the caller no longer has a loop to Port A. Instead, the caller connects to the CLEC network. The CLEC in this example has limited geographic scope. It therefore delivers the call to trunk-side Port C at LEC Central Office 1. The LEC switches the call from Port C to an interoffice trunk. It then transports the call to Central Office 2 and switches it to Port D.

In this example, as before, the LEC enjoys little or no cost savings in terms of call handling. The LEC must still switch the call twice and transport it between central offices. The only difference is that the call comes through a different port on the same switch. Also, as before, the CLEC provides little or no value in terms of call handling.[14] Interconnection prices should reflect the costs incurred by the LEC.

Figure 3.3 illustrates a type of call wherein the CLEC's offering does significantly reduce the resources the LEC needs to deploy. In this figure, the caller and the called party are served by different LEC central offices, and the CLEC serves a large geographic area. Prior to competitive entry, the call proceeded exactly as described in Fig. 3.2. After competitive entry, the caller no longer has a loop to Port A. Instead, the caller connects to the CLEC network. The CLEC in Fig. 3.3 has large geographic scope and delivers the call to trunk-side Port E at LEC Central Office 2. The LEC switches the call from Port E to Port D.

In this example, the LEC has substantial cost savings associated with call handling. It needs to switch the call only once, instead of twice, as before competition. It does not need to transport the call from Central Office 1 to Central Office 2. These cost savings are over and above any savings in loop costs and connection costs. Interconnection pricing should be based on the cost incurred by the incumbent LEC (ILEC) which will provide incentives for CLECs to handle the type of call depicted in Fig. 3.3.

Figure 3.3
CLEC Serves a Large Geographic Area

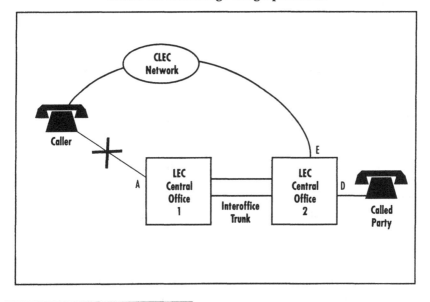

---

14. The CLEC may well be providing value elsewhere—for example, contract terms, complementary services, and so on. The CLEC's customers may be willing to pay significant amounts for such value. We focus here only on call-processing efficiency, which is the proper focus of traffic-sensitive interconnection charges.

We also need to consider the possibility that a call originates on the LEC network and terminates on the CLEC network. In this case, it turns out that the LEC's cost savings are virtually identical to those in our previous discussion. One need only reverse the direction of the call in Fig. 3.1, 3.2 and 3.3. We get the same result. There is little, if any, cost saving if the caller and called party are served by the same LEC central office. There is little, if any, cost saving if the CLEC has limited geographic scope. The cost savings are, however, substantial if the CLEC transports the call from the area served by one LEC switch to the area served by another LEC switch.

As before, interconnection pricing should provide accurate, cost-based price signals for CLECs to handle the types of calls depicted in Fig. 3.1 and 3.2 with reverse directions. It should provide incentives for CLECs to handle the type of call depicted in Fig. 3.3 (with reverse direction).

LEC INTERCONNECTION PRICING

In this section, we consider a variety of approaches for call-termination charging: bill-and-keep, symmetrical (or reciprocal) pricing schemes, pricing interconnection slightly above incremental cost, and the ECPR.

Bill-and-Keep

Under bill-and-keep, each carrier has an obligation to complete, at no charge, calls that originate on other carriers' networks. Each carrier keeps all the revenues that it collects from its own customers—both usage charges and fixed monthly charges.

The regulator can administer bill-and-keep without setting any rates or estimating any costs. Bill-and-keep is therefore administratively simple, though it may provide incorrect price signals. Furthermore, the administrative ease is largely illusory. Difficult pricing issues will still need to be resolved in other interconnection proceedings that will entail significant administrative costs.

The rationale for bill-and-keep is that each carrier contributes value by terminating other carriers' calls on its network. In exchange, each carrier enjoys the benefit of terminating calls on other networks for free. But this rationale breaks down if a carrier actually contributes little added value in terms of call handling.

Symmetrical Pricing

Under symmetrical (or reciprocal) pricing, the LEC and the CLEC each charge the same price for terminating each other's calls. Bill-and-keep is thus just a special case of symmetrical pricing — with the price set equal to zero. Symmetrical pricing does not require balance between originating and terminating traffic. As with bill-and-keep, symmetrical pricing may afford little incentive for the CLEC to expand its facilities. Symmetrical pricing thus has the same fundamental problem as bill-and-keep. The CLEC may receive rewards, but not add corresponding value in terms of handling calls.

The disabilities of symmetrical pricing are much reduced if the policy is construed as "equal pay for equal work." Under that pricing structure, there could be separate prices for each switching occurrence and for interoffice transmission. CLECs would therefore pay considerably less to the LEC to terminate the type of call depicted in Fig. 3.3 than the type of call depicted in Fig. 3.2. CLECs would have incentives to carry calls across town themselves, rather than drop each call off at the nearest LEC central office.

Even so, symmetrical pricing does not lead to efficient outcomes. If a CLEC handles a call, the call may need to be switched more times than if the LEC handled it. For example, in Fig. 3.1 the LEC can handle the call by switching it only once. In contrast, the call must be switched at least twice if it is handled by a CLEC. As previously noted, CLEC entry in this example does not reduce the LEC resources utilized to handle calls. Symmetrical pricing, even with equal pay for equal work, provides an artificial incentive for a CLEC to handle calls of this type. The CLEC is paid for each switching occurrence, regardless of whether the switching occurrence could be avoided by having another carrier (viz., the LEC) handle the call. More generally, the CLEC has incentives to handle calls, even if doing so increases total costs—LEC plus CLEC. The likely consequence of affording such incentives is that total industry costs will, indeed, increase unnecessarily.

Consider a specific example. Suppose that a Bell Atlantic customer in Bethesda, MD, calls a nearby MFS Communications Company Inc. (MFS) (i.e., a wireline CLEC) customer in Bethesda, MD. In this case, Bell Atlantic saves little, if any, costs, relative to the alternative of (Bell Atlantic's) delivering the call to the called party. Unless MFS provides some special value to the Bell Atlantic customer (over and above what Bell Atlantic would provide by completing the call), MFS's compensation should be minimal.[15] Under these circumstances, the compensation should not be based on the costs that MFS may incur by hauling the call to its switch in Virginia and then back to Maryland.

The disabilities of symmetrical pricing are ameliorated if the incumbent's prices for call termination are based on costs (and the incumbent has flexibility in setting retail rates, as discussed later). In that case, the call termination charge may be an adequate (if not perfect) proxy for the incumbent's cost savings.[16]

---

15. If MFS provides special value to its own customers, it can charge those customers higher rates, apart from the compensation scheme. Such pricing would be economically efficient, because the party that benefits from the special value (MFS's customer) pays the cost of providing it.

16. A still better approach may be to deploy symmetrical pricing, together with considerable regulatory flexibility with regard to LEC interconnection pricing. This idea (*inter alia*) is explored in detail in J. Haring and J. H. Rohlfs (1997). *Efficient competition in local telecommunications without excessive regulation.*

Pricing Interconnection Slightly Above Incremental Cost

Entrants sometimes argue that the incumbent should price interconnection slightly above long-run incremental cost (LRIC). In this section, we examine that proposal. It can be applied to the LEC's connection charges and its call-termination charges. We find that the viability of pricing interconnection slightly above LRIC depends on the incumbent's freedom to rebalance and geographically deaverage rates to end users. In particular:

- Pricing interconnection slightly above LRIC may be reasonable if the incumbent is allowed to rebalance and deaverage end-user prices—as long as rates remain significantly above incremental costs.
- Otherwise, pricing interconnection slightly above LRIC may encourage inefficient competition (i.e., competition that increases costs without compensating advantages) and entail a regulatory giveaway.

Freedom to Rebalance and Deaverage Rates. Competition is naturally attracted to markets where prices far exceed costs. Incumbent carriers would likely rebalance rates so as to be more competitive if they were allowed to do so. They would lower rates where they faced elastic demand and simultaneously raise rates in other markets. The incumbent would likely raise residential rates and lower business rates and also deaverage local rates to some extent to reflect geographic differences in costs.

The rate increases would presumably be limited by an overall price cap or revenue requirement. In addition, price increases could be limited by band constraints to avoid rate shock. Price decreases might be limited by a requirement that the new prices significantly exceed LRIC.

Appropriate rebalancing and deaveraging rates will improve economic efficiency and are desirable in their own right. Furthermore, rebalancing and deaveraging are necessary in order to promote efficient competition. After rates are rebalanced and deaveraged, a competitor will be able to enter profitably if it is more efficient than the incumbent (as long as the incumbent's rates for competitive services remain at or above LRIC). However, an inefficient competitor will have greater difficulty operating profitably.

Regardless, the proper price signal is sent. The entrant is encouraged to enter where it is more efficient than the incumbent and discouraged where it is less efficient. In both cases, output ends up being produced in the most efficient manner.

No Freedom to Rebalance and Deaverage Rates. We now consider the possibility that the incumbent is not free to rebalance rates. Regulators may insist that the current rate structure be maintained, even after the advent of competition. In that case, pricing interconnection slightly above LRIC invites inefficient competition where markups are high. It transfers substantial wealth from customers and/or shareholders of the LEC to customers and stockholders of the entrant. This combination of policies—no freedom to rebalance/geographically deaverage rates and pricing interconnection slightly above LRIC—cannot reasonably be regarded as sound public policy.

The Efficient Component Pricing Rule

The Efficient Component Pricing Rule (ECPR) was developed by W. J. Baumol and R. D. Willig on behalf of New Zealand Telecom for its litigation with Clear Communications. Baumol and Sidak subsequently elaborated and refined the analysis.[17] The ECPR is a guide for evaluating the economic reasonability of charges for essential inputs—either connection or call-termination charges.

Under the ECPR, interconnection prices consist of two components: the incremental cost of interconnection, and the contribution that the incumbent loses (in end-user markets) as a result of selling interconnection services. Baumol and co-workers have shown that, under idealized assumptions, use of the ECPR encourages efficient entry and discourages inefficient entry. The idealized assumptions do not necessarily obtain in the real world.[18] Even so, prices based on the ECPR have desirable properties and may be reasonable. In particular:

- If entrants produce the same services as the incumbent, the ECPR encourages entry if and only if the entrants' costs are lower than the incumbent's. The outcome is an industry structure in which the costs of production are minimized.
- Entrants and their customers appropriate the entire value of the entrants' efficiency improvement, relative to the incumbent. However, the incumbent and its customers are not forced to incur any additional costs or sacrifice profits.

For these reasons, the ECPR deserves serious consideration, especially if the LEC is afforded little or no freedom to rebalance and/or geographically deaverage rates.

COMMINGLING OF DIFFERENT TYPES OF TRAFFIC

As previously discussed, CLECs are likely to be full-service providers. They may be interexchange carriers (IXCs) and wireless carriers, as well as local wireline

---

17. See, W. J. Baumol and J. G. Sidak (1994). *Toward Competition in Local Telephony* (pp. 93–116). Cambridge, MA: MIT Press.

18. See, for example, J. J. Laffont and J. Tirole (1986). Using cost observations to regulate firms *Journal of Political Economics*, *614* 94; and N. Economides and L. J. White (1995). *Access and interconnection pricing: how efficient is the "efficient component pricing rule"?* EC-95-04, March. Baumol and Sidak (see note 17) acknowledged the limitations of the ECPR, when all the assumptions are not satisfied.

carriers. CLECs will therefore carry a mix of types of traffic. Suppose, for example, that AT&T delivers a call to a LEC for termination. AT&T may have handled the call in its capacity as an IXC, wireless carrier or CLEC. Today, LEC interconnection prices for these various types of usage differ substantially. Consequently, interconnection prices to CLECs must also differ, depending on the type of usage. Otherwise, users could avoid paying the highest charges (long-distance access) by using a CLEC as an intermediary. Such a policy would encourage inefficient entry and would ultimately collapse the LEC price structure.

Unfortunately, the policy of having different interconnection prices for different types of usage has its own problems. The LEC may not be able easily to distinguish among the different types of traffic, such as long-distance access, wireless access, or local usage. Consequently, CLECs would have to report the amounts of each type or usage. Their interconnection cost would then depend on their reporting. Under such a regime, CLECs have an obvious incentive to evade interconnection costs by misreporting the types of usage.[19] These incentives would be especially great for small CLECs, whose financial condition may be marginal (and the temptation to cheat on interconnection payments correspondingly strong).

The first best way to avoid these problems is to have a uniform retail charge for all these types of usage of the LEC network. For example, the uniform charge might be $0.02 per minute for residential usage, business usage, long-distance access, or wireless interconnection. Such uniform pricing can be justified, even apart from competitive considerations, because the LEC's costs of providing these various services are similar.[20]

With uniform retail pricing, the interconnection price could be the same for all types of usage without creating arbitrage opportunities. Consequently, the CLEC would not need to report type of usage, and there would be no problems of misreporting.

Suppose, however, that regulators insist on maintaining different interconnection prices for different types of usage—notwithstanding the resulting need for different interconnection prices. In this world of second best, effective procedures for monitoring and enforcement are critical elements of a sound interconnection policy. Metering the interconnection traffic of each CLEC would be necessary in order to monitor the accuracy of the CLEC's reporting. Thus, the use of bill-and-keep for local calls does not obviate the need to meter traffic.

CONCLUSION

The ground rules governing economic intergration of complementary supply capabilities in a "network of networks" will play a very important role in determin-

---

19. This issue already arises, and has been actively debated, with respect to percentage of interstate use (PIU) of facilities that carry both interstate and intrastate traffic.

20. One can, however, easily justify charging more for calls that are switched twice (with interoffice transmission) than for those that are switched once.

ing whether competition actually enhances productive efficiency and consumer welfare. In economics, it is accepted that "prices reflect value." When prices do not accurately signal the value of marginal resource productivity, distortions of economic activity are likely to occur; resource misallocation and inefficiency are the predicted consequences. Even when prices do fairly reflect value, there is a danger that arrangements for economic integration of complementary supply capabilities may engender inefficient resource deployments. Supply efforts must be efficiently coordinated to ensure beneficial results from competition in an operating environment where different productive capabilities exist in both a substitute and complementary relationship to one another.

Spectrum Sharing Without Licenses: Opportunities and Dangers

Durga P. Satapathy
Carnegie Mellon University

Jon M. Peha
Carnegie Mellon University

The primary method of spectrum allocation today is based on licenses granted by the Federal Communications Commission (FCC) that give recipients exclusive rights to spectrum for a limited duration. It has also been suggested that these temporary licenses be replaced with permanent deeds, like property (Pressler, 1996). This too is a spectrum management policy in which users are given exclusive rights to spectrum. Generally, those with exclusive rights to a given block of spectrum have exclusive access at all times.

There are wireless applications that cannot be supported efficiently under a system based on permanent exclusive access to spectrum, but would be well served with real-time access to spectrum, even if that spectrum is shared (Peha, 1994-1995). Such applications include mobile wireless applications that would desire the ability to access spectrum anywhere within a wide area, but require much smaller coverage at any given time, such as a mobile wireless local area network (LAN), or a wireless private branch exchange (PBX). Granting permanent exclusive access to such a device for any location at which it might ever be operated would be grossly inefficient. Other applications that are poorly served with exclusive access are those that need only sporadic access to spectrum and can tolerate widely varying access delays, like a wireless electronic mail service. These applications can share spectrum with minimal penalty. For any application, sharing spectrum is inherently more efficient (Peha, 1997; Salgado-Galicia, Sirbu, & Peha 1995, in press). For example, consider the case where eight PBXs are sharing enough spectrum to support 32 simultaneous calls. Calls arrive according to a

49

Poisson process, and their duration is exponentially distributed. If they share the spectrum, they can sustain a traffic load of 68.9% while blocking only 1% of the calls (Peha, 1997), where a call is blocked when no channels are free at the time the call is attempted. In contrast, if each of the eight PBXs is given exclusive access to four channels, the 1% blocking probability is not achievable with a load over 21.7% (Peha, 1997). Thus, sharing makes it possible to carry over three times as many calls. Finally, there are also novel and rapidly evolving applications and technologies for which experimentation is important. Such applications are poorly served in a system where access to spectrum involves long administrative delays.

Consequently, there is a need for shared spectrum that allows real-time access. Real-time sharing is made difficult by three problems. The first is that because devices do not have exclusive access, they may interfere with each other's transmissions. To deal with this mutual interference, a set of rules are required that dictate when, where, and how devices may transmit. The second problem is that, because shared spectrum with real-time access would be valuable for a wide variety of applications and devices, there is motivation to create bands supporting diverse applications rather than to create many different bands of shared spectrum. These applications may vary greatly in terms of average data rate, transmission duration, or even the technology used. Such variations make it difficult to enforce efficient utilization for all applications. The third problem is that because spectrum is shared, there is no inherent incentive to use the spectrum efficiently, which may result in a tragedy of the commons (Hardin, 1968; Hundt & Rosston, 1995; Peha, 1997). This problem made the citizens band radio service highly inefficient and undependable in crowded regions, where users wasted spectrum with high-power transmitters. This problem can occur if too many devices are deployed in shared spectrum, or if individual devices waste spectrum. In the former case, the problem is relatively easy to avoid by requiring a fee to be paid for each device deployed, and for a number of reasons, this is a good policy (Peha,1996). However, in the latter case, the problem is more difficult.

There are three possible approaches to offer real-time access for shared spectrum. One is unlicensed spectrum. This spectrum is under FCC control, and any device is allowed to use it—a public park for wireless devices. The other possible approaches are appropriate for either licensed spectrum or spectrum for which there are property rights. In these cases, the profit-driven license - holder or spectrum - owner would demand compensation for the use of spectrum, and there are two ways to do this. The first is usage-based pricing, where each device is charged a fee that depends on how much spectrum it uses (Noam, 1995). The second option is that a one-time fee be charged for each device deployed, independent of how much spectrum the device actually uses (Peha, 1997).

In a system with usage-based pricing, a centralized authority must monitor, control, and regulate usage. Devices must explicitly obtain permission from this centralized authority before transmitting, and where there is conflict, those that are willing to pay more will gain access. As in wired networks like the Internet (Wang, Peha, & Sirbu, 1997; Wang, Sirbu, & Peha, 1996), this approach has two advantages. First, because fees depend on the amount of spectrum resources consumed, there is an incentive to conserve spectrum. Thus usage-based pricing can prevent a tragedy of the commons. Second, when there is conflict, resources go to applications whose value exceeds the price. However, there are many complications in implementing usage-based pricing for real-time access to shared spectrum. One problem is that it is difficult to avoid mutual interference in a system with centralized control, unless each device has the ability to convey signal measurements to the central controller. Another problem is that supporting a diversity of devices increases the system complexity, because the central controller needs to support a variety of wireless communication interfaces to enable communication with different devices. Most importantly, in order to have mobile applications communicate with the access provider over a wireless link from anywhere in the access provider's region of coverage, a highly complex and massive infrastructure is needed. This raises the transaction costs so high that usage-based pricing for spectrum access is impractical.

The other two schemes for spectrum sharing, unlicensed spectrum and one-time deployment fees to a license-holder, are quite similar. Both employ decentralized control; that is; both require the devices to follow an etiquette to determine when, where; and how they may transmit. An example is the etiquette set forth by the FCC for unlicensed operation in the 2-GHz personal communication services (PCS) band (FCC, 1994; Steer, 1994). An etiquette solves the problem of mutual interference, and its possible that a well-designed etiquette can help alleviate a tragedy of the commons. The difference between the two schemes lies in the motivation of the entity in control: The FCC is motivated to serve the public good, and a license - holder is motivated by profit. Profit-seeking entities tend to be more efficient, but there are two dangers to be addressed. One is that, because revenues come from devices at deployment time, a profit-seeking entity has less incentive to protect the performance of devices that have already been deployed. It may therefore change the etiquette to favor new devices over those already deployed. The other danger is that a profit-seeking entity that controls a critical resource may engage in anticompetitive behavior. For example, it may exclude devices from some manufacturers in return for suitable compensation from their competitors, or it may overcharge. This danger may be mitigated through competition, but only if multiple identical bands of this type can be created and controlled by different profit-seeking entities. Besides these differences, the two schemes are essentially the same, so we discuss the problems of real-time sharing with distributed control in the context of unlicensed spectrum for the rest of the chapter.

Given the significant advantages and potential problems of real-time sharing in unlicensed spectrum, it is important to evaluate whether allocating additional unlicensed spectrum is justified. The opportunity costs of allocating spectrum for unlicensed use are high, as shown by the billions of dollars fetched by the recent PCS auctions (Pressler, 1996). The FCC has already allocated 30 MHz of unlicensed spectrum in the 2-GHz PCS band (1910–1930 MHz, 2390–2400 MHz), and is now considering additional allocations in the 5-GHz range (FCC, 1996) and at 59–64 GHz (Marcus, 1994, 1996). These allocations would not be justified if the utilization of unlicensed spectrum is likely to be highly inefficient. Although many spread spectrum devices have been developed for unlicensed use in the industry, science and medicine (ISM) bands (902–908 MHz, 2.4–2.48 GHz; and 5.725–5.85 GHz), it remains to be seen whether or not a diverse group of unlicensed systems can coexist efficiently, even with an etiquette. It is therefore important to evaluate whether the provisions present in the etiquette are sufficient to prevent the tragedy of the commons. It is equally important to evaluate whether these provisions are necessary, because an overly restrictive etiquette can both reduce the spectral efficiency and increase the cost of unlicensed devices. In this chapter, we demonstrate that there is a potential risk of the tragedy of the commons occurring in unlicensed spectrum, and suggest possible techniques to avoid the problem through modifications to the spectrum etiquette if required.

In the second section we discuss proposals for real-time access to shared spectrum with distributed control that have been previously suggested. The third section discusses the strategies that designers may adopt for unlicensed devices that might result in a tragedy of the commons. In the fourth section we demonstrate the potential risk of a tragedy of the commons through analysis and simulation. In the fifth section we suggest etiquette modifications that may be used to deal with a tragedy of the commons if required. Finally, we present our conclusions.

PREVIOUS PROPOSALS FOR REAL-TIME SHARING WITH DISTRIBUTED CONTROL

In this section we look at four previous proposals for real-time sharing with distributed control and examine their potential to avoid a tragedy of the commons. The most far-reaching plan is based on the premise that emerging technology will eliminate spectrum scarcity, thereby eliminating both the need for licenses for exclusive access, and the potential for a tragedy of the commons. Instead of getting licenses, users will access spectrum through frequency-nimble devices that allow them to find spectrum as needed without having a band exclusively reserved for them (Gilder, 1994). This approach is inappropriate, at least at this time, for two reasons. One is that the technology required for this plan to work is still too expensive (McGarty & Medard; 1994), and the other is that even though technical innovations are constantly increasing the effective availability of spectrum, demand for spectrum is also increasing rapidly. There are no signs that supply will surpass demand.

Spectrum sharing with real-time access is possible without such radical change. There are existing unlicensed bands within the current licensed system. The industrial, scientific, and medical (ISM) bands are the simplest, where unlicensed devices are allowed secondary access, and they have to share the bands with licensed devices that have primary access. Unlicensed devices must not cause harmful interference to licensed devices, and must accept any interference from licensed devices as well as other unlicensed devices. Unlicensed operation in this band is restricted to devices using spread spectrum modulation only, and the maximum transmission power of these devices is restricted to 1 W. The low signal energy density of spread spectrum modulation and the limit on power reduce interference to some extent, but devices are always at risk of unmitigated interference from licensed devices and unavoidable interference from unlicensed devices. There is obviously no scope for diversity, because only spread spectrum devices are allowed access. The only provision to deal with the risk of the tragedy of the commons is an upper limit on power, which is a solution with limitations. Although an upper limit prevents excessive use beyond the limit, it provides no incentive to use only what is necessary. Devices may therefore choose to always transmit at the maximum power to achieve a high signal-to-noise ratio (SNR), even if lesser power would be adequate. In this case the power limit is low. This solves some problems, but creates inefficiency for other reasons, because devices cannot transmit at greater power even when no interference or spectrum contention would result.

The other form of unlicensed band involves the use of an etiquette, as used in the 2-GHz PCS band. This etiquette (FCC, 1994) uses a "listen before talk" (LBT) approach, which requires devices to first sense the channel for a specified time and determine whether there is a transmission underway. If the received power is sufficiently low that they are unlikely to experience or cause interference, they can transmit. The LBT feature inherently provides much better protection from interference than a system of power limits where there is no sensing. The etiquette has no restriction on the technologies that devices may use except that they must follow the etiquette, and therefore it supports more diversity. (However, the etiquette has an inherent limitation that it cannot distinguish between applications of low value and of high value, and restricts all applications equally.) The etiquette also includes many provisions to improve spectrum efficiency. For example, under one such provision, devices requiring bandwidth less than 625 kHz in the isochronous band must search the band from left to right, which reduces inefficiencies due to spectrum fragmentation. A few features of the FCC etiquette were also designed to inhibit excessive use of spectrum, thereby reducing the potential for a tragedy of the commons. For instance, the etiquette encourages operation at low power by raising the threshold used to determine whether a channel is free or not for low power transmissions, thereby reducing the access delay. Features that inhibit excessive use of spectrum often complicate access strategy design, like the requirement that transmission in the isochronous band must cease if an acknowledgment has not been received within the last 30 sec. Whether the restrictions in the etiquette are necessary for efficient operation, or are sufficient for efficient operation, remains to be seen.

The issue of promoting efficient spectrum sharing with real-time access has been addressed using an approach similar to that of the unlicensed PCS band, but it does so in the more limited case of common carriers offering the same service using cellular infrastructure (Salgado-Galicia et al., 1995, 1997). As with the FCC etiquette for unlicensed PCS, all devices are required to follow a specific protocol in acquiring access to spectrum. In this case, the etiquette is based on dynamic channel allocation (DCA), through which channels are dynamically shared amongst the competing operators on a call by call basis. All devices search channels in the same order, and the first channel that meets the minimum carrier-to-interference (C/I) ratio is selected. This approach was found to offer far more efficient spectrum utilization than a traditional licensing approach based on exclusive access to spectrum, provided that all of the firms were willing to invest in equipment that would enhance spectral efficiency. Unfortunately, this sharing also led to a tragedy of the commons. The reason is that as channels saturate due to increased usage, cells should be made smaller to increase system capacity through frequency reuse. However, as the spectrum is shared, the operator investing in a new cell site bears all the costs, but the benefits are shared by all. Consequently, there is little incentive to increase effective capacity of a given spectrum block. The authors of this approach addressed this problem by limiting the number of transceivers deployed per broadcast tower. This helps to alleviate the risk of a tragedy of the commons in this particular scenario, but is not applicable for all types of devices, so another approach is needed in unlicensed spectrum supporting greater diversity. The fifth section of this chapter focuses on suitable provisions for the unlicensed spectrum etiquette.

WHAT CAUSES THE TRAGEDY OF THE COMMONS?

In all wireless systems, design decisions are exclusively based on the self-interest of the users of the device being designed. The design of the access strategy involves a trade-off between competing goals and interests. One goal is to conserve spectrum; others might be to reduce equipment and operating costs, or to optimize some measure of performance like access delay or reception quality. In licensed spectrum, where the spectrum consumed is typically the exclusive domain of the users of the device, the goal of conserving spectrum is important to device designers. What sets unlicensed spectrum apart is that, although conservation of spectrum is no less important from a system perspective, there is considerably less incentive to design individual devices to conserve the shared spectrum, as mentioned in the introductory section. Thus, in unlicensed spectrum, it is more likely that the best design decision from the selfish perspective of the designer of a given device is also a "greedy" approach, where the more a device is designed to waste shared spectrum unnecessarily in favor of its own goals, the more we consider it to be greedy. The amount of resources a device consumes with a transmission depends on three factors: the transmission duration, bandwidth, and coverage area, of which the last is a function of transmission power. Thus, the transmissions of a greedy device have greater duration, bandwidth, or power than is necessary. We refer to these three factors as the three dimensions along which devices may manifest greedy behavior. We now present examples where designers of unlicensed devices have motivation to be greedy along each of these three dimensions and the trade-offs involved, beginning with transmission duration.

If a device could always access spectrum within an acceptable delay, there would be no reason for it to transmit longer than necessary. However, in unlicensed spectrum there can be no such guarantees and the access delay may vary considerably. Devices may therefore be designed to transmit longer just to avoid the access delay whenever they have a message to transmit again. For example, consider a wireless bridge operating in the PCS unlicensed band that connects two wired local area networks (LANs). Whenever a packet must be forwarded from one LAN to the other, the bridge has to wait for a given monitoring time before it may begin transmission as per the "listen before talk" (LBT) protocol. Instead of releasing the channel at the end of packet transmission, the bridge may be greedy by continuing to transmit even if it has nothing to send. This way all packets that arrive after the first one are spared the access delay imposed by the LBT rule. Essentially, the device is hoarding spectrum that it may or may not need later. The bridge may continue to hold the channel for as long as the etiquette permits. However, there is a cost in doing so. The greedy bridge prevents other unlicensed devices from using the channel, and the queue of packets awaiting transmission at the other devices would grow. Thus, when the bridge finally releases the channel, it may take much longer to reclaim it. Consequently, this form of greed can be beneficial to the user, but isn't always.

We now consider greed in the bandwidth dimension. Application designers may be attracted to higher bandwidth choices because of the cost advantage of inefficient modulation schemes, or for improved performance such as better video quality. However, the disadvantage of using more bandwidth is that there would be a low probability of finding the needed bandwidth free because spectrum is shared. Consequently, as with the duration dimension, greed in the bandwidth dimension may or may not benefit the user.

Devices may likewise be designed to transmit at a higher transmission power than necessary in order to improve the signal-to-noise ratio (SNR), which in a digital transmission decreases the bit error rate. Transmitting at a higher power also reduces the frequency reuse in the system. However, for a given device for which power consumption is not an issue, there would be no disadvantages to high power unless those disadvantages are imposed by the etiquette. As mentioned in the preceeding section, under the current FCC etiquette, the LBT noise threshold by which a device determines whether a channel is busy or free is a function of the transmit power. Consequently, if a device increases its transmit power, its noise threshold would be lower, resulting in greater access delay for the device. Of course, battery-powered devices would have the additional disadvantage of a reduced battery life if they transmit at high power.

To evaluate the potential for a tragedy of the commons, we need to determine whether unlicensed applications will be designed to be greedy, and if so, to what extent and how greed would be manifested by different applications. If the potential greedy behavior of devices significantly degrades system performance, there is cause for concern, and etiquette modifications to discourage greedy behavior would be necessary.

AN EXAMPLE OF A TRAGEDY OF THE COMMONS

In this section we consider a practical scenario where devices may be designed to be greedy in the duration dimension. First we provide a description of this scenario and specify how devices may be greedy. Unfortunately, this is a difficult system to analyze. The next sub-section gives a closed-form analysis that approximates the greedy behavior model well in many scenarios, particularly when the devices are greedy. Simulations are then used to demonstrate the accuracy of this approximation. The final sub-section discusses the implications of our findings.

Our Scenario

In this scenario, two devices compete for access to a wireless channel in the 2-GHz unlicensed PCS band. Devices are sufficiently close together that each device receives the other's transmissions, and no frequency reuse is possible. The devices follow the isochronous band etiquette, and their transmissions require the same bandwidth. Transmission power and bandwidth would therefore have no impact on device performance. We assume that messages arrive at the devices for transmission according to a Poisson process, and that the message lengths are exponentially distributed. Messages are queued in a buffer of infinite length until they can be transmitted. Devices may be greedy in the duration dimension by holding on to the channel longer than necessary, as discussed in the preceeding section. Non-greedy devices release the channel as soon as they transmit the last message in their queue, whereas a greedy device with a greed of duration T holds the channel for duration of at least T even if it has no messages to send. At the end of duration T, the greedy device releases the channel once its queue becomes empty.

If greed can improve performance, then equipment designers will select their greedy strategy based on their projections of the extent to which they will share spectrum with competing devices, and the extent to which those competing greedy devices will also be greedy. Because the strategies employed in one device can influence the optimal design for another device, a useful measure of resulting behavior is the *reaction function* $r_i(T_j)$ for each device i, which gives the optimal greed for a device in response to that of the other device. If device 1 has a greed of T_1, then $r_2(T_1)$ is the greed that minimizes the device 2 delay. Such dynamic reactions to another device's greed may occur in one of two ways. It is possible that greed on some devices can be adjusted by a system administrator, the way one might change the maximum packet length in a LAN. In other systems, the extent of greed will be fixed when the device is manufactured. However, greed can change over time when equipment is replaced. For example, if most CB radios in use have a power of P, then someone buying a new radio will buy one with power $r(P)$, and generation after generation, the power levels will change.

How Greed Affects Performance

We use analysis and then simulation to demonstrate the potential tragedy of the commons in this scenario. We show that if one device is designed to be greedy, it increases the average queuing delay for the messages transmitted by the other device. Furthermore, this other device can always reduce its delay by increasing its own greed. This degrades performance for the first device, and it can reduce its own delay by increasing its greed again, thereby continuing the process whereby each device always responds with more greed than the other. To make analysis tractable, we use a fluid-flow model (Anick, Mitra & Sondhi, 1982; Tucker, 1988) for message arrivals where messages arrive at a constant rate. More precisely, the amount of data received in any period of duration τ is exactly $\rho\tau$. In practice, the arrival rate fluctuates, causing the delay calculated by the fluid-flow model to have an error in the range of a few message transmission times. Therefore, whenever the delay caused by waiting for the other device to release the channel is much greater than the delay caused by the arrival rate fluctuations, the error percentage would be small, and the fluid-flow model would therefore be a good approximation. As a result this model should be more accurate when devices are greedy.

We now define the variables we use for our analysis. Let ρ_i be the load of device i. We assume $\rho_1 > 0$; $\rho_2 > 0$, and $\rho_1 + \rho_2 < 1$; that is, the total message arrival rate at both devices does not exceed the total capacity. Device $i : i \in \{1,2\}$ has a greed of duration T_i, and it holds the channel for a period of duration H_i, where $H_i \geq T_i$. For a period of $X_i : X_i \leq H_i$, device i has messages queued and is transmitting at the maximum rate possible. Devices are required by the FCC etiquette to start transmission only if they find the channel to be free for a duration M. We assume devices sense the channel with persistence; that is, devices continuously monitor the channel until they detect it to be idle (Vukovic & McKown, 1996). Figure 4.1 shows the device 1 *unfinished work* as it varies over time, where unfinished work is the amount of time it would take to transmit all currently queued messages. We define time 0 to be the time when device 2 starts monitoring the channel. At that time, device 1 has just relinquished the channel, and presumably has emptied its queue of messages. Device 2 finds the channel free and begins transmission at time M. It then releases the channel at time $H_2 + M$. The device 1 queue increases from time 0 until $H_2 + 2M$, when Device 1 begins transmission after monitoring the channel for duration M. Its unfinished work built up at this time is given by $\rho_1(2M + H_2)$. The device 1 queue length then starts decreasing and reaches zero at time $2M + H_2 + X_1$. The device continues to hold the channel, transmitting messages as quickly as they arrive until time $2M + H_2 + H_1$. The process is then repeated.

Figure 4.1
Device 1 Unfinished Work as a Function of Time Under
a Fluid-flow Model for Message Transmissions

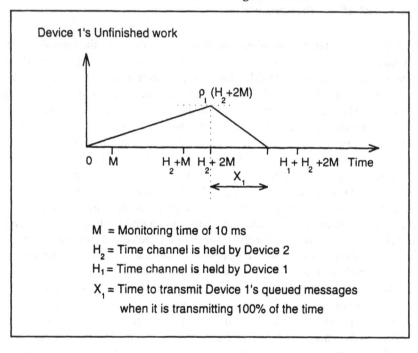

The following theorems characterize the potential greedy behavior in this system. (See Appendix for proofs.) Theorem 1 shows the holding times H_1^* and H_2^* for devices 1 and 2 respectively, when neither device is greedy. Note that device i is never greedy if $T_i \leq H_i^*$, because the device always holds the channel for at least H_i^* anyway. Therefore, delays are identical for any $T_i \leq H_i^*$. Theorem 2 shows the general reaction functions. Theorem 3 shows that even if device i is nongreedy, if $H_i^* < 2M$, then the other device is better off being greedy. To determine the ultimate results of these reaction functions, let device 2 select an initial greed $T_2^{(0)}$. Devices 1 and 2 will then take turns responding to each other's greed, that is, for $i > 0$, $T_1^{(i)} = r_1(T_2^{(i-1)})$ and $T_2^{(i)} = r_2(T_1^{(i)})$. Theorem 4 shows that if one device is greedy, it will cause both devices to escalate their greed to infinity. Theorem 5 shows that whenever a device's greed is increased, delay increases for the other device.

Theorem 1: If $T_1 = T_2 = 0$, then

$$H_1^* = \frac{2M\rho_1}{1 - \rho_1 - \rho_2}$$

and

$$H_2^* = \frac{2M\rho_2}{1 - \rho_1 - \rho_2}$$

Theorem 2:

$$r_1(T_2) = \max\{T_2, 2M\}\frac{1 - \rho_2}{\rho_2} - 2M$$

$$r_2(T_1) = \max\{T_1, 2M\}\frac{1 - \rho_1}{\rho_1} - 2M$$

Theorem 3: If $T_1 < H_1^*$ and $H_1^* < 2M$ then $r_2(T_1) > H_2^*$.

If $T_2 < H_2^*$ and $H_2^* < 2M$ then $r_1(T_2) > H_1^*$.

Theorem 4: If $T_2^{(0)} > H_2^*$ then $\lim_{i \to \infty} T_2^{(i)} = \infty$ and $\lim_{i \to \infty} T_1^{(i)} = \infty$.

Theorem 5:

$$\frac{\partial D_1}{\partial H_2} > 0$$

and

$$\frac{\partial D_2}{\partial H_1} > 0$$

All together, these theorems show the potential for a tragedy of the commons. If $H_1^* < 2M$ or $H_2^* < 2M$, which occurs when $\rho_1 + \rho_2 + \min\{\rho_1, \rho_2\} < 1$, then it will inevitably lead to an escalation of greed until both devices hold the channel as long as possible, and neither has adequate performance.

Results from Simulations

Of course, as described in the preceding sub-section, the results of a fluid-flow approximation are accurate only when the delay caused by waiting for the other device to release the channel significantly exceeds the delay caused by fluctuations in the arrival rate of the messages. This is certainly the case when the greed of both devices is significant, but may not be when devices are not greedy. We test these results through simulation.

Figure 4.2

Delay of each device as a function of device 1 greed. Device 2 is nongreedy.

(Device Load = 40%, Average Message Transmission Time = 0.5 msec, and Monitoring Time = 10 msec.)

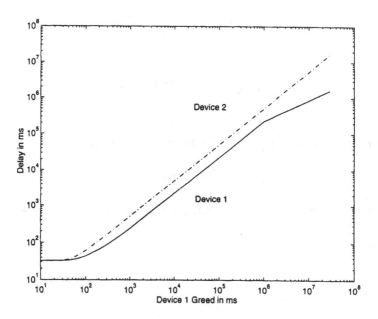

We first present the case when each device has a load of 40% and an average message transmission time of 0.5 msec. Device 1 then varies its greed from zero to 8 hrs. whereas device 2 remains nongreedy. Figure 4.2 shows that the greed of device 1 results in increased delay for both devices. As predicted by the analytic approximation, the device 1 delay is minimized at a greed of 0, so there is no incentive to make it greedy. If this were always true, there would be no risk of a tragedy of the commons. However, when the devices are less heavily loaded at 10% load, each having an average message transmission time of 0.5 msec, we see in Fig. 4.3 that the device 1 performance does benefit from being greedy. Again, this result was predicted through analysis. The delay of device 1 is minimized at 160 msec, whereas the delay of device 2 increases monotonically with the greed of device 1. Thus, there is incentive to make device 1 greedy in this case, and Device 2 performance will suffer as a result. Indeed, as Fig. 4.4 shows, in some cases a device can decrease its delay by an order of magnitude through greed.

Figure 4.3
Delay of each device as a function of device 1 greed. Device 2 is nongreedy.
(Device Load = 10%, Average Message Transmission Time = 0.5 msec, and Monitoring Time = 10 msec.)

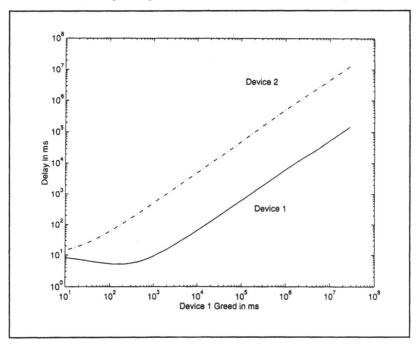

We saw in the preceding sub-section that in a fluid-flow model, if one device is made greedy, the other device will as well, and greed will escalate. Fig. 4.5 shows that this phenomenon also occurs as predicted. This figure

shows the reaction functions when both devices have a load of 10% and the
average message transmission length is 0.5 msec. When the greed chosen for
device 1 significantly exceeds monitoring time, the device 2 delay is reduced
by making it more greedy, by about $(1 - p_1)/p_1 = 9$ times as much, as predicted in Theorem 2. For example, if the devices start with greed indicated
by point A on Fig. 4.5, they will progress to points B, C, D, and so on. Eventually they reach point H, where both hold the channel for 8 hr at a time,
which is the maximum transmission duration allowed by the FCC etiquette.
Note that this is the only point where the reaction functions intersect, so it is
the only equilibrium. If there were no such upper limit imposed by the etiquette, our analysis predicts that the greed of devices would tend to infinity.
Figure 4.6 shows that as the greed of devices escalates from zero to 8 hr, the
average delay increases monotonically. This is a tragedy of the commons.
We have similarly observed this phenomenon in other reaction functions that
were derived through simulation.

Figure 4.4
Delay of Each Device as a Function of Device 1 Greed. Device 2 Greed = 10,000 msec.
(Device Load = 10%, Average Message Transmission Time = 0.5 msec, and Monitoring Time = 10 msec.)

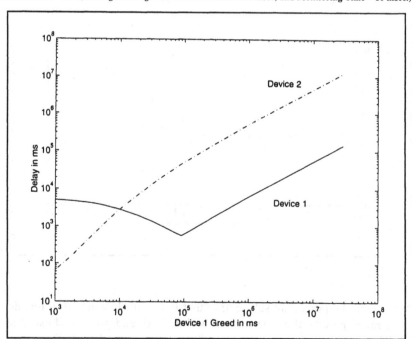

Figure 4.5
Reaction Function for Two Devices at 10% Load Each.

(Average Message Transmission Time = 0.5 msec, and Monitoring Time = 10 msec.)

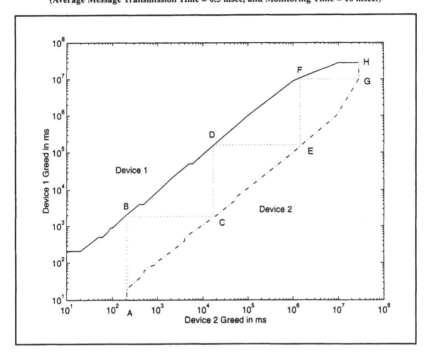

Implications

As described earlier, the reaction functions just employed are applicable when there are two devices competing for spectrum, and greed for each device is chosen independently. Initially, there will be few unlicensed devices competing for spectrum. Consequently, the first set of unlicensed devices produced by the industry is less likely to be designed greedy. As usage of the unlicensed PCS bands increases, and more devices compete with each other for access to spectrum, their performance would decrease. The reaction of equipment designers would depend to a large extent on the diversity of devices using unlicensed spectrum, and the corresponding industry structure. At one extreme, consider the case where a single manufacturer produces all of these unlicensed devices. This manufacturer would only produce nongreedy devices, as there is nothing to be gained from a tragedy of the commons between two or more of its own customers. As we see in Fig. 4.7, if there are two devices that have the same greed, their delay increases monotonically with their greed. However, the situation is very different if all manufacturers have so little market share that two devices from the same manufacturer rarely compete for spectrum. In this case, greedy devices will emerge and spectral efficiency will be reduced. Of course, reality is somewhere in between, and it will be explored in future work, as will the impact of greed when there are more than two competing devices.

Figure 4.6
Device 1 Delay at Its Optimal Greed vs. Device 2 Greed.
Devices Are Each at 10% Load.
(Average Message Transmission Time = 0.5 msec, and Monitoring Time = 10 msec.)

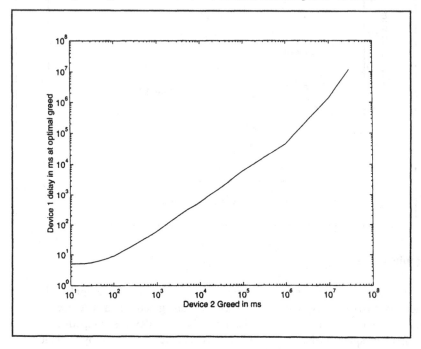

The way to prevent manufacturers from marketing greedy devices would be to discourage greedy behavior through modifications to the FCC etiquette. We provide examples of modifications that may be used to discourage greedy behavior in the following section.

WAYS TO AVOID THE TRAGEDY OF THE COMMONS

One way to control greed is to set upper limits on each of the dimensions of greedy behavior; that is, one might impose a maximum duration, power, or bandwidth. However, there are problems with this approach. For example, setting a time limit on a voice conversation would not be desirable as phone calls may then be prematurely terminated. Also, an upper limit may result in spectrum inefficiency unless it is chosen appropriately. As discussed in the second section, a device may be designed to use as much of a resource as is allowed by the upper limit even if it is not necessary for adequate performance. Hence a higher than optimal limit would be inefficient. A lower limit may restrain the use of resources to the extent that spectrum is unnecessarily made unusable for some applications. Another option is to have a slightly more flexible form of upper limit. As mentioned in the third section , the maximum transmitted power allowed by the FCC etiquette is a function of bandwidth. Narrow-bandwidth applications are permitted higher power spectral density, which provides an incentive for devices to not use excessive bandwidth. Although there is some flexibility in the dimension of resource consumption a device chooses, this is still an upper limit, and there is no incentive to do better than the limit requires. Upper limits are indeed a blunt instrument to deal with such a delicate problem.

Figure 4.7
Delay of Each Device As a Function of Greed. Devices Have Equal Greed.
(Device Load = 10%, Average Message Transmission Time = 0.5 msec, and Monitoring Time = 10 msec.)

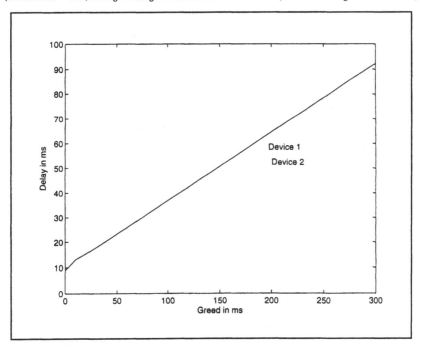

To produce further incentives, we might give devices that consume less spectrum some form of priority in accessing spectrum. For incentives to be effective, they must be based on parameters that strongly affect the device performance. We now provide examples of such parameters. In the current etiquette, a device must monitor the channel before transmitting to ensure that detected power remains below a threshold throughout the monitoring period. Altering either monitoring time or power threshold would affect a device's chances of accessing spectrum. For example, if two devices begin monitoring at the same time, the one with the smaller monitoring period will get access. (Currently, monitoring period is fixed in both bands, but threshold is a function of transmit power.) Thus, factors like monitoring time and power threshold can be used to provide incentives; if devices seeking a large bandwidth were assigned a large monitoring time, there would be more incentive to use efficient modulation. Another factor that influences access to spectrum is *interburst gap*, which is the minimum amount of time a device must wait to transmit after completing a transmission. For example, if a need is discovered to induce a device to end its transmission early, the interburst gap following a long transmission could be made large compared to the monitoring times of other devices, whereas a short transmission can be rewarded with short interburst gaps. In the current etiquette, interburst gap also does not depend on resources consumed. It is 10 msec in the isochronous band, and it is selected randomly in the asynchronous band, where the distribution depends only on the number of previously unsuccessful attempts to access the spectrum. Another parameter that affects a device's performance is the *back-off period*, which is the minimum time a device has to wait before attempting to access a channel again once the channel is detected busy. Making the back-off period depend on the spectrum resources consumed by the devices might provide an incentive to design devices that conserve spectrum. Our future work will address and quantify the extent to which such parameters affecting access to spectrum can be used to induce efficient utilization of the spectrum resource.

CONCLUSION

Some wireless applications are not well-served under a system of exclusive rights to spectrum, and are better off in shared spectrum supporting real-time access. Real-time sharing has several advantages. It supports mobility of wireless applications, allows spectrum sharing, and facilitates experimentation and innovation. Unlicensed spectrum offers the potential to realize these benefits. However, because there is little inherent incentive for individual devices to be designed to use unlicensed spectrum efficiently, they may engage in greedy behavior; that is, device designers may sacrifice the goal of spectrum efficiency to meet other design goals. In this chapter we have demonstrated, in a simple scenario, that greedy behavior is sometimes rewarded, and it can lead to a tragedy of the commons. The severity of this problem will be determined in future work, which will consider more complex models. For example, we will consider more complicated wireless applications, other forms of greed, the result of competition for spectrum among more than two devices, and the impact of industry and market structure. If the tragedy of the commons proves to be significant, the most practical way to reduce the risk of a tragedy of the commons is offered by the etiquette for unlicensed spectrum. We have suggested examples of etiquette modifications that may prove effective in discouraging greedy behavior. These will be addressed in more detail in future work.

Meanwhile, demand for more unlicensed spectrum is high, and the FCC is moving forward. At this point, there is still little evidence that provisions in the current FCC etiquette are both necessary and sufficient. It is possible that a tragedy of the commons will lead to poor performance in this band, and it is also possible that many existing restrictions in the etiquette could be relaxed without penalty, thereby simplifying designs. Little has been published on this issue, and the industry groups backing unlicensed spectrum have kept their work mainly proprietary. Consequently, before industry invests significantly in the manufacture of unlicensed devices, and before the FCC releases more large blocks of unlicensed spectrum, caution is advised until more is known.

APPENDIX

Theorem 1

If $T_1 = T_2 = 0$, then

$$H_1^* = \frac{2M\rho_1}{1 - \rho_1 - \rho_2}$$

and

$$H_2^* = \frac{2M\rho_2}{1 - \rho_1 - \rho_2}$$

Proof for Theorem 1

When $T_1 = T_2 = 0$, both devices are nongreedy, so $H_1^* = X_1$ and $H_2^* = X_2$. Device 1 has $\rho_1(2M + H_1^* + H_2^*)$ unfinished work built up during the period from 0 to $2M + H_1^* + H_2^*$, which it transmits in time H_1^*. Therefore we have $H_1^* = \rho_1(2M + H_1^* + H_2^*)$ and $H_2^* = \rho_2(2M + H_1^* + H_2^*)$ from symmetry.

The solutions to these equations are given by

$$H_1^* = \frac{2M\rho_1}{1 - \rho_1 - \rho_2} \qquad (1)$$

and

$$H_2^* = \frac{2M\rho_2}{1-\rho_1-\rho_2}$$

Theorem 2

$$r_1(T_2) = \max\{T_2, 2M\}\frac{1-\rho_2}{\rho_2} - 2M$$

$$r_2(T_1) = \max\{T_1, 2M\}\frac{1-\rho_1}{\rho_1} - 2M$$

Proof for Theorem 2

Device 1 responds with the greed $T_1 = r_1(T_2)$ that minimizes the device 1 delay. The total unfinished work $\rho_1(2M + H_2 + X_1)$ built up by device 1 during the period from 0 to $2M + H_2 + X_1$ is transmitted in duration X_1, so $X_1 = \rho_1(2M + H_2 + X_1)$, that is,

$$X_1 = \rho_1\frac{2M + H_2}{1-\rho_1}$$

and similarly   (2)

$$X_2 = \rho_2\frac{2M + H_1}{1-\rho_2}$$

The average delay for the device 1 messages is its average unfinished work divided by ρ_1. As Fig. 4.1 showed, the device 1 average unfinished work during the period from 0 to $2M + H_2 + X_1$ when it has queued messages is $0.5\rho_1(2M + H_2)$. The fraction of time device 1 has queued messages is

$$\frac{2M + H_2 + X_1}{2M + H_2 + H_1}$$

so

$$D_1 = 0.5(2M + H_2)\frac{2M + H_2 + X_1}{2M + H_2 + H_1}$$ (3)

We consider two cases: $T_2 > X_2$ and $T_2 \leq X_2$.

Case 1: $T_2 > X_2$, so the device 2 holding time is $H_2 = T_2$, independent of T_1. Consequently,

$$\frac{\partial D_1}{\partial H_1} = -0.5(2M + T_2)\frac{2M + T_2 + X_1}{(2M + T_2 + H_1)^2}$$

which is always negative. Thus, as long as

$$T_2 > X_2 = \rho_2 \frac{(2M + H_1)}{(1 - \rho_2)}$$

device 1 would choose to increase H_1, which it can do by increasing T_1.

Case 2: $T_2 \leq X_2$, so $H_2 = X_2$ and the device 1 average delay is

$$D_1 = 0.5(2M + X_2)\frac{2M + X_2 + X_1}{2M + X_2 + H_1}$$

where

$$X_1 = \rho_1 \frac{2M + X_2}{1 - \rho_1}$$

and

$$X_2 = \rho_2 \frac{2M + H_1}{1 - \rho_2}$$

Differentiating, we get

$$\frac{\partial D_1}{\partial H_1} = \frac{0.5(2M + H_1\rho_2)[\rho_2(H_1 + 2M) - 2M(1 - \rho_2)]}{(1 - \rho_1)(1 - \rho_2)(2M + H_1)^2}$$

Then

$$\frac{\partial D_1}{\partial H_1} > 0$$

if and only if $\rho_2(H_1 + 2M) > 2M(1 - \rho_2)$, or equivalently, if

$$X_2 = \frac{\rho_2(2M + H_1)}{(1 - \rho_2)} > 2M$$

Thus, in the case where $T_2 < X_2$, device 1 would choose to increase H_1 (and thus T_1) if $X_2 < 2M$, decrease H_1 if $X_2 > 2M$, and leave H_1 unchanged if $X_2 = 2M$.

Considering both cases, we see that device 1 would choose an H_1 at which $X_2 = \max\{T_2, 2M\}$ if possible. (It may be that H_1 will be larger than this even if device 1 is nongreedy.) Consequently, we set T_1 equal to this value of H_1, and by Equation 2, we have

$$T_1 = r_1(T_2) = \max\{T_2, 2M\} \frac{1 - \rho_2}{\rho_2} - 2M \qquad (4)$$

By symmetry,

$$T_2 = r_2(T_1) = \max\{T_1, 2M\} \frac{1 - \rho_1}{\rho_1} - 2M \qquad (5)$$

Theorem 3

If $T_1 < H_1^*$ and $H_1^* < 2M$ then $r_2(T_1) > H_2^*$. If $T_2 < H_2^*$ and $H_2^* < 2M$ then $r_1(T_2) > H_1^*$.

Proof for Theorem 3

Let $T_1 < H_1^*$ and $H_1^* < 2M$. From Theorem 2, we have

$$r_2(T_1) = \max\{T_1, 2M\} \frac{1 - \rho_1}{\rho_1} - 2M = 2M \frac{1 - 2\rho_1}{\rho_1} \qquad \text{as } T_1 < 2M$$

From Theorem 1, $H_1^* = 2M\rho_1/(1 - \rho_1 - \rho_2)$. As $H_1^* < 2M$, $2M\rho_1/(1 - \rho_1 - \rho_2) < 2M$. Multiplying both sides by $(1 - \rho_1)/\rho_1$ and subtracting $2M$, we get

$$\frac{2M\rho_2}{1 - \rho_1 - \rho_2} < 2M \frac{1 - 2\rho_1}{\rho_1}$$

that is, $H_2^* < r_2(T_1)$.

Hence if $T_1 < H_1^*$, and $H_1^* < 2M$ then $r_2(T_1) > H_2^*$. By symmetry, if $T_2 < H_2^*$ and $H_2^* < 2M$ then $r_1(T_2) > H_1^*$.

Theorem 4

If $T_2^{(0)} > H_2^*$ then $\lim_{i \to \infty} T_2^{(i)} = \infty$ and $\lim_{i \to \infty} T_1^{(i)} = \infty$.

Proof for Theorem 4

Let $S_2^{(0)} = T_2^{(0)}$. We define

$$f_1(S_2) = S_2 \frac{1-\rho_2}{\rho_2} - 2M \text{ and } f_2(S_1) = S_1 \frac{1-\rho_1}{\rho_1} - 2M$$

Note that $\partial f_1/\partial S_2 > 0$ and $\partial f_2/\partial S_1 > 0$, that is, that f_1 and f_2 are monotonically increasing functions. Let

$$(S_2^{(i)}) = f_1(S_2^{(i-1)}) \text{ and } S_2^{(i)} = f_2(S_1^{(i)}) = f_2(f_1(S_2^{(i-1)}))$$

Likewise, let

$$T_1^{(i)} = r_1(T_2^{(i-1)}) \text{ and } T_2^{(i)} = r_2(T_1^{(i)}) = r_2(r_1(T_2^{(i-1)}))$$

By Theorem 2,

$$r_1(T_2) = \max\{f_1(T_2), 2M\frac{1-\rho_2}{\rho_2} - 2M\} \geq f_1(T_2) \tag{6}$$

and

$$r_2(T_1) = \max\{f_1(T_2), 2M\frac{1-\rho_2}{\rho_2} - 2M\} \geq f_1(T_2) \tag{7}$$

We first prove three propositions, which we use to prove Theorem 4.

Proposition 1:

$$T_2^{(i)} \geq S_2^{(i)} \ \forall i \geq 0$$

Proof: We prove proposition 1 by induction.

Basis: Let $i = 0$. $T_2^{(i)} \geq S_2^{(i)}$ becomes $T_2^{(0)} \geq S_2^{(0)}$ which is true as $T_2^{(0)} = S_2^{(0)}$.

Inductive step: If $T_2^{(i)} \geq S_2^{(i)}$ for a given i, we show $T_2^{(i+1)} \geq S_2^{(i+1)}$.

From Equation 7 we have $T_2^{(i+1)} = r_2(r_1(T_2^{(i)})) \geq f_2(r_1(T_2^{(i)}))$. From Equation 6 we have $r_1(T_2^{(i)}) \geq f_1(T_2^{(i)})$. Since f_2 is monotonically increasing, $f_2(r_1(T_2^{(i)})) \geq f_2(f_1(T_2^{(i)}))$. Therefore $T_2^{(i+1)} \geq f_2(f_1(T_2^{(i)}))$. Because f_1 is also monotonically increasing, and $T_2^{(i)} \geq S_2^{(i)}$, we have $f_1(T_2^{(i)}) \geq f_1(S_2^{(i)})$ and therefore $f_2(f_1(T_2^{(i)})) \geq f_2(f_1(S_2^{(i)}))$. We thus have $T_2^{(i+1)} \geq f_2(f_1(T_2^{(i)})) \geq f_2(f_1(S_2^{(i)}))$, that is, $T_2^{(i+1)} \geq S_2^{(i+1)}$.

Hence $T_2^{(i)} \geq S_2^{(i)} \ \forall i \geq 0$.

Proposition 2:

$$S_2^{(i)} = (1 + \frac{1 - \rho_1 - \rho_2}{\rho_1 \rho_2})^i (S_2^{(0)} - H_2^*) + H_2^* \ \forall i \geq 0$$

Proof: We prove this by induction.
Basis: When $i = 0$,

$$S_2^{(i)} = (1 + \frac{1 - \rho_1 - \rho_2}{\rho_1 \rho_2})^i (S_2^{(0)} - H_2^*) + H_2^*$$

becomes $S_2^{(0)} = S_2^{(0)} - H_2^* + H_2^* = S_2^{(0)}$, which is true.

Inductive step: If proposition 2 is true for a given i then we show it is true for $i + 1$. Assume that

$$S_2^{(i)} = (1 + \frac{1 - \rho_1 - \rho_2}{\rho_1 \rho_2})^i (S_2^{(0)} - H_2^*) + H_2^* \text{ for some } i$$

By definition,

$$f_1(S_2) = S_2 \frac{1 - \rho_2}{\rho_2} - 2M \text{ and } f_2(S_1) = S_1 \frac{1 - \rho_1}{\rho_1} - 2M$$

Therefore, we have

$$S_2^{(i+1)} = f_2(f_1(S_2^{(i)})) = (f_1(S_2^{(i)})) \frac{1 - \rho_1}{\rho_1} - 2M = \frac{1 - \rho_1}{\rho_1} \frac{1 - \rho_2}{\rho_2} S_2^{(i)} - \frac{2M}{\rho_1}$$

From Equation 1 we have

$$\begin{aligned}
S_2^{(i+1)} &= (1 + \frac{1 - \rho_1 - \rho_2}{\rho_1 \rho_2}) S_2^{(i)} - H_2^* \frac{1 - \rho_1 - \rho_2}{\rho_1 \rho_2} \\
&= (1 + \frac{1 - \rho_1 - \rho_2}{\rho_1 \rho_2})(S_2^{(i)} - H_2^*) + H_2^* \\
&= (1 + \frac{1 - \rho_1 - \rho_2}{\rho_1 \rho_2})\left[(1 + \frac{1 - \rho_1 - \rho_2}{\rho_1 \rho_2})^i (S_2^{(0)} - H_2^*) + H_2^* - H_2^*\right] + H_2^* \\
&= (1 + \frac{1 - \rho_1 - \rho_2}{\rho_1 \rho_2})^i (S_2^{(0)} - H_2^*) + H_2^*
\end{aligned}$$

Hence

$$S_2^{(i)} = (1+\frac{1-\rho_1-\rho_2}{\rho_1\rho_2})^i(S_2^{(0)} - H_2^*)+H_2^* \ \forall i \geq 0$$

Proposition 3: If $S_2^{(0)} > H_2^*$, then $\lim_{i\to\infty} S_2^{(i)} = \infty$ and $\lim_{i\to\infty} S_1^{(i)} = \infty$.

Proof: From proposition 2, we have

$$S_2^{(i)} = (1+\frac{1-\rho_1-\rho_2}{\rho_1\rho_2})^i(S_2^{(0)} - H_2^*)+H_2^* \ \forall i \geq 0$$

Because $\rho_1+\rho_2 < 1$, $[1 + (1-\rho_1 - \rho_2)/\rho_1\rho_2] > 1$. Also, $S_2^{(0)} - H_2^* > 0$. Consequently,

$$(1+\tfrac{1-\rho_1-\rho_2}{\rho_1\rho_2})^i(S_2^{(0)} - H_2^*)$$

approaches ∞ as i approaches ∞ and so does $S_2^{(i)}$. We have $S_1^{(i)} = f_1(S_2^{(i-1)}) = S_2^{(i-1)}[(1-\rho_2)/\rho_2] - 2M$. Thus if $S_2^{(i-1)}$ goes to ∞ as i goes to ∞, $S_1^{(i)}$ must do the same. Hence if $S_2^{(0)} > H_2^*$ then $\lim_{i\to\infty} S_2^{(i)} = \infty$ and $\lim_{i\to\infty} S_1^{(i)} = \infty$.

We now prove Theorem 4. From proposition 3 we see that if $S_2^{(0)} > H_2^*$ then $\lim_{i\to\infty} S_2^{(i)} = \infty$. From proposition 1 we have $T_2^{(i)} \geq S_2^{(i)} \ \forall i \geq 0$. Therefore, $\lim_{i\to\infty} T_2^{(i)} = \infty$. We have $T_1^{(i)} = r_1(T_2^{(i-1)}) = \max\{T_2^{(i-1)}, 2M\}$ $1-\rho_2/\rho_2- 2M$. Thus if $T_2^{(i-1)}$ goes to ∞ as i goes to ∞, so does $T_1^{(i)}$. As $T_2^{(0)} = S_2^{(0)}$, we get if $T_2^{(0)} \geq H_2^*$ then $\lim_{i\to\infty} T_2^{(i)} = \infty$ and $T_1^{(i)} = \infty$.

Theorem 5

$$\frac{\partial D_1}{\partial H_2} > 0 \text{ and } \frac{\partial D_2}{\partial H_1} > 0$$

Proof for Theorem 5

As shown in the proof of Theorem 2, the device 1 delay is given by

$$D_1 = 0.5(2M + H_2)\frac{2M+H_2+X_1}{2M+H_2+H_1} \text{ where } X_1 = \frac{\rho_1}{1-\rho_1}(2M + H_2)$$

Therefore

$$D_1 = 0.5(2M + H_2)\frac{2M+H_2+\frac{\rho_1}{1-\rho_1}(2M+H_2)}{2M+H_2+\max\{T_1,\frac{\rho_1}{1-\rho_1}(2M+H_2)\}}$$

If $T_1 < X_1$ then $D_1 = 0.5(2M + H_2)$ and $\partial D_1/\partial H_2 = 0.5$.

If $T_1 > X_1$ then

$$D_1 = 0.5(2M + H_2)\frac{2M+H_2+\frac{\rho_1}{1-\rho_1}(2M+H_2)}{2M+H_2+T_1}$$

and

$$\frac{\partial D_1}{\partial H_2} = \frac{0.5}{1-\rho_1}\frac{(2M+H_2)(2M+H_2+2T_1)}{(2M+H_2+2T_1)^2}$$

Because $M > 0$, $H_2 > 0$, and $T_1 > 0$, we have $\partial D_1/\partial H_2 > 0$. By symmetry, the same is true for $\partial D_2/\partial H_1$. This shows that greed always hurts the other device.

ACKNOWLEDGEMENT

We gratefully acknowledge the support of Bellcore through Carnegie Mellon University's Information Networking Institute. However, the views expressed in this chapter are those of the authors, and do not necessarily reflect the views of Bellcore.

REFERENCES

Anick, D., Mitra, D., & Sondhi, M. M. (1982). Stochastic theory of a data-handling system with multiple sources. *Bell System Technical Journal, 61* (8), 1871–1894.

Federal Communications Commission. (1994). *Amendment of the Commission's Rules to Establish New Personal Communications Services.* Memorandum Opinion and Order, Federal Communications Commission. 94-144, Gen. Docket No. 90-314.

Federal Communications Commission. (1996). *NII/SUPERnet at 5 GHz.* Notice of Proposed Rulemaking, Federal Communications Commission. 96-193, ET Docket No. 96-102.

Gilder, G. (1994). Auctioning the airways. *Forbes ASAP Supplement, 153* (8), 99–112.

Hardin, G. (1968). The tragedy of the commons. *Science, 162*, 1243–1248.

Hundt, R. E., & Rosston, G. L. (1995). Spectrum flexibility will promote competition and the public interest. *IEEE Communications, 33* (12) 40–43.

Marcus, M. J. (1994). Millimeter wave spectrum management. *Applied Microwave & Wireless, 6,* 98–106.

Marcus, M. J. (1996). Recent progress in U.S. millimeter wave spectrum management policy. *IEEE MTT-S International Microwave Symposium Digest, 2,* 505–507.

McGarty, T. P., & Medard, M. (1994). Wireless architectural alternatives: current economic valuations versus broadband options, The Gilder conjectures. *Paper presented at the 22nd Telecommunications Policy Research Conference.*

Noam, E. M. (1995). Taking the next step beyond spectrum auctions: Open spectrum access. *IEEE Communications, 33* (12) 66–73.

Peha, J. M. (1994-1995). Seamless wireless networks: Motivation and challenges. *International Engineering Consortium Annual Review of Communications, 48,* 553–556.

Peha, J. M. (1996). *A proposed new spectrum management policy* [Online]. http://www.ece.cmu.edu/afs/ece/usr/peha/peha.html.

Peha, J. M. (1997). Developing equipment and services for shared spectrum: Is it a good gamble? *International Engineering Consortium Annual Review of Communications, 50.*

Pressler, L. (1996). Spectrum reform discussion draft. *Congressional Record (Senate)* S4928–S4936.

Salgado-Galicia, H., Sirbu, M., & Peha, J. M. (June 1995). Spectrum sharing through dynamic channel assignment for open access to personal communications services. *Proceedings of the IEEE International Conference on Communications (ICC)* 417–422.

Salgado-Galicia, H., Sirbu, M., & Peha, J. M. (1997). A narrow band approach to efficient pcs spectrum sharing through decentralized DCA access policies. *IEEE Personal Communications 4* (1) 23–34.

Steer, D. G. (1994). Wireless operation in the unlicensed band. *IEEE Personal Communications, 1* (4) 36–43.

Tucker, R. C. F. (1988). Accurate method for analysis of a packet-speech multiplexer with limited delay. *IEEE Transactions on Communications 36* (4) 479–483.

Vukovic, I., & McKown, J. (November 1996). Spectrum sharing under the asynchronous UPCS etiquette: The performance of collocated systems under heavy load. *Proceedings of ACM Mobicom.* 67–72.

Wang, Q., Peha, J. M., & Sirbu, M. (1997). Optimal pricing for integrated-services networks with guaranteed quality of service. In J. Bailey & L. McKnight (Eds.), *Internet economics.* Cambridge, MA: MIT Press, 353–376.

Wang, Q., Sirbu, M., & Peha, J. M. (1996). Dynamic pricing for integrated services networks. In G. W. Brock & G. L. Rosston (Eds.), *The Internet and telecommunications policy: Selected papers from the 1995 Telecommunication Policy Research Conference* (pp. 65–78). Mahwah, NJ: Lawrence Erlbaum Associates.

II

INTERNET GROWTH
AND COMMERCE

Motivations for and Barriers to Internet Usage: Results of a National Public Opinion Survey

James Katz
Bellcore, Morristown, New Jersey

Philip Aspden
Center for Research on the Information Society, Pennington, New Jersey

There has been an explosion of interest in and surveys on the Internet. Most interest focuses primarily on the commercial potential and secondarily on the demographics of users, often ignoring the perceptions of nonusers—many of whom will eventually become users (Dataquest, 1995; Nielsen Media, 1996; Public Perspective, 1996). Thus, despite the plethora of surveys, significant subjects appear to be unexplored, both among users and especially relative to the general population. Hence our purpose is to complement existing studies by looking at both users and nonusers.

We believe gaining a systematic understanding of the "social" issues is a worthwhile objective for both intellectual and policy reasons. According to many observers, there are important societal and political consequences if a "digital divide" should long endure. For example, Anderson, Bikson, Law, and Mitchell (1995) highlight the range of services becoming available via the Internet, and the equity implications if certain segments of the population are excluded from them. The issue becomes increasingly urgent as more governmental functions become available through the World Wide Web (WWW) and the Internet (Edmondson, 1996). Those without computers and Internet accounts can neither access nor participate in "point-and-click" government. Highlighting this inequity is the fact that the 1996 political conventions of the Republic and Democratic parties were "carried" over the Internet, and included opportunities for interaction and feedback. Perhaps most troubling of all aspects of this "information rich-information poor" issue is that the gap of computer ownership between the "rich" and the "poor" has widened since the mid-1980s, so that now higher proportions of the wealthier portions of the population have computers relative to the poorer portions (Kraut et al., 1996). Further, increasing political resources made available over the Internet (Woodward, 1995; Trounstine, 1995) and their exploitation by Internet denizens (Miller, 1996a) exacerbate the situation.

To the extent that it actually exists, a digital divide has troubling implications for those without ready access to the Internet that extend well beyond political empowerment and go to the heart of issues concerning economic participation and equity. For example, help wanted offerings are increasingly being carried on the Internet, sometimes on an exclusive basis (Galifianakis, 1996). Internet usage provides both a growing opportunity for home-based entrepreneurs (Gupta, 1995) and a business status symbol (Maney, 1995), and has been described as being increasingly central to the operations of small businesses (Clark, 1996). To the extent that any demographic group becomes excluded from and underrepresented on the Internet, the group will also be excluded from the economic fruits that such participation promises. Likewise, to the extent that Internet participation is educationally beneficial, those locales that do not have appropriate Internet access will suffer, hobbling the academic competitiveness of children and diminishing the brightness of their future (*New York Times*, 1996). (This is not withstanding President Clinton's recent call to wire all U.S. schools to the Internet by the year 2000, because, to paraphrase the First Lady, it takes more than a wire to make a scholar.)

The Internet is so new that various national or random surveys and what they portend have not, by and large, appeared in the scholarly literature. Instead, this material has become the daily grind of mass and trade media, which frequently offer a variety of informed speculations concerning individual psychology of users, social equity, economic impacts, commercial aspects, and cultural consequences of the Internet (and from which perforce we derive some of our citations). This is not to say that academics have ignored the Internet, because there is a vigorous flurry of publishing activity from them (e.g., Batteau, 1996). Rather academics tend to rely on fragmented evidence, case studies, and personal experience without benefit from national random surveys. Hence we hope our research will not only have merit on its own but also provide grist for subsequent work by our scholarly colleagues.

Our purpose with this chapter is largely normative: We seek to describe behavior, opinions, and experience as told to us by Internet users and nonusers with an eye toward summarizing them in a way that will have relevance for policy and praxis. We anticipate later developing a more theoretical analysis in which we can test a variety of hypotheses about the social processes involved with the Internet. Yet for now, we believe that an analysis of a random national sample of individuals can help clear up many misconceptions that have been based on narrowly gathered data. Specifically, this portion of our study is directed at answering the following questions:

- What are the demographic characteristics of users and nonusers?
- What are the motivations for using the Internet? How do these vary across current users, former users, and nonusers?
- How are users introduced to the Internet, and where do they seek support?
- What are the barriers to Internet usage?
- What are the least attractive features of the Internet and what changes would users like to see?

Finally, we sought to explore via statistical models Internet awareness and usage based on demographic variables to identify the key predictors.

METHOD

Data for our study were taken from an October 1995 national random telephone sample, surveyed by a commercial firm working under our direction. The survey yielded 2500 respondents, of whom 8% reported being Internet users, 8% reported being former Internet users, 69% reported having heard of the Internet but not being users, and 15% reported not having heard of the Internet. The sample of Internet users was augmented by a further national random telephone sample of 400 Internet users. Of the total of 600 Internet users, 49% reported being "long-time" Internet users—current users who started using the Internet prior to 1995, as distinct from "recent" Internet users—current users who started using the Internet in 1995. Both surveys asked about social and personality attributes, and demographic and occupational characteristics.

Our survey sample of 2,500 respondents has a close match on socioeconomic variables compared with the U.S. population as a whole. Based on comparisons with 1990/1991 U.S. Census data, respondents in our sample are similar to the national average in gender, ethnic mix, and age composition, but slightly wealthier and better educated. In addition to the higher income and educational levels, our data are subject to all the inherent limitations of phone surveys (Katz, Aspden, & Reich, 1997).

RESULTS

Demographics of Users and Nonusers Are Markedly Different

The demographics of the five categories of respondents (current long-time and recent Internet users, former users, nonusers who have heard of the Internet, and nonusers who have not heard of the Internet) are very different:

- Current long-time Internet users were more likely to be male (76% of our sample were male), somewhat younger than average, very well educated

(58% of our sample had a bachelor of science degree or better), and very much better off (59% had a household income over $50,000).

- Current recent Internet users were more likely to be male (55% of our sample were male), somewhat younger than average, considerably better educated (50% of our sample had a bachelor of science degree or better), and better off (49% had a household income over $50,000).
- Former Internet users were more likely to be male (58% of our sample were male), very much younger than average (60% were 34 years and under), and slightly better educated with average household incomes.
- Nonusers who have heard of the Internet were more likely to be female (58% of our sample were female) but otherwise close to average in terms of age, educational skills, and household income.
- A disproportionate number of black and Hispanic respondents reported not being aware of the Internet (21% were black, and 10% were Hispanic). In addition, this group was more likely to be female (64% of our sample were female), older (41% were 50 years and over), less well educated, and less well off (58% had a household income below $25,000).

Interest in Internet Fueled by Sociopersonal Development

Ranking of Reasons for Becoming Internet Users. We asked respondents (both users and nonusers who had heard of the Internet) to comment on the importance of various reasons for joining the Internet. Based on our survey, important reasons appear to be communicating with people using e-mail, getting information of both general and special interest, and keeping up-to-date—43% of respondents reported that sending and receiving e-mail is a very important reason; 41% reported the same for finding information about special interests; 36% for finding out general information of importance; and 35% for staying up-to-date with what is new. Collectively these reasons point to sociopersonal development being the major driver of interest in the Internet among our survey respondents.

Business, commercial reasons, and contacting new people appeared to be lesser reasons for using the Internet—28% of respondents reported that the Internet providing a business opportunity to make money is a very important reason for becoming an Internet member, whereas only 17% reported the same for the Internet being a convenient way to do banking, 16% for having contact with new people, and only 10% for the Internet being a good way to shop for needed items. These data are summarized in Table 5.1.

We examined how these views vary across Internet experience categories—nonusers, former users, and recent and long-time current former users. Broadly, long-time users, recent users, and former users share the same priorities, whereas nonusers have a different set of priorities. For each Internet experience category, we ranked the reasons on the basis of the proportion who answered "very important." These responses are shown in Table 5.2.

TABLE 5.1
Reasons for Becoming Internet Users (*N*=1,296)

How important are these reasons for becoming an Internet user? (in %)	Very Important	Important	Not Important at All
Send and receive e-mail	44	38	18
Find out information about special interests	41	46	13
Find out general information of importance	36	53	11
To stay up-to-date with what is new	35	48	17
Have customers/clients be able to reach you	32	31	37
Contact people sharing your special interests	29	45	27
Read messages and bulletin boards	28	50	22
Business opportunity to make money	28	38	33
A convenient way to do banking	17	30	53
Have contact with new people	16	40	44
Nowadays it is just a good thing to do	14	44	42
A good way to shop for needed items	10	33	57

The rankings by long-time users, recent users and former users are almost identical. The most important reasons appear to be communicating with people using e-mail (ranked 1st for long-time users, 1st for recent users, and equal 2nd for former users), getting information of both special (ranked 2nd, 2nd, and 1st) and general (ranked 3rd, 3rd, and equal 2nd) interest, and keeping up-to-date (ranked equal 4th, 5th, and 4th). Less important reasons for using the Internet are business opportunities (ranked 8th, 8th, and 5th), contacting new people (ranked 9th, 10th, and 10th), and providing a good way to shop (ranked 12th, 12th, and 12th).

The ranking of nonusers who had heard of the Internet is somewhat different—they rank staying up-to-date first, more highly than the other experience categories whose ranks are equal fourth, fifth, and fourth. Nonusers rank business opportunity to make money equal second, more highly than the other experience categories whose ranks are eighth, eighth, and fifth. Finally, nonusers rank sending and receiving e-mail sixth, much less highly than the other experience categories whose ranks are first, first, and equal second.

84

TABLE 5.2
Reason Ranking by Respondents Answering "Very Important"

How important are these reasons for becoming an Internet user—ranking by percent who answered "Very Important"	Long-time Internet User	Recent Internet User	Former User	Nonuser Heard of Internet
Send and receive e-mail	1	1	=2	6
Find out information about special interests	2	2	1	=2
Find out general information of importance	3	3	=2	=4
To stay up-to-date with what is new	=4	5	4	1
Have customers/clients be able to reach you	=4	4	7	=4
Contact people sharing your special interests	6	6	6	7
Read messages and bulletin boards	7	7	9	9
Business opportunity to make money	8	8	5	=2
A convenient way to do banking	11	11	8	8
Have contact with new people	9	10	10	10
Nowadays it is just a good thing to do	10	9	11	11
A good way to shop for needed items	12	12	12	12

These results suggest that sociopersonal development is the major driver for current and former users, whereas nonusers perceive the reasons for joining the Internet as somewhat different. Nonusers appear to be more strongly drawn to the Internet for the business reasons and the opportunities for staying up-to-date.

Ethnic Groups Vary on Priority of Business/Commercial Opportunities. We also examined how views on the importance of reasons for becoming Internet users varied by ethnic group (Table 5.3). We consider four categories: White and Asian users, Black and Hispanic users, White and Asian nonusers, and Black and Hispanic nonusers.

White and Asian users rate most highly sociopersonal development reasons—sending/ receiving e-mail, finding information, and staying up-to-date. Each of the other groupings believe sociopersonal development is important, but also see business/commercial opportunities as much more important than the White/Asian grouping.

Black and Hispanic users rate most highly having customers/clients be able to reach you, followed by sending/receiving e-mail and finding general information of importance. Interestingly, this group rates finding specialized information much less highly than White/Asian users.

TABLE 5.3
Ethnic Groups and Priorities of Opportunities

How important are these reasons for becoming an Internet user—ranking by percent who answered "Very Important"	Rankings			
	White/Asian User	Black/ Hispanic User	White/Asian Nonuser	Black/ Hispanic Nonuser
Send and receive e-mail	1	2	6	7
Find out information about special interests	2	7	1=	3
Find out general information of importance	3	3	4	=4
To stay up-to-date with what is new	4	4	1=	2
Have customers/clients be able to reach you	5	1	5	4=
Contact people sharing your special interests	6	9	7	8=
Read messages and bulletin boards	7	6	9	10
Business opportunity to make money	8	5	3	1
A convenient way to do banking	11	8	8	6
Have contact with new people	9	10=	10	11
Nowadays it is just a good thing to do	10	10=	11	12
A good way to shop for needed items	12	12	12	8=

White and Asian nonusers rate most highly finding information about special interests jointly with staying up-to-date, followed by providing a business opportunity to make money.

Black and Hispanic nonusers rate most highly providing a business opportunity to make money, followed by staying up-to-date and finding information about special interests. For this grouping sending/receiving e-mail is much less important than for White/Asian users.

Social/Work Networks Important for Stimulating Interest/ Providing Support

Routes Into the Internet. We then asked current and former users how they were originally introduced to the Internet (Table 5.4). Responses to our question indicated that social and work networks are important for introducing people to the Internet. Over half the respondents reported that they were introduced to the Internet either by learning at work or by being taught by friends or family. Only about a quarter reported being introduced to the Internet through a university or other formal course. The remaining quarter of respondents reported being self-taught.

TABLE 5.4
Routes Into the Internet

How were you originally introduced to the Internet (in percent)?	University Course	Other Formal Course[a]	Self-Taught	Learned at Work	Taught by Friends or Family	Number of Respondents
Former user	16	10	13	15	46	144
Recent Internet user	9	6	27	28	30	308
Long-time Internet user	19	7	22	35	18	293

[a] Other formal courses: school in general (21%), magazines/brochures (13%), through e-mail (10%), other (44%), don't know (10%); 61 respondents.

The proportions attributed to the various introduction routes varied across Internet user category. For long-time users, the most popular route to the Internet was via work, with 35% reporting being introduced this way. The next most popular route was being self-taught, with 22% of long-time users reporting being introduced to the Internet this way. Eighteen percent of the long-time group reported being introduced by friends or family—a much smaller proportion than for the recent and former users. About a quarter (26%) of long-time users reported being introduced to the Internet via a university or other formal course.

For recent users, the three main introduction routes were roughly equally popular: taught by friends or family (30%), learned at work (28%), and self-taught (27%). Fifteen percent of recent users reported being introduced to the Internet via a university or other formal course.

For former users, the most popular route into the Internet was being taught by friends or family (46% reported being introduced to the Internet this way), much more popular than the other routes—26% via a university or other formal course, 15% learned at work, and 12% self-taught.

Sources of Internet Advice. We also asked current Internet users to identify their sources of Internet advice. Again, we found that that it was social and work networks that were important. When asked "When you have a problem using the Internet, where do you first turn for help?" 35% of current users reported asking a personal friend and 24% a professional colleague. About 40% reported turning to formal help services—on-line help (14%), support person (10%), book or manual (10%), and phone service or technical support line (8%). There appeared to be no difference in the reported responses of long-time and recent users.

Regarding the last person that helped the respondent, we asked what was the helper's relationship to the respondent. In two-thirds of cases, users sought help from a coworker (39% of the time), friend (16%), and relative (14%), whereas in only 16% of cases was help sought from a specialist support person. Long-time users were more likely than recent users to use a colleague or a coworker or a classmate, and recent users were more likely to use a relative.

Online Support—Most People Do Not Use It. Earlier we reported that only a seventh of users turned first to online services when they had a problem. We probed the use of online help services further and found that most people reported not using them at all. Only 31% of current users reporting using online services at any time. Their use was distributed as follows: online reference files (14%), online list-servs or chatgroups (6%), online consultants (5%), and other online services (5%). Long-time users were more likely to use online services—34% of them reported doing so as compared to 28% for recent current users.

Even Experienced Users Perceive Significant Barriers to Getting Started

Our survey also explored the barriers to Internet usage as perceived by users and nonusers alike. Respondents were asked, "For someone who has not tried the Internet before, how difficult would you say it is to get started?" Surprisingly, the responses from former, recent and long-time current users were approximately the same: very difficult, 16%; a little difficult, 59%; and not at all difficult, 25% (791 respondents). Thus three-quarters of respondents believed that getting started on the Internet represents some degree of difficulty. We next probed how respondents perceived a number of potential obstacles to using the Internet.

Ranking of Barriers. We asked each respondent "How much of an obstacle is [rotated list of obstacles] this to you" on a scale "very much an obstacle," "an obstacle," "not an obstacle at all," or "don't know." The list of obstacles was: (1) no idea about how to do it, (2) costs too much, (3) no way to get access, (4) too complicated, and (5) uncomfortable sitting at a computer (e.g., eye strain or back strain). The results of our survey (Table 5.5) show that these five obstacles can be clustered in three groups in decreasing order of importance:

- Group I: cost (59% reported this to be an obstacle or very much an obstacle).
- Group II: no idea how to do it (48% reported this to be an obstacle or very much an obstacle); no way to get access (43% reported this to be an obstacle or very much an obstacle): too complicated (42% reported this to be an obstacle or very much an obstacle).
- Group III: discomfort using computers (21% reported this to be an obstacle or very much an obstacle).

We examined how these views vary across Internet awareness/usage categories—nonusers, former users, and recent and long-time current former users. Cost is a major issue for all, particularly for former users, nearly three-quarters of whom reported it being an obstacle. Having "no idea how to do it" is, perhaps not surprisingly, more of an obstacle to people who have never used the Internet (53% of whom reported it being an obstacle or very much an obstacle). Lack of access is, again as might be expected, more of an obstacle of for former users (of whom 51% reported it being an obstacle or very much an obstacle) and nonusers (of whom 47% reported it being an obstacle or very much an obstacle) than for currents users. However, users and nonusers are largely in agreement about the degree to which they see the complexity of using the Internet as an obstacle.

TABLE 5.5
Ranking of Barriers

How much of an obstacle is this to you? (in %)	Very Much an Obstacle	An Obstacle	Not an Obstacle at All	Don't Know	Number of Respondents
Costs too much	20	39	36	5	1,277
No idea of how to do it	15	33	50	3	1,283
No way to get access	14	29	54	3	1,280
Too complicated	9	33	55	3	1,282
Uncomfortable sitting at a computer	5	16	77	2	1,290

Users Want Traffic and Navigation Problems Addressed

Closely allied to the barriers to Internet usage are inhibiting factors that make usage difficult, even discourage usage and in the long-term might lead to people stopping using the Internet. Accordingly, we asked respondents, "What do you like least about the Internet?" Results indicate that the key concerns of users are primarily traffic and navigation problems, with cost a distant third.

About a quarter of respondents expressed concerns about traffic problems: "A lot of traffic/too slow" was cited by 18% of respondents and "delays/connection problems" was cited by 8% of respondents. About a fifth of respondents expressed concerns about navigation problems: "Difficult to find things/complicated" was cited by 15% of respondents, "difficulty in finding out what is there" was cited by 3% of respondents, and "not having a guide to the Internet book" was cited by 3% of respondents. Only 8% of respondents said "cost" was the least attractive aspect of the Internet, perhaps reflecting the relatively higher wealth of current Internet users, alluded to earlier.

Most Desirable Improvement—Make It Easier to Use. When asked, "If there could be one improvement in the Internet, what would that be?" respondents' focused on making the Internet easier to use (Table 5.6). Two-fifths of respondents said "make the

Internet more user friendly" or "easy/improved access" or "having a map address" or "more powerful search commands." These improvements address the navigation problems mentioned earlier. One in nine users wanted "quicker speed in accessing information," addressing the traffic problems also mentioned earlier.

TABLE 5.6
The Most Desirable Improvements to the Internet (in percent)

Make more user-friendly	15%
Easy/improved access	12%
Quicker speed in accessing information	11%
Having a map address	10%
Lower rates	7%
More security against theft or fraud	4%
More powerful search commands	2%
Other	10%
Need no improvement	5%
Have no idea/still learning	4%
Don't know	20%

STATISTICAL MODEL ANALYSIS

Although there is of course much value in examining frequency and contingency tables to explore our data, we also wanted to create a few simple models of our data that could be readily interpreted by the nonexpert. Hence we turned to log-linear models, which can provide a straightforward way to see the relative contribution of each of the variables while "controlling" for the others. In the models that follow, the chi-square statistics show the magnitude of effect (larger = greater a variable's explanatory power), whereas the significance level in essence estimates the likelihood such a result would be due to chance (smaller = less likely result due to chance). Adhering to our desire for easy interpretation, we dichotomized all variables and did not examine possible higher order interactions. In particular, we believe it important to dichotomize all variables at this point in light of the established biases of mixing levels of measurement (Blalock, 1971) and that some of the targets of our research seem inherently dichotomous, such as gender, African-American versus White race/ethnicity.

Model of Internet Awareness

We began by examining the relative impact of various demographic variables in explaining, in a statistical sense, who might be aware of the Internet in the first place versus those who were not. The log-linear model had the following dichotomous dependent variable:

Awareness or knowledge of the Internet's existence

with the following independent variables (also dichotomized): (a) gender; (b) age of respondent; (c) highest educational level of achieved by respondent; (d) household income of respondent; (e) ethnic /racial group of the respondent; and (f) number of children in the respondent's household.

As can be observed in Table 5.7, race/ethnicity was the most important explanatory variable in predicting who would be aware of the Internet, with people of White or Asian ethnicity/race being more aware than those of African-American or Hispanic ethnicity/race. Education, income, and age were also important explanatory variables. Children in the household was mildly predictive of knowledge of the Internet. Females were very little different than males in their likelihood of Internet awareness, controlling for the other demographic variables.

TABLE 5.7
Those Who Had Heard of the Internet Versus Those Who Had Not

Variable	Maximum Likelihood Estimate	Standard Error	Chi-Square	Significance Level
Ethnic background	−0.6784	0.0808	70.45	<.000
Age	−0.4313	0.0827	27.22	<.000
Educational level	0.5690	0.1104	26.55	<.000
Household income	0.5857	0.1147	26.06	<.000
Children in household	−0.2229	0.0764	8.52	.004
Gender	0.1248	0.0725	2.96	.085

Model of Internet Usage

To determine which variables were relatively more important in potentially explaining Internet usage we again applied a log-linear model. This model had the following dependent variable:

Internet usage in the group consisting of nonusers who had heard of the Internet and current users, both recent and long-time users

with the independent variables as specified earlier.

The significance levels of the parameter estimates and related results for the model are given in Table 5.8.

TABLE 5.8
Nonusers Who Had Heard of the Internet Versus Current Users

Variable	Maximum Likelihood Estimate	Standard Error	Chi-Square	Significance Level
Educational level	0.6322	0.0621	103.7	<.000
Age	−0.6744	0.0853	62.6	<.000
Household income	0.3925	0.0632	38.7	<.000
Gender	0.2602	0.0604	18.6	<.000
Children in household	−0.0697	0.0618	1.3	.260
Ethnic background	−0.1058	0.1022	1.1	.300

Judging by the chi-square and significance levels of the estimates of the independent variables, the key predictors of Internet usage, in decreasing order of importance, are education level, age, household income, and gender. Race and number of children in the household are not significant predictors. It should be noted that race/ethnicity differences show up strongly in relation to awareness or nonawareness of the Internet. However, race/ethnicity differences are not important in distinguishing users from nonusers in the subsample that had heard of the Internet.

DISCUSSION AND CONCLUSION

Our survey, carried out in October 1995, revealed that 8% of respondents reported being current Internet users. Somewhat surprisingly, our survey also revealed almost an equal number of former users, indicating considerable churn among Internet users. Lost access appeared to be the key reason for stopping using the Internet. Former users, a group hardly recognized in the published literature, warrant further study because analysis of their Internet experiences could point to ways of improving the Internet.

We found the demographics of users and nonusers to be very diverse. A disproportionate number of Black and Hispanic respondents reported not being aware of the Internet. Further, current users of the Internet were more likely to be male, younger, better educated, and wealthier. These results, if confirmed by other studies, point to a worrying "digital divide" based on race, gender, educational attainment, and wealth.

Interest in the Internet appears to be fueled by the desire for sociopersonal development. The main motivations for using the Internet are communicating with people, getting information on special interests, and keeping up-to-date. These appeal most strongly to long-time Internet users. Recent Internet users find them more attractive than nonusers. Business/commercial reasons (banking, shopping) were considered to be lesser reasons for using the Internet, and had more appeal to nonusers. Social (using the Internet to contact new people) and educational aspects were hardly valued at all.

Social/work networks appear to be important for stimulating interest in the Internet and providing users with support. In our survey over half of the users reported that they were introduced to the Internet either by learning at work or by being taught by friends or family. Surprisingly, only a fifth of current users had been introduced to the Internet through a university or other formal course—for recent users the proportion was even less. The picture is the same when Internet users are seeking advice. About 60% of current users report seeking advice from friends and professional colleagues. Very few people reported seeking advice from online support services. These findings have implications for the digital divide alluded to earlier. Through social/work networks the digital "haves" will be able to bolster their position, whereas the digital "have nots," being generally poorer and less well educated, may not have access to these important social/work networks. Thus, the digital divide may be perpetuated and perhaps even strengthened.

Even experienced users perceive significant barriers to getting started with the Internet. For users and nonusers alike cost is the main barrier. Other significant barriers are lack of understanding of how to use the Internet, lack of access to the Internet, and the complexity of navigating the network. Our finding concerning difficulty in getting started coincides with research done by the Home Net Project at Carnegie Mellon University (Kraut et al., 1996).

When asked about the least attractive features of the Internet, users' answers overlap with the preceding barriers. The key concerns of users are primarily traffic and navigation problems, with cost a somewhat distant third. It is clear users want an Internet that is easier to use, particularly with regard to tools for navigating the Internet and finding information.

Our simple statistical models strongly suggest that the fears of a digital divide may be well founded. Age, education, and income all help distinguish Internet aware and users from their counterparts, despite recent broadening in Internet participation (Husted, 1995; Miller, 1996b). However, the existence of a racial divide in Internet awareness is especially disconcerting. Certainly individuals, regardless of their social situation, should be free to decide whether they want to attempt to surmount the not inconsiderable barriers to become Internet users. But our initial model suggests that those of African-American or Hispanic heritage are much less likely to even be aware of the Internet in the first place, independent of age or education, relative to Whites. Lack of awareness of the Internet in the first place would of course make it difficult for any individual to elect to participate in this technology. The model of Internet use suggests that a gender differential continues to exist, despite the much greater participation of women in the Internet over the past few years (Swisher, 1995). Finally, the role of children in the household, which some have attributed to being an important driver to Internet participation

(Kraut et al., 1996), does not seem important in our simple model analyzing user/ non-user patterns. However, it is important relative to being Internet aware.

In sum, our preliminary results suggest that although the race/ethnicity divide among users and nonusers who are aware of the Internet is not highly significant, there continues to be a gender differential. Perhaps most disturbingly, there appears to be a prominent racial/ethnic divide between Internet aware and nonaware respondents (Wynter, 1996). Moreover, the bias of our survey (like similar phone surveys of the American public) in favor of richer and more educated people leads to an underestimation of the already significant magnitude of this divide. Consequently, the enduring concern over the inequities of access to the important social, economic, political, and personal resources—made possible by the Internet—appears to be justified in terms of what our research seems to be suggesting.

ACKNOWLEDGEMENTS

The authors thank the Markle Foundation for funding some of the research reported in this chapter. They also thank two former colleagues at Bellcore, Leora Lawton and Ming Ye, for helping, respectively, with early design and subsequent analysis of the questionnaire. Prof. Robert K. Merton has contributed helpful ideas throughout the course of this investigation.

REFERENCES

Anderson, R., Bikson, T. K., Law, S. A., & Mitchell, B. M. (1995). *Universal access to e-mail: Feasibility and societal implications*. RAND Corporation.
Anonymous. (1996, June/July). Caught in the net: What to make of user estimates. *Public Perspective, 6*, 37–40.
Batteau, A. (1996, Spring). Symposium on social aspects of the national information infrastructure. *Social Science Computer Review, 14:1*.
Blalock, H. M. (1971). Aggregation and measurement error. *Social Forces, 50* (2), 151–165.
Clark, D. (1996, June 7). A third of small businesses benefit from Internet. *American Banker*, p. 16.
Dataquest. (1995, May 22). *Internet users: Who they are; What they want*. Dataquest report OLST-WWW-UW-9501. San Jose, CA.
Edmondson, B. (1996, July/August). The point-and-click government. *Marketing Tools Magazine, 3:5*, p. 68.
Galifianakis, N. (1996, March 14). Internet help wanted. *USA Today*, p. 14 B-1.
Gupta, U. (1995, October 13). Home-based entrepreneurs expand computer use sharply, study shows. *Wall Street Journal*, p. B-2.
Husted, B. (1995, December 17). Average age of users declining. *Atlanta Journal Constitution*, p. B-7.

Katz, J., Aspden, P., & Reich, W. (1997). Public attitudes toward voice-based electronic messaging technologies in the United States: A national survey of
opinions about voice response units and telephone answering machines. *Be-
haviour and Information Technology, 16,* p. 125–144.

Kraut, R., Scherlis, W., Mukhopadhyay, T., Manning, J. & Kiesler, S. (1996, December). The HomeNet field trial of residential Internet services. *Communications of the ACM, 39:12,* p. 55–63.

Maney, K. (1995, June 20). Doing business on the Web is new status symbol. *USA
Today,* p. E5-1.

Miller, L. (1996a, June 10). Web surfers keen on politics and privacy. *USA Today,*
p. D-1.

Miller, L. (1996b, August 14). Net surfers becoming more mainstream, survey
shows. *USA Today,* p. D-1.

New York Times. (1996, February 27). AT&T makes belated move on Internet, p.
D-1.

Nielsen Media. (1996). Commerce Net survey [Online] http://www.nielsonme-
dia.com

Swisher, K. (1995, October 31). Internet's reach in society grows, survey finds.
Washington Post, p. A-1.

Trounstine, P. J. (1995, June 18). GOP candidates decide '96 campaigns will be
regulars on the Internet. *Detroit News & Free Press,* p. F-6.

Woodward, C. (1995, December 18). Candidates open up in Vote Smart survey.
Atlanta Constitution, p. A-9.

Wynter, L. E. (1996, March 6). Business & race: Survey of black hackers shows
an elite audience. *Wall Street Journal,* p. B-1.

6

Combining Sender and Receiver Payments in the Internet

David D. Clark
MIT Laboratory for Computer Science

There have been a number of proposals for usage-sensitive payment schemes for the Internet. This chapter considers one aspect of payment for service in the Internet, the problem of allocating payment between the sender and the receivers of traffic. It proposes a scheme, called *zone-based cost sharing*, in which the sender and the receiver can pay some proportion of the cost for use of the Internet. The chapter proposes that any practical scheme must permit cost sharing among the participants, and also proposes that this sort of pricing should be applied only to enhanced services, such as real-time traffic or predictable capacity under congestion, rather than the basic packet transport service of the Internet.

THE CURRENT INTERNET PRICING MODEL

The current payment model for the Internet might be called a *regional payment* scheme. As illustrated in Fig. 6.1, users attached to the Internet make payments to their access provider, who in turn makes payments to the regional or wide area provider who serves them, and this pattern has the effect that each user contributes to the payments for the part of the Internet that is "nearby." In the "center" of the network, the various wide area service providers interconnect at revenue neutral points, at which no money flows in either direction. The assumption is that these points represent a form of equilibrium in which the providers on each side have been adequately and, (by some approximation) equivalently compensated by the flow of money in from the edges, and thus there is no residual imbalance at these points in the "middle" of the Internet.

Figure 6.1
Flow of Money in the Current Regional Payment Scheme

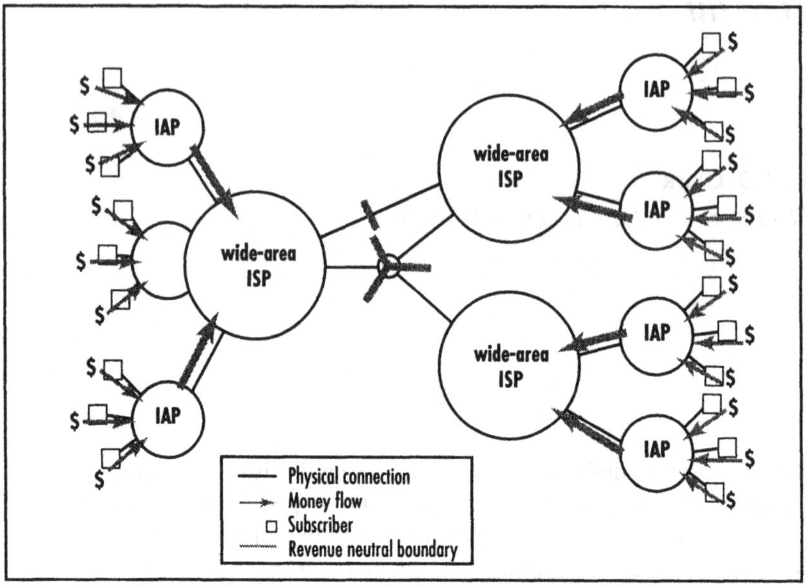

This scheme seems to work today, to the extent that it has allowed a healthy competitive market for Internet service providers to come into existence. However, we argue in this chapter that there are a number of limitations to this scheme.

What Is Wrong with Today's Scheme?

The basic issue with the regional payment scheme is that it provides no explicit way to measure or determine from user behavior what value each customer actually associates with the various uses of the network, and what the customer's willingness to pay for service is. There are three specific problems with the regional payment scheme in use today.

First, there is no way of demonstrating that the revenue-neutral boundaries are actually in the correct place in the Internet. The assumption that this boundary is at the points of connection among the major wide area service providers is just a simplifying approximation negotiated by the various providers in order to avoid disputes.

Second, there is no effective way to rationalize the payment for expensive components in the "middle" of the network. A specific example is a transoceanic link between different Internet service providers on opposite sides of the ocean. If this were a terrestrial connection, it would be assumed to be a revenue neutral point. But a transoceanic link is a very expensive facility, and this expense triggers debate as to how the cost should be shared. In the absence of any explicit way to determine the value of this link to either side, the providers are again forced to make a negotiated assumption. For example, if a small country connects to a large country, it is often assumed that most of the value is to the smaller country (and that the small country has little bargaining power), so the small country pays for the whole link.[1] Whatever the resolution, the balance of payment is not derived from actual indication of value to the different users.

Third, the regional payment scheme forces all users to fit into the same assumed balance of values. There is no way to identify those users who would pay more to have better service and serve them appropriately. There might be substantial additional revenues to be extracted from customers if there were a mechanism to serve and bill them according to their individual service objectives, but the static regional scheme today, which imposes a uniform assumption of payment flow, precludes serving each customer according to need.

There are some benefits to the current static scheme, as well as problems. First, of course, it is simple to implement, requiring no data gathering, traffic monitoring, or control inside the network. Second, it forces each region to be as cost-effective as possible, because all the costs of each region must be recovered from that region. However, there may be circumstances in which there is a benefit if funds can flow across some of the presumed revenue-neutral boundaries. For example, if there is a particular region of the Internet that has higher intrinsic costs to serve, that region is forced to carry all the costs of provisioning that region. There is no way for funds to flow in from outside, even if the collective users of the network would see value in having this happen.

DIFFERENT USERS—DIFFERENT VALUE

A good example of different value presumptions can be found by looking at World Wide Web servers on the Internet. One Web server might be a commercial site that is very interested in attracting users, and be willing to pay substantial fees to the network provider if this successfully subsidizes customers who thus are willing to explore the site. A second commercial Web site may believe that the content provided there is of such intrinsic value that the user should pay all the network costs of accessing

---

1. This cost appears to be a noticeable component of the higher prices seen for some overseas Internet service.

it.[2] In contrast, a third class of Web server, offering content on the network essentially as a public service, may not be able to afford to pay any network fees to support access to this data by users. The assumption in this case is that if a user assigns sufficient value to the information to pay for its retrieval, then the transfer is justified.

From the perspective of the user, rather than the provider of content, there may be a class of potential users who assign so little value to the overall use of the Internet that under the static regional payment scheme, they refrain from subscribing. However, having these customers on the network may be of sufficient marginal value to some commercial content providers that they would be willing to see their payments flow toward these users to subsidize their basic access.

Today, all of these customers, both users and content providers, are required to purchase service within the context of the same regional payment scheme. There is no real way (except by attaching to different physical networks with different service features and pricing) for the various classes of users to purchase different sorts of services; there is no current way for one service provider to serve all of them with the same Internet infrastructure. One of the goals of the mechanisms described in this chapter is to permit this flexibility in service offering.

There are parallels to this range of assumptions in other media. Television broadcast, at least in the United States, is free to consumers, under the assumption that the advertisers derive sufficient value to pay for the transmission. Cable and satellite companies challenged this assumption, and proved that they can extract substantial payments from consumers for access to television. Public broadcasting attempts to fit into this structure without advertising revenues, and thus has to step outside the normal framework of revenue flows, and ask the consumers to contribute directly to the payment for the production and transmission in a rather ad hoc way.

The telephone system reflects this variation in willingness to pay, with callers paying for certain of their long-distance calls, whereas merchants offer 800 service, in which they pay the per-call costs, in order to attract customers to call. The 800 service has added substantially to the total revenues derived from phone usage, and by analogy, a service with similar objectives might add to the total revenues derived from the Internet.

ASSIGNING VALUE TO FLOWS

Although a service that captures the same objectives as the telephone 800 service might be useful in the Internet, one cannot carry the details of the 800 service directly into the Internet. The Internet has no concept of a call, packet flows in each

---

2. This chapter is concerned with the structure of payments for Internet service itself, not with payment for use of higher level services such as for-fee Web sites. A suitably designed payment mechanism might be extended to cover the latter function as well, but that is not the primary focus of this chapter.

direction are not linked, and today there is no indication in a packet as to how value should be associated with it. A scheme with different structure will have to be devised to match the nature of the Internet.

To begin, let us introduce the term *flow*, which is often used to describe a sequence of packets, going from a specific source to a specific destination (which could be a multicast address describing multiple physical destinations), that share a common set of service requirements and value assumptions. A flow could be associated with the retrieval of a Web page, a file transfer, a video stream, one direction of a telephone call, and so on. Although there are often packets flowing in both directions as a part of a network interaction, we model this as two flows, one in each direction. The goal of a pricing scheme must be to deduce how much value is associated with a flow by each *participant* (sender and set of recipients), and charge the various parties accordingly.

One approach to offering and procuring service is that the end nodes will perform a separate real-time determination of their willingness to pay for each flow of packets that they deal with (like dealing separately with each collect call). This degree of dynamic interaction and control is probably not what the customers want. Instead, they probably want to enter into some long-term agreement that they will pay in some proportion for flows of various sorts (like an 800 service contract), perhaps in the form of a fixed fee for a certain level of service. This requires that there be some way for the customers to describe the sorts of flows they wish to pay for.

Assume that the users of the network—the sources and destinations of flows—had the capability to declare explicitly what value they associated with their flows. What information would such an expression of value contain?

In general, it would capture two aspects of the value equation: first, the performance parameters of the service that the flow should receive, and second, the price that the user will pay. It might be that the price is not expressed directly, but is derived as a function of the requested service. However note that, both predictable price and predictable service must be part of the service offering. Because there are at least two participants to a flow—the sender and one receiver, and in the case of multicast there can be many receivers, it should be possible for each of these participants to share in paying the cost of the flow. These points are discussed later.

It Must Be Possible to Specify the Desired Service for a Flow

In contrast to the telephone system, where all calls require the same network capacity and the same quality of service, flows in the Internet can differ widely in their need for capacity, control of delay, or other features. Especially in the context of associating value with enhanced services, it must be possible for the user to describe the service desired.

There have been a number of proposals for how services should be characterized. Some proposals, such as the committed information rate (CIR) of frame relay, allow the subscriber to specify a service in terms of a minimum bandwidth
(Stallings, 1992). Some, such as the discard eligible (DE) bit of Frame Relay or
the cell loss preference (CLP) bit of ATM (ATM Forum Technical Committee;
1996), allow the user to specify a loss preference for various packets in time of
overload. The expected capacity scheme described in Clark (1995) allows a user
a wide variation in the burstiness and bandwidth of the service that the user requests. The guaranteed service (Shenker, Partridge, & Guerin, 1996) of the Internet, intended for real time flows, allows the user to specify the average rate and
maximum burst size of traffic and then find out the end-to-end worst-case delay
for the packets in the flow. In contrast, the controlled load service of the Internet
(Wroclawski, 1996), also intended for real-time traffic, allows the user to specify
the average rate and maximum burst size, but, although trying to reduce the delay
of the packets, does not provide to the user any explicit indication of the maximum
delay to anticipate.

This chapter is not concerned with which sorts of services are available, or
which approach to specification is used, as long as the definition is consistent
across the Internet and conforms to certain requirements as discussed later. The
point is that any proposal for usage-sensitive payment schemes must be expressed
in the context of payment for something. The range and nature of the services that
might be invented in the future is difficult to predict. A general pricing scheme
must thus be able to control the use of enhanced services, without knowing in detail what these services are. What is needed is to understand enough about the general characteristics of these services to build suitable cost allocation mechanisms.

Benefit of Shared Payment Schemes

Consider again the current payment practice in the Internet, the static regional
payment plan. The consequence of this pattern is that the origin and destinations
of a flow share the cost of moving the data.

The long-distance component of telephone service has a different pattern: Either the sender pays all the costs, or (for collect or 800 calls) the receiver pays all.
This "all-or-nothing" pattern has a drawback, which only occasionally becomes
apparent. The 800 numbers can only be called from the region of the telephone
system within which the receiver agreed to make full payment. If the caller is further away than this, the call cannot be completed. It is necessary, in that case, for
the caller to find an alternative number for the receiver, a number associated with
the "sender pays all" pattern, and call that number instead.[3]

---

3. This feature has caused callers in Europe a degree of frustration, because many U.S.
companies only put their 800 number in their ads.

A more general payment structure would be for the sender and the receiver to share the service-specific costs, just as with today's static regional payment scheme they share the overall network costs. This requirement is probably even more important in the Internet than in the phone system, because, as noted earlier, there is a wide range of services that the Internet offers, not just a phone service. This suggests that there will probably be a much wider variation in willingness to pay, depending on the service. Therefore, this chapter proposes that any payment scheme proposed for the Internet should be a shared payment scheme, in which all the participants in a flow can potentially contribute to payment, each according to their willingness to pay.

How to Define Scope of Payment

There are a number of ways that the sender and receiver of a flow could express their partial willingness to pay. The scheme proposed in this chapter is that each end of the flow (source and destinations) declares the region of the network in which it will pay the service-specific costs. For example, a Web site might agree to pay all costs for flows within the United States. Another might only pay costs local to a city, whereas another might not pay any costs beyond their local site.

This pattern for service-specific payments is similar to the static regional payment scheme, and has the following features:

- At each point in the network, either the sender or some set of receivers is paying the cost. It is thus clear which way the compensation should flow at any point in the net, in particular, at interprovider connection points. (The scheme must be implemented in a way that makes this information available; discussed later.)
- The cost of agreeing to a certain service can be determined in advance. The cost of a service within a certain region is determined only by the prices charged within that region.
- There is an obvious way to deal with the case when the total willingness to pay from the sender and receivers is "not enough" to cover the total cost. In a regional payment scheme, the manifestation of insufficient willingness to pay is that a region of the network "in the middle" is uncompensated for providing service. A reasonable way to deal with this would be to give the flow only the lowest class of service across that region. For example, the traffic could be the first to be discarded in case of congestion. This is discussed later.

Preserving the Basic Service of the Internet

As noted earlier; if there is insufficient total willingness to pay, the consequence should be reasonable and predictable. One feature of the current regional payment scheme is that it provides complete connectivity. In almost all cases, any two parties on the Internet are assumed to be able to exchange packets. Anything that disrupts this feature is considered a serious flaw. Per-user boundaries for payment run the risk of having this consequence. If individual senders and receivers need to fully cover the value of a flow before it can be accomplished, then either a complex negotiation must occur before packets can flow—which is completely at odds with the current design of the Internet, which has no "setup" phase before starting to send—or there is a risk that if a particular sender and receiver do not happen to associate enough value to a flow, it is simply not carried across the Internet.

One way to deal with this issue is to assume that there will be a range of network services—a basic service and enhanced services. If we assume that the basic service is similar to what the Internet offers today, with universal connectivity, a simple subscription pricing scheme, and no assurances of service quality, then users can assume that the Internet of tomorrow will at least be similar to the Internet of today. Then more advanced pricing can be associated with more advanced services, which might, for example, give assured access during times of congestion and so on.

The proposal here is that the willingness of the user to pay should control the willingness of the network to offer enhanced services to the flow of packets.

IMPLEMENTING THIS PROPOSED SCHEME

The previous section proposed a specific approach to assigning value and collecting costs for packet flows in the Internet. This section discussed how this might actually be implemented, given the existing details of the Internet design.

A Framework for Implementation

Each router inside the net, when forwarding the packet, must be able to determine two things easily: what enhanced service is requested for the packet, and whether the sender or some set of receivers is willing to pay for that service. Because routers are high-performance devices, this determination must be easy to make, and must not require any computational complexity.

In order to determine which participant is paying, each router, on receiving a packet, must be able to determine whether the router is within the sender's region of willingness to pay for that packet, the receiver's region, or neither. This information has to come from somewhere. Mechanically, there are only two ways that a router can have access to information about a flow. It is stored either in the router, or in the packets of the flow. There is no other option.

Information about a flow stored in a router, called *per-flow state*, is in general not a preferred design approach for the Internet. Either it is put in place in the routers as it is needed, which requires a dynamic protocol for setting up the state, or it is stored as long-term tables, which might get very large, because there are tens of millions of endpoints today, and the net is growing rapidly. The only viable approach to the long-term storage of state information in the routers of the Internet is to group users with similar willingness to pay into large aggregates, rather than describing each end-node separately.

The alternative to the per-flow state is to put the information in the packet. This has the advantage that the flow is self-describing, but has the disadvantage that the amount of information that can be carried in a packet is very limited, because as a practical matter there are at most a very few bits in the header that could be assigned to this purpose, This implies that any practical scheme must involve some creative approximation, which allows for a reasonable flexibility in expressing willingness to pay, although still reducing the information that must be in both the routers and the packets. However, if a reasonable approximation can be devised, the avoidance of per-flow state is very desirable. So a starting point for a design should be to maximize the use of packet-borne information.

One approximation for expressing a region of willingness to pay is a zone system. The Internet could be divided into zones within which service is provided at a uniform, distance-insensitive way. A possible set of zones might be:

- Within site
- Within city
- Within region
- Across wide area in country
- End-to-end within country
- Transocean
- End-to-end

Many systems use zone approximations to provide cost tiers, and it seems a reasonable approach here. In fact, the current regional payment scheme could be viewed as a very coarse static zone system. The zones, however defined, would have to have meaning that is globally agreed, but in different parts of the world there could be different bindings for different zones. This chapter uses this proposal, which we call *zone-based cost sharing*, as an example of a scheme that takes very few bits to express but potentially has enough flexibility to express a reasonable range of user options.

The Crux of the Implementation Problem

Proposing a zone system does not solve the implementation problem. There is an essential asymmetry in the network: Packets flow from sender to receivers. This very obvious point has important implications. When a packet leaves the sender, the sender can put information in the packet that directly expresses that sender's willingness to pay. For example, the packet could contain an indication of the payment zone of the sender. But it is much harder for the packet to carry information about the receiver, for several reasons. The most obvious reason is that the sender may not know about the zone within which the receiver is willing to pay.

In the case of unicast, it is possible in principle to carry this information. A control packet could be specified that carried information about the receiver back to the sender, and this information could then be put into the packets headed toward the receiver. However, for multicast this approach does not work. There is an unknown number of receivers, potentially in the thousands. The packet does not have room in the header to carry information about each of them, and the control information necessary to carry the information about all of these receivers back to the sender could be overwhelming. In fact, it is a design goal of Internet multicast that the sender does not need to know the identity of the receivers. So, at least in the case of multicast, information about the receiver's willingness to pay cannot be carried in the packet from the source. Instead, for multicast, the information about the willingness of the receiver to pay must be stored as a state within the routers along the multicast tree, and the control messages will have to flow to these points to install this information, thus creating per-flow state.

At worst, this would imply that there would be a control message sent for every enhanced service flow in the network, and a state stored for every node in every multicast tree in the network. However, there is a technique that can be used to reduce the need for control information. Once the packet leaves the region in which the sender is willing to pay, it initially carries no information about the payment zone of the receiver. However, it still carries the indication that it requires enhanced service. As it passes through the net toward the receiver, it encounters a sequence of routers, each of which attempts to provide that service. If the router is congested, as the packet is forwarded according to the request for enhanced service, the router marks the packet with the zone in which congestion is first encountered. The packet then makes its way to the final receiver. At the attachment point for the receiver, there is a new component in the network, called a *receiver traffic meter*, which has been set up to know the receiver's willingness to pay. It examines the incoming packets, and sees if, for the zone and the enhanced service indicated in the packet, there is willingness to pay. If there is, the packets in the flow are allowed to continue unmodified. If, on the other hand, the receiver is not willing to pay for this enhanced service in the indicated zone, a *zone exception* control message is sent back, either to the source or to the suitable point on the multicast tree, reporting the zone within which the receiver is willing to pay for this enhanced service. This fact can then be remembered in routers as needed, or (in the case of unicast) written into future packets at the source. Thus, a control message need be sent, and a state entry recorded in a router, only in the case where there the zones of the sender and receivers do not properly overlap, or where there is a mismatch in the enhanced service that the various participants are willing to pay for.

This zone-based scheme makes use of both the per-flow state in the router and the information contained in the packets. How poorly does this scale?

In the case of unicast, the sender (once it has been informed by the zone exception message) can note the receiver's zone directly in future packets. So only the sender (or the traffic meter at the source) needs to remember the receiver zone, and the complexity there scales as the number of ongoing transfers from the sender within some time period. This is not an excessive amount of information to remember. In the cast of multicast, information must be remembered at each point in the multicast tree where the tree branches, and recipients down each branch have different zones of willingness to pay. So the zone exception control message must make its way backward up the multicast tree until it reaches a branch point, and trigger the installation of a piece of information. Although this is perhaps more complex than would be desired, it turns out that the multicast protocols already have a message with exactly this behavior, the *prune* message. And in fact the zone exception message is exactly a form of prune; it does not prune the receiver from the tree, but it prunes it from the enhanced service along (part of) that tree. So adding this new control message and router state information does not add a new form of overhead to multicast.

INTERPROVIDER PAYMENTS

As already described, the preceding scheme provides a crucial piece of information that is needed at interprovider points of connection. If a packet requests an enhanced service, then it carries, in the zone fields, an explicit indication of which end is willing to pay the service-specific costs for this flow at this point in the net. The providers can look at these indications to determine whether there is an imbalance in the expressed value flowing in each direction. The providers will still have to negotiate the price each charges the other to carry various sorts of enhanced services to/from specific endpoints, but the raw information as to the direction of the value flow is now present in the packets.

The cost of carrying a packet will depend on how far it is going, and thus a simple count of packets in each direction is not adequate. However, the information needed for accounting is very simple. For a packet in the region of the sender, where the compensation is flowing in the direction of the packet, the total cost of the packet depends on how much further it is going, which is determined solely by the sender zone. In the region of the receiver willingness to pay, where the compensation is flowing counter to the packet, the total cost depends on where the packet came from, which again can be determined looking only at the receiver's zone. Thus, what the two providers need to keep track of is how many packets (or bytes) marked with which zones have crossed the interface. More detailed information, such as the source and destination addresses, is not required.

MARKET FORCES

If we had a scheme like this in place, how might the market actually evolve? Consider a particular example to illustrate the issues that might emerge.

Determining the Costs

The region scheme, in principle, isolates the costs for each user to the region of willingness to pay. But it does not necessarily make it easy to determine what the total cost will be. Consider the earlier zone example of "end-to-end within country", illustrated in Fig. 6.2. From the sender's perspective, this would cover the Internet access provider (IAP) of the sender, some number of wide area Internet service providers (ISPs), and the remote IAP serving the receiver of the flow at the "other end" of the network. Using the United States as an example, there are hundreds of IAPs serving customers, and each might set a different price to deliver a flow to an attached user. One of these distant IAPs could set a very high fee for incoming service, and the sender would then discover that he or she has just paid this very high fee for a transfer to that region. How can an arrangement be established, such that a sender of information across the Internet could agree to pay for service-related charges across a distant service provider, and still have some assurance of what the cost would be?

Figure 6.2
Sender Pays, End to End. Can the Remote IAPs Set Independent Prices,
or Will the ISPs Dictate a Common Price?

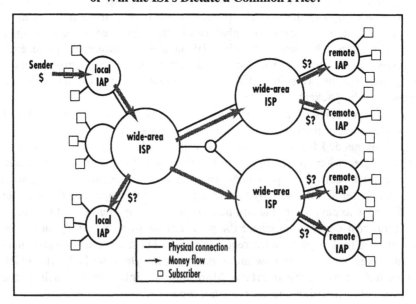

One possibility is that the wide-area ISPs will find themselves in the role of rate setters, in order to create the market. An ISP will explore the price structure of the various of the IAPs, and then set some fixed amount it will be willing to pay to any distant IAP to carry flows with a given enhanced service. If a distant IAP will not carry the traffic for the rate that the ISP will pay, then the ISP will turn off the bits in the packet requesting the enhanced service. The distant IAP will then receive no payment, and the flow will not receive enhanced service across that portion of the network. In this scenario, the ISPs will have to pick a target price that reflects some norm of the price asked by all the potential distant ISPs, and the process of picking this price will be a major factor in the evolution of the overall pricing structure. If the ISP sets too low a price that it will pay distant IAPs to carry sender-compensated traffic, then many distant IAPs will be excluded from the payment region because their costs are too high. If it sets too high a price, potential customers may seek out alternative ISPs to carry their traffic.

Just How Far Should the Money Flow?

Today, the assumption of most IAPs is that they must recover all their costs from the subscriber payments. They make no assumptions of any other income, either from regulatory subsidy or from some free market structure. This assumption, if sustainable, will greatly help the market, because it removes the need for substantial payments to flow across the IAP interface. But the question of whether the market will "stall" is an interesting one. Additional sources of revenue into the IAP, from remote participants with high willingness to pay, or from some other source, such as advertising, might provide a means to lower attachment costs and bring a larger pool of users onto the network.

In the extreme, an IAP could set itself up with a price structure that had very low attachment costs for users, hoping to recover most of its costs from the income derived across the network from distant participants willing to pay. This pattern would match somewhat the provision of "free" broadcast television, and might permit pricing that would attract potential users with very low willingness to pay. It is not clear, given the pricing structure of the current market, how this model could come into existence. But some sort of pricing based on expressed value, together with a means for funds to flow among ISPs, would be essential.

OTHER ROLES OF PRICING

Another role of pricing is to shape user behavior in socially desirable ways. Attaching some incremental cost to network usage tends to discourage low-value usage. This is viewed as a disadvantage when it discourages exploration of new uses for which the value is not clear. It is viewed as an advantage if it discouraged annoying usage such as junk e-mail. In this context, it is worth noting that some usage is sender-driven, and some receiver-driven. Junk e-mail is sender driven, so the pricing scheme would have to include a "sender-pays" mode to discourage mail sending. Casual Web browsing, which can clog servers, is receiver initiated, and could only be controlled by a "receiver-pays" mode. So if we anticipate the use of payments as a social control, both sender and receiver modes must be supported. This leave unanswered how the correct payment mode would be forced upon the user for the particular uses.

WHAT HAPPENS IF NOTHING HAPPENS?

This zone-based cost-sharing scheme is rather complex, with new fields in packets and new protocols to allow the receivers to express their willingness to pay. This may not come to pass, and certainly will not come to pass unless there is enough anticipated benefit compared to the consequence of taking no action.

How might pricing evolve if no architectural extensions are put in place? The most obvious possibility is that service providers will offer separate networks (how the separation is achieved, and at what level of the protocol stack, is not relevant), and offer different levels of service on these different nets. Because there would be no specific flags in packets to select the enhanced service, all users of a given net would see the same service. The most obvious version of "enhancement," which is already beginning to show up, is a lightly loaded network, which simply makes "things work better." Because these networks will have different addresses, the addresses become an indirect way of expressing the scope and nature of the service and willingness to pay. All the subscribers have to accept the same scope of willingness to pay, and the same service objectives, because there is no field in the packet to convey individual preference.

For example, as a way to offer an "enhanced" service, a provider might construct a network that was designed to offer uncongested access from the United States to points of attachment on the major continents. Local IAPs in other countries could then connect to this net (presumably at low cost) and the users in those countries would discover that they had very responsive access to sites (such as commercial Web servers) attached to this high value network in the United States. However, the network would not offer transit service back to sites on other networks in the United States, so another (perhaps much more congested) path would have to be used to reach sites that would not pay to attach to this high-performance network.

The addresses on this "enhanced network" function somewhat similarly to 800 addresses in the phone system. Users connecting to these addresses would observe that connections "cost less" or "work better" than connections to other sites. Users might thus be tempted to use them preferentially. It might be possible to make this feature easily visible to users to convey this appeal.[4] These addresses have the additional advantage, relative to 800 numbers, in that they support a (static form of) copayment, so that they can be used from any part of the Internet.

This sort of specialized ISP offering is bound to happen in the short run. The interesting speculation is whether this approach will provide enough ability to allocate costs so that no further mechanism is demanded. With the simpler scheme, there is no way for a user to change the service it purchased from the network except by changing to a new network provider, which implies renumbering all the hosts of the user. There is no way to select among a range of enhanced services on a dynamic basis, so services such as real-time telephony and bulk transfer cannot be differentiated. There is little likelihood that payment flow can be provided that reaches all the way from the sender to the receiver, to facilitate the class of "cross-subsidy" discussed earlier. However, the scheme might still provide enough service and price differentiation to succeed in the marketplace.

CONCLUSIONS

Summary of Mechanism

This chapter proposes a set of mechanisms that together permit flexible pricing and cost allocation for services on the Internet. The key assumptions of the proposal are the following.

- Both sender and receivers should be permitted to share in the payment for services. This chapter proposes a zone-based cost-sharing model, in which each user identifies a region of the network within which the user will pay. This matches the existing static payment scheme for basic network service.
- Usage sensitive payments should be associated with enhanced services, such as real-time flows or flows that are assured some expected capacity in times of congestion, and not the basic service of today, which provided universal connectivity without much assurance of service level.
- Each participant in the flow should be able to predict that user's component of the costs.
- It should be possible, at interprovider connection points, to determine easily the direction in which compensation for the flow should go.

---

4. In contrast to 800 numbers, where the meaning of the 800 prefix is well known, Internet addresses are not usually seen by users. Instead, users see higher level names, such as domain names and World Wide Web names. Those do not directly indicate the address to be used. So the sender's willingness to pay for the transfer needs to be made evident in some way other than the address prefix. Some service icon on the Web page pointing to the object is one approach.

- The scheme should minimize the need for dynamic, distributed protocols to compute total end-to-end costs, or cost allocation formulas. Local costs should be locally determinable.
- Multicast is an important service, and any pricing scheme should be
  compatible with multicast. Multicast is relevant because it implies that
  there may be more than one recipient of a flow.
- An advanced pricing scheme must not break the current operation of the
  Internet.

The prices being discussed in this chapter are not necessarily dynamic per-flow
prices. What is more likely is that providers will enter into long-term contracts that
set the rates that they will pay each other to offer various enhanced services. These
interprovider arrangements will specify such things as the peak loads or long-term
limits on the total amounts of enhanced service traffic, and other parameters, such
as time of day. These expectations about service levels and prices will then become part of the overall context in which individual participants in flows express
a willingness to pay for service across different regions of the net. In most cases,
those individual arrangements will also be long-term agreements.

Other, simpler approaches to supporting users with different willingness to pay
are possible. The most simple approach is the construction of different Internets
with different degrees of overcapacity and different regions of coverage. These
will attract different sets of customers. This approach can be understood as an approximation to the scheme being proposed here, with major limitations but the potential to be "good enough."

Support for Explicit Pricing Schemes

The zone-based scheme assumed that the packet carried, in addition to the zone
information, the indication of the enhanced service that the packet requested.
There is another class of pricing scheme, in which the packet carries a bid for service, rather than describing the service parameters. These schemes include the
smart market proposal of MacKie-Mason & Varian (1993) and the priority scheme
of Gupta, Stahl, & Whinston (1997). Some adjustment of these schemes may be
required in order to fit them into a zone-based scheme.

The first issue is that, to be compatible with a cost-sharing model, any pricing
scheme must come in two forms, which express respectively the willingness of the
sender and the receivers to pay. The schemes mentioned earlier were described in
terms of packet fields that express the sender's willingness to pay. The second issue is that those schemes, by intention, compute the actual price to be paid for service in a very dynamic way at the points of congestion. This implies that there
must be some means to reflect these charges back to the sender and forward to the
receiver, and thus to relate the bid carried in the packet to the total cost, which depends on the price set at each point of congestion. In contrast, the zone scheme

assumes that the cost for any specific enhanced service can be computed in advance for each zone, so that the network nodes can just count the packets or bytes going to and from each zone, and compensate each other based on the aggregated traffic flows. The sender and receiver can know in advance the cost of agreeing to a particular enhanced service across some particular zone. No per-flow accounting need be done inside the network.

Research Agenda

This speculative design of a cost-sharing model raises a number of questions, mostly related to user requirements, economics, and business structure. The first step is to validate the need for a copayment model in which the receiver as well as the sender can express willingness to pay. Following that, the task is to explore the utility of a region scheme, and a zone scheme as an approximation to allocate shared payments, and consider what granularity of zones/regions will meet the need of the market.

Another set of questions is whether this market of providers compensating each other based on value of flow is similar to other markets that have useful or known properties. Is this a stable market with proper incentives? Will competition at the various levels cause the market to converge to a useful operating point?

Finally, a critical question is how much better this zone-based scheme is than the "do-nothing" alternative. That model, with different ISPs building different networks with different service levels and different regions of coverage, is the likely outcome if no extensions to the Internet are undertaken.

ACKNOWLEDGEMENT

This research was supported by the Advanced Research Projects Agency of the Department of Defense under contract DABT63-94-C-0072, administered by Ft. Huachuca. This material does not reflect the position or policy of the U.S. government, and no official endorsement should be inferred.

REFERENCES

ATM Forum Technical Committee. (1996). *Traffic management specification 4.0.*
Clark, D. (October 1995). Adding service discrimination to the Internet. *Proceedings of the 23rd Annual Telecommunications Policy Research Conference (TPRC)* Solomons, MD. Also appeared in *Telecommunications Policy, 20,* 169–182.
Gupta, A., Stahl, D., & Whinston, A. (1997). Priority pricing of integrated services networks. In L. McKnight and J. Bailey (Eds.), *Internet economics.* (pp. 323–352). Cambridge, MA: MIT Press.

MacKie-Mason, J. & Varian, H. (1993). *Pricing the Internet*. Presented at Public Access to the Internet, JFK School of Government. Available [Online] ftp:/ /ftp.econ.lsa.umich.edu/pub/Chapters/Pricing_the_Internet.ps.Z

Shenker, S., Partridge, C., & Guerin, R. (1996). *Specification of guaranteed quality of service*. Available [online]: ftp://ftp.ietf.org/internet-drafts/draft-ietf-intserv-guaranteed-svc-06.txt.

Stallings, W. (1992). *ISDN and broadband ISDN with Frame Relay and ATM*. Englewood Cliffs, NJ: Prentice Hall.

Wroclawski, J. (1996). *Specification of the controlled-load network element service*. Available [online]: ftp://ftp.ietf.org/internet-drafts/draft-ietf-intserv-ctrl-load-svc-03.txt

Evaluating and Selecting Digital Payment Mechanisms

Jeffrey K. MacKie-Mason
University of Michigan

Kimberly White
Price Waterhouse, Washington, DC

Money is not an invention of the state. It is not the product of a legislative act.
The sanction of political authority is not necessary for its existence.
— Carl Menger

The Internet is growing rapidly as a marketplace for the exchange of both tangible and information goods and services. Numerous payment mechanisms suitable for use in this marketplace are in various stages of development.[1] A few have been implemented; most have been merely proposed or are undergoing trials.

Potential participants in electronic commerce are having difficulty evaluating and selecting payment mechanisms because the field is in constant flux. To begin with, there are many methods for making payments. Consider the tremendous variety of familiar mechanisms: coin, bills, personal checks, cashier checks, money orders, credit cards, debit cards, and so forth.[2] Each of these has multiple digital analogues. Information about the alternatives is limited and costly to obtain; even

---

1. For a listing of dozens of proposals and links to further information about them, see the Net Commerce page on the Telecom Info Resources Directory (URL: http://www.spp.umich.edu/telecom/net-commerce.html). Payment protocols in use at the time of this writing include DigiCash, NetCash, NetCheque, First Virtual, CyberCash, NetBill, and traditional credit cards using PGP and other encryption.

implemented mechanisms have been in use for only a short period of time. Further, all mechanisms, including those already in use, are being continually modified. With conventional money we may "know it when we see it,"[3] but most decision makers have only just begun to see digital money, and they don't know much about it yet.

We offer both information and a decision strategy to assist decision makers evaluate and select digital payment mechanisms. The information is in the form of two matrices that characterize numerous payment mechanisms according to about 30 characteristics. We have prepared this characterization for 10 leading digital mechanisms, and also, for comparison and benchmarking, for 7 different conventional payment mechanisms. In the main text we focus on the method and its application; the complete definitions of the characteristics are provided in Appendix 1.

The decision strategy we discuss is not very sophisticated. However, it is fully consistent with rational choice theory, feasible, and generalizable. It also has the advantage of focusing costly information gathering on those aspects of the mechanisms that are relevant to the decision, thereby limiting wasteful efforts.

We apply the decision method to the problem of selecting payment mechanisms for the University of Michigan Digital Library (UMDL) project.[4] This application supports our views that:

- User-centric mechanism evaluation may lead to different preferred mechanisms for different users, even within a single project,
- The simple selection approach can reach a conclusion quickly,
- The information burden is modest.

SELECTION PROCESS

A payment mechanism is only as good as its users perceive it to be. Therefore, we take a decision-maker-centric approach to comparison and selection. This may seem tautological, but in fact other authors have followed an ad hoc approach, which may lead to unproductive infrastructure investment, frustration of potential clients, damage to the organization's reputation, and thus foregone sales of goods and services.

Each of the bewildering array of possible mechanisms can be identified by its performance on a number of characteristics that affect the decision maker. Thoughtful

---

2. In the 19th century multiple banks issued their own cash, and cotton bills (invoices) were used as currency (Temin, 1969). The variety of alternatives increases the further back we look.

3. Apologies to Justice Potter Stewart.

4. The UMDL is a research project to develop an agent-based distributed digital library that facilitates commercial document transactions, as well as cost recovery for system resources. A digital payment mechanism is an essential component of its architecture. See the third section for more detail.

selection of a digital money mechanism requires a way to compare and evaluate different systems in terms of these characteristics. Previous authors select an arbitrary subset of the possible criteria on which to evaluate their proposed mechanism. In fact, the selection and application of criteria is the decision maker's prerogative, and thus should reflect the decision maker's circumstances and preferences.

After identifying the relevant characteristics, the decision maker needs a method for selecting one or more mechanisms based on their performance on the characteristics. We show that a generalized form of prioritizing characteristics provides a workable method for selection. In many cases, prioritization will be a simple ranking of characteristics. In more complex cases the method is consistent with rational choice theory (maximizing a utility function).

The problem of selecting from among many novel mechanisms based on a large number of characteristics is not trivial. However, in practice, decision-maker-centric selection may be simplified for two reasons:

- Different attributes may be important to various parties in a transaction, but only those that affect the decision maker (either directly or indirectly) are relevant to selection,

- Only a few high-priority attributes may be sufficient for making a selection from available mechanisms.

We find it useful to distinguish between two types of mechanism characteristics: those that affect the decision maker directly, and those that affect other users of the mechanism directly, and thus affect the decision maker indirectly. For example, suppose the person (or organization) implementing a mechanism is a seller: The seller might care directly if the mechanism is *nonrefutable* (loosely, the transaction details can be verified by a third party; see Appendix 1 for detailed definitions of characteristics). In addition, the seller's potential customers might care about the one-time setup costs to be able to use a mechanism (e.g., account creation and software installation). Thus, the seller as decision maker cares about buyer fixed costs indirectly, because they may affect the number of potential buyers.

The approach is quite general. A user, and thus a decision maker, could be a seller, a buyer, or an intermediary. All parties relevant to a decision are either users or decision makers. This simple dichotomy is exhaustive: Everyone who matters is either the decision maker, or someone whose preferences the decision maker cares about.[5] For our UM Digital Library examples in the next section, the decision maker is an intermediary (the library), and the users are buyers and sellers (and possibly other intermediaries, as digital libraries may interoperate with each other).

---

5. To be truly complete, the decision maker may also care about regulators or others with authority to impose constraints on the decisions. The legal environment is underdeveloped currently, so we ignore regulators in this chapter.

We now proceed through the steps of the selection method. First we define the master vector of possibly relevant characteristics. We then narrow the vector of characteristics to those relevant to the decision maker. The final step is to identify a set of sufficiently well-ordered preferences, and to apply them to the characteristics matrix to reach a decision.

Characteristics of Payment Mechanisms

The idea of evaluating digital payment mechanisms based on a set of desirable characteristics is not new. Several papers on electronic currency include a discussion of characteristics the authors consider desirable.[6] However, none of the papers recommend the same characteristics; our literature review yielded about 30 different desirable characteristics. These characteristics are defined in Appendix 1. Although long, the vector of characteristics is not exhaustive, and illustrates how complex the task of characterizing and comparing payment mechanisms may be.[7] Compiling such a list is an important first step in making a selection between competing proposals, because it forces the decision maker to explicitly think about how to distinguish mechanisms, and about which characteristics are important from his or her perspective.

To test the usefulness of the criteria for discriminating between payment mechanisms, and to offer a familiar point of departure, we first evaluated seven forms of conventional money. The results are shown in Appendix 2. Because these payment mechanisms are quite familiar, it is easy to understand the table and to observe how the vector of characteristics can be used to discriminate.

We now turn to digital payment mechanisms. We evaluated 10 popular payment mechanisms on 30 characteristics.[8] See Appendix 3.[9]

---

6. These papers include: XIWT (1995), Camp (1994), Chaum (1987), Neuman and Medvinsky (1995), Matonis (1995), Neuman and Medvinsky (1993), Mao (1995), Manasse (1995), Neuman (1995).

7. The task is further complicated because some characteristics themselves have several dimensions. As an example, we collapse a number of considerations into a single characteristic: "privacy." In fact, there are a number of different pieces of information about each transaction that may or may not be private (e.g., identity, value of transaction, item transacted), and there are a number of different potential observers from whom the information may be obscured (e.g., buyer, seller, law enforcement, third party). Camp (1994) explored the privacy attributes of payment mechanisms in depth; she suggested about 20 distinct privacy characteristics that might independently matter to decision makers.

8. We evaluated each mechanism on all 30 characteristics as a public service. One of the main points of our approach is that a decision maker not need to go through the entire time-consuming, costly exercise. The cost-conscious decision maker will first narrow the list of relevant characteristics, as we describe later, and then evaluate the candidate payment mechanisms on the reduced list of characteristics.

Narrowing the List: The Decision Maker's Perspective

To evaluate and compare every possible characteristic before selecting a mechanism would be expensive and impractical. The literature to date lacks any reasoned argument about on which subset of characteristics to base the selection and evaluation of digital payment mechanisms. In fact, using the characteristics advocated by any particular author may be quite misleading. Several of the discussions of desirable characteristics are conducted by developers of payment mechanisms that of course possess all of the characteristics advocated. Instead, we recommend selecting from the master list by focusing on the needs of the actual decision maker (or the organization the decision maker represents).

Consider two characteristics for digital payment mechanisms. Low financial risk (due to, e.g., low default risk) may be important, for example, to a digital library if it takes a direct interest in the transactions (i.e., if it takes and makes payments).[10] Monetary value is also frequently cited as a desirable characteristic (i.e., are balances or tokens convertible to cash by a bank on demand; see, e.g., XIWT, 1995. However, for its own transactions, the digital library may care about monetary value only insofar as it reduces default risk; it is but one of several aspects of a payment mechanism that may affect financial risk.[11] The library may be willing to accept a mechanism that does not have monetary value (e.g., a "credit") if for other reasons the mechanism has sufficiently low financial risk. Therefore, only low financial risk, not monetary value, should be retained in the vector of decision characteristics (for this particular decision maker).

To illustrate, we have applied this dimension reduction approach for an example decision maker: a commercial digital library, for infrastructure transactions in which it takes a direct financial interest.[12] Obviously, the procedure is subjective, as is any model-building exercise.[13] We first eliminated irrelevant characteristics, and then determined which characteristics were subsumed by others. We then evaluated a variety of digital payment mechanisms for their performance on these characteristics. The reduced set of characteristics is presented in Table 7.1.

---

9. All of the digital mechanisms we examine are still under development, and limited information is available. Thus, at any time our characterizations are likely to have some inaccuracies. Readers are responsible for determining the validity of a particular characterization.

10. Different decision makers will tolerate different degrees of risk. Generally in market economies there is a trade-off between risk and expected return. Some decision makers will choose to bear higher risk in exchange for higher expected returns.

11. Others include the risk that the digital money or other sensitive financial information will be stolen or "overheard" by parties outside the transaction and used fraudulently; risk that digital money is being double -spent; and risk of nonpayment or other customer fraud.

12. To make this example realistic, we conferred with UMDL researchers (including ourselves) to determine the appropriate reductions of the characteristics space.

TABLE 7.1
Reduced Evaluation of Digital Payment Mechanisms
for Example Decision Maker

	First Virtual	NetBill	Millicent	Ecash	Cyber Cash
Easily exchangeable	N/A	Yes	Yes	Yes	N/A
Locally scalable	Yes	Limited	Yes	Yes	Yes
Acceptable to users	Yes	?	?	Yes	Yes
Low transactions delay	No	No	Yes	No	No
Low transactions cost:					
For micro transactions	No	No	Yes	No	No
For large transactions	Yes	Yes	No	Yes	Yes
Low fixed costs (for seller)	Yes	No	?	No	?
Nonrefutable	No	Yes	No	No	Yes
Transferable	No	Limited	No	No	No
Financial risk:					
Buyer subject to risk?	No	Low	Low	Low	Low
Seller subject to risk?	Yes	Low	Low	Low	No
Unobtrusive	No	Yes	No	Yes	Yes
Anonymous:					
For buyer	No	No	No	Yes	No
For seller	No	No	No	No	No
Immediately respendable	No	No	No	No	No
Privacy	Some	High	Some	High	Some
Two-way	No	No	No	Yes	No

Prioritize Characteristics and Select a Mechanism

We now have a reduced set of characteristics on which we have scored the various alternatives. We propose an axiomatic approach to mechanism selection. By axiomatic selection we mean that a mechanism will be acceptable (to the decision maker) only if it satisfies certain a priori conditions, taken as given. This approach

---

13. It is not as obvious to us, however, that the process is more subjective than is forming the initial master list of objectives, for example, or than any other aspect of the decision-making process. The problem is not the subjectivity of decision-making—indeed, we are trying to focus the decision process on the responsible subject, or decision maker—but the extent to which the decision maker forms explicit objectives, methods, and assumptions at each stage of the process.

is trivial and noncontroversial if there exist one or more mechanisms that perform adequately on all of the characteristics important to the decision maker.[14] We are interested in the harder case, for which no mechanism exhibits all of the desired characteristics (but there is at least one mechanism that performs well enough overall that the decision maker prefers to choose such a mechanism rather than forego digital payment altogether). In this case, the decision maker must make a judgment about which characteristics are axiomatically required and which are not. In a simple world, the decision maker will be comfortable prioritizing the relevant characteristics, and establishing independent thresholds for each.[15] The decision procedure is then straightforward, and can be quite illuminating. Each characteristic serves as a screen: The set of candidate mechanisms is passed through each screen in priority order until the desired number remain. A useful by-product of this procedure, especially for digital payment designers, is guidance on which absent or inadequate characteristics are critical for particular mechanisms (at least for the decision maker in question).

Of course, decision makers may not have a strict preference ordering across all characteristics; for example, there may be "ties." Then the screens can be applied in sets. A problem arises if applying a set of "equally important" screens eliminates all remaining mechanisms. This could indicate one of three different states:

- None of the considered mechanisms performs sufficiently well, and the decision maker chooses to forego digital payment.
- The decision maker is willing to forego all of the equally important characteristics in the subset, and accepts the set of mechanisms remaining before applying the set.
- The decision maker realizes on further reflection that the criteria in the subset are not equally important, and arrives at a partial ordering that leads to a successful selection.

The last possibility suggests the usefulness of the approach in helping the decision maker to refine the selection method efficiently. If it is difficult (costly) to strictly order a subset of criteria, it makes sense to first see whether such an ordering is necessary. If it is necessary because no mechanisms survive the subset of screens, then the decision maker can choose whether to invest the extra effort required to refine the ordering. Thus, rather than an inflexible decision support tool, the approach is interactive and flexible: The decision maker is confronted with the need to make decisions that trade off certain characteristics against others only

---

14. If too many mechanisms perform "adequately" on all of the relevant characteristics (perhaps too many because the costs of implementing each incremental mechanism outweigh the benefits of having multiple mechanisms), then the decision maker needs either to tighten the sense of "adequate," or develop some way to otherwise (partially) rank order the surviving mechanisms.

15. In the simple case of an independent rank-ordering of characteristics that must be satisfied, the axiomatic approach is lexicographic.

when such trade-offs are critical to the selection. We show an example later in which such a trade-off is explicit, and is reached surprisingly quickly (that is, after only two screening characteristics are applied).

There may be cases in which the decision maker does not have a simple preference ordering with well-defined, independent thresholds for each characteristic. The method admits the use of a much more complex class of preferences. Any independent sequence of criteria that can be formed as logical operations on the decision maker's list of characteristics can be applied. For example, the following could be a (sub)sequence of screens:

1. Exchangeable
2. (Financial risk ≤ "medium" AND anonymous) OR immediately respendable
3. Low transactions delay, and so on.

By the use of generalized screens, the axiomatic approach is made fully general. Indeed, generalized screens support a continuum of metrics, from the simplest lexicographic prioritization to a complete cardinalization that selects the mechanism with the maximum weighted score.[16]

The basic insights from our formulation of the decision process are that:

- Hard decision problems necessarily involve trade-offs between multiple desirable features.
- Good decision making requires that the trade-offs be made explicit.
- A good decision support method assists the decision maker in finding which screening criteria can be applied at low cost, and which must be more carefully defined in order to discriminate between acceptable and unacceptable choices.

APPLYING THE SELECTION APPROACH IN A DIVERSE ENVIRONMENT

We use the University of Michigan Digital Library (UMDL) to motivate and illustrate the usefulness of our method. The UMDL architecture is built on multiple specialized information agents. Each agent can reason about its resources and objectives, communicate and negotiate with other agents, and choose its actions autonomously (Birmingham, 1995; Wellman, Durfee, & Birmingham, 1996). Agent types include user interface agents, collection interface agents, task planners, information service agents (providing, e.g., thesaurus look-ups or searches), and system service agents (providing, e.g., registration, notification, and auctions). These agents agree on terms of exchange—intermediated through decentralized

---

16. Maximum weighted score could be implemented to any desired degree of precision. Call the weighting function $F(x)$. Construct a sequence of screens of the form: (1) $F(x) > e_1$, (2) $F(x) > e_2$, and so on, for e_i an increasing sequence with small increments.

auctions—before providing services to each other, and thus are engaged in continuous, real-time electronic commerce.

A system with many autonomous, goal-oriented agents acting on their own behalf is an excellent setting in which to emphasize the importance of approaching the payment mechanisms selection problem from the decision maker's viewpoint. Each agent is an autonomous decision maker, and may have different preferences over payment mechanisms than do other agents. Further, the UMDL architecture is self-consciously modeled on a distributed, human agent economy, so this example also illustrates our method for the broader context of free-market electronic commerce in general.

Within the UMDL, various agents may prefer different digital payment mechanisms for at least two reasons: diversity of transactions, and diversity of trust. We first consider transaction diversity. A distributed digital library requires the performance of a wide variety of functions. With a distributed agent architecture, these functions are provided in the form of transactions between autonomous agents. For an example, a collection agent will register with the registry agent; the registry agent will notify users of the availability of a collection; and a user and a collection agent may negotiate and contract to exchange a digital document for monetary compensation. Transactions may be large (document delivery) or small (notification); they may be occasional (bibliographic search) or frequent (registrations and notifications); and so on. Agents may prefer different money mechanisms for different transaction types, much as we each do in familiar daily transactions: Sometimes we use cash, sometimes checks, credit cards, bank checks, and so forth. Frequent, small transactions are best served by a low-overhead, lightweight mechanism; irregular, large transactions may benefit from a more secure, but higher overhead, mechanism.

User preferences for a payment mechanism may also vary due to trust. UMDL is designed to be an open system, in that autonomous human agents may participate merely by providing their own software interface agents that meet the interface requirements of the UMDL.[17] Each autonomous agent in an open distributed system will trust other agents to a varying degree. For example, many agents in the initial implementation of the UMDL will be "owned" by the UMDL itself, and although each seeks to maximize its own objective function, they are known to each other and ultimately are part of a system designed to accomplish a unified goal.[18]

---

17. For example, a publisher unaffiliated with the university may offer access to its collection through a collection interface agent. This is quite different from some computational market agent systems in which all of the software agents are designed and owned by the same organization, and thus can trust each other.

18. The fact that self-interested individual agents can achieve a result that maximizes an overall, social objective follows from the First Welfare Theorem in economics (see, e.g., Varian, 1992). This result (and the Second Theorem) motivates the use of a distributed agent architecture based on economic exchange for solving complex system objectives.

When trusted agents engage in commerce, they may be willing to rely on a light-weight protocol with low peer-to-peer security, analogous to merchants who accept checks only if drawn on a local bank. When these same agents transact with unrelated agents (e.g., with the agent of a commercial book publisher), they may prefer a stronger payment mechanism, despite the higher overhead cost (cf. checks, credit cards, with immediate authorization but a 3–4% service charge).

The diversity of preferences for a payment mechanism, together with differences in mechanism overheads, may have important implications for the participant configuration in a distributed agent architecture. In the UMDL, for example, there are many infrastructure services that must be provided continuously, resulting in a heavy transactions load (examples include registry, notification, auction setup, and operation). The architecture is open, so any participant could write a competing agent to provide some of these infrastructure services. However, frequent low-value transactions may not be feasible if monetized with a heavyweight protocol. Thus a single provider of a cluster of different infrastructure agents may have a decided competitive advantage over a configuration with multiple competing providers, each offering only one or two infrastructure agents, because the single provider can use a lightweight protocol for exchanges between its chain of agents.[19]

For less frequent, higher value transactions, such as an extensive bibliographic search, the value of the transaction may be sufficient to support greater payment mechanism overhead. Then we would expect to see multiple, competing agents with different (untrusted) owners who offer search. The case is clearer still for the content providers: The overhead of a secure payment mechanism is small relative to the value of content, and thus we should expect to see a diversity of content providers, even though a single integrated provider of content and library services would have a lower overhead cost.

A similar effect of payment mechanism overhead in conventional commerce can be observed in near-border trade. Retail establishments situated near national borders often accept currency from both sides of the border at approximate exchange rates, bearing some risk of exchange rate fluctuation and the transactions cost of later performing an exchange, in order to lower the overhead costs to the customers of changing currency for frequent, small transactions. However, this behavior is more likely to be seen for exchanges of modest value (e.g., a newspaper or restaurant meal), and almost never for purchases of consumer durables (e.g., home appliances, automobiles).

---

19. The question of which functions are most efficiently provided as clusters of trusted agents is closely related to the question of which functions should be provided by a single agent, and which should be distributed to separate agents. The boundaries of agent functionality and trust relations involve a trade-off between the efficiencies of more centralized control and the efficiencies of market-mediated transactions. See, for example, Williamson (1975).

Therefore, we expect that in UMDL the selection of acceptable payment mechanisms will affect the configuration of participants and the agents they offer, and that there is likely to be considerable demand for more than one payment mechanism within the system. Diversity in transaction types and in trust are two reasons for choosing one payment mechanism over another, and these reasons are likely to diverge significantly within an open, distributed agent economy. However, as we see later, there are a number of other considerations in choosing a digital payment mechanism, and the selection problem can be quite complex.

EXAMPLES: PAYMENT IN A DIGITAL LIBRARY

To illustrate our method, and some of the helpful by-products of following the process, we present two simplified examples of how payment mechanisms might be selected for different transaction types in the UMDL.[20]

Example 1

Many transactions in the UMDL distributed agent environment will have low value, will occur frequently, and will require low transaction delay. For example, a notification agent will poll the registry to check for new events (such as new collections being added), and a new auction service will register with the registry agent. Some of these high-frequency, low-value transactions will take place between trusted agents, such as agents that are owned by a single entity. We suppose that for these transactions, the agents require a payment mechanism that satisfies three properties (in priority order; see Appendix 1 for definitions):

1. Low transactions delay,
2. Low transactions cost for microtransactions,
3. Unobtrusive.

By referring to the characteristics matrix in Table 7.1 we conclude that only Ecash, NetBill, Millicent, CyberCash, NetCheque, Mondex, MicroMint, and PayWord pass through the "low transactions delay" screen.[21] Millicent, NetCheque, MicroMint, and PayWord also pass the "transaction cost" screen. Of these four, only PayWord and MicroMint also pass the "unobtrusive" screen (Fig. 7.1). If the UMDL wishes to select only one mechanism, it can now determine which further criteria are important, and start applying those until either PayWord or MicroMint

---

20. The UMDL has not reached any final decisions about digital payment mechanisms. The examples presented here should not be viewed as an endorsement or criticism of any particular mechanism.

21. To save space we did not include columns for all of the payment mechanisms in Table 7.1; the underlying information is available in Appendix 3.

is eliminated. In the alternative, the UMDL might decide to accept both forms of digital money if the costs of setting up and handling two types are sufficiently low.

Figure 7.1:
Payment mechanism selection example

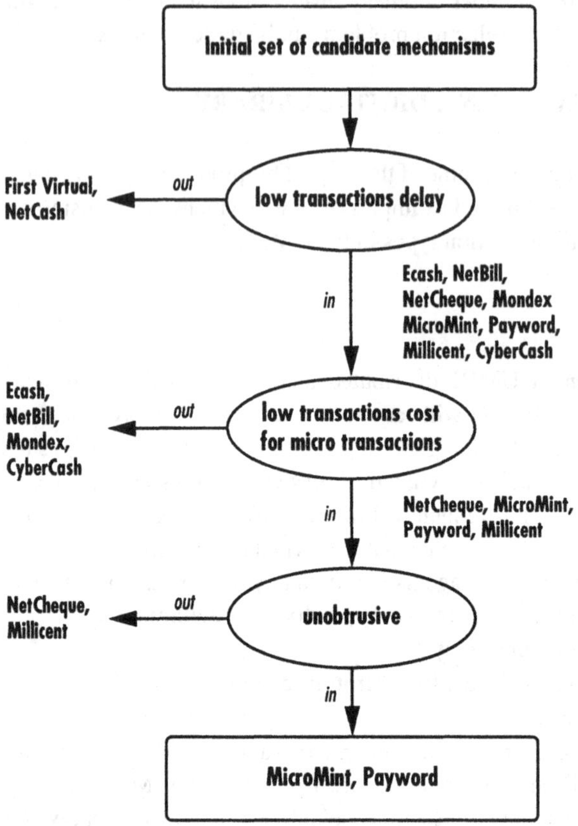

Example 2

To illustrate a different type of outcome, consider now a document exchange transaction in which a collection agent delivers a substantial document to a user agent. We presume that these agents do not fully trust each other (any more than a conventional bookstore fully trusts its customers and vice versa). Suppose that for this low-trust, high-value, relatively infrequent transaction agents require a payment mechanism that satisfies the following properties (in priority order):

1.   Low financial risk for buyer and seller,
2.   Low transactions cost for large transactions,
3.   Nonrefutable,
4.   Atomic transactions.

By again applying our evaluation matrix, we find that NetBill, Millicent, Ecash, CyberCash, NetCash, MicroMint, PayWord, and Mondex all pass the first screen. Of those, NetBill, Ecash, CyberCash, Mondex, and NetCash also pass the "transactions cost" screen. Only NetBill, CyberCash, and Mondex also pass the "nonrefutable" screen. When we further impose "atomicity," only NetBill survives.

One striking observation from this example is that the bewildering variety of digital money characteristics (and the size of the decision matrix) is not very important, at least in this application. The decision makers needed to identify only the top four criteria in order to eliminate all but two of the options. The next dozen or more criteria of interest don't affect the decision, and thus no effort need be devoted to evaluating those criteria or attempting to determine their strict ordering or interdependent rates of trade-off.[22] This is not to say that the decision is easy; but the decision method limits the effort and focuses it on the hard part: determining the priority of the most important characteristics.[23]

DISCUSSION

We have emphasized that when the analysis of payment mechanisms proceeds from the decision maker's viewpoint, only a very few of the many possible characteristics may be needed to make a selection. We expect, however, that some readers may be uncomfortable with what may seem to be too intensive an emphasis on individual decision makers, making independent decisions. Have we missed the most important consideration: that digital money users will want to adopt a mechanism that others are adopting? Won't the only characteristic that matters to most decision makers be that they adopt the "standard" mechanism?

We have no quibble with this prediction, but our purpose is not to predict which characteristics will be important to users. If widespread adoption is important to a decision maker, then that can be a characteristic in that person's vector, and he or she can assign it a high rank when the vector is prioritized. Indeed, if this characteristic is dominant for some users, then one of our main points is reinforced: The list of characteristics needed for any particular decision maker to select a mecha-

---

22. Indeed, if it is expensive to fill out the characteristics matrix, it would make sense to fill it out in priority order, researching the distinctions between the mechanisms only up to the point necessary to make a decision.

23. A second observation is that if our sole purpose in this example is to select a mechanism, then the third criterion was redundant. That is, if we apply only screens 1, 2, and 4, we arrive at the same conclusion: NetBill.

nism may be very short, and a careful quantification and weighting of numerous factors may be irrelevant when some characteristics are axiomatic (e.g., "must be a widely adopted standard").

On the other hand, it is well to remember that to date, no payment mechanism has been widely adopted as a standard. Unlike the speed with which Web browsers and HTML language modifications have become de facto standards, some payment mechanisms have been in use for more than 2 years without critical mass gathering behind any of them. This should not be surprising: There are a dozen or so forms of payment in regular use for nonelectronic commerce, with no single standard winning after a few thousand years of experience. Evidently the diversity of decision maker and user needs is greater than the benefits of a single standard.

In any case, every adopter is an early adopter, and by definition early adopters are those for whom selecting the successful standard is not as important as the value of selecting some mechanism and moving forward. We think it is almost surely true that payment mechanisms will exhibit *positive network externalities*—that is, that users will value a mechanism more, the more other users who also use the mechanism—and thus that payment mechanisms will be subject to the phenomena of critical mass and "tipping" (rapid standard adoption) that has characterized so many other software applications (see, e.g., Katz & Shapiro, 1994, for a discussion). (This does not preclude multiple branding, as we see today for the "standard" credit card.) For now, each individual decision maker needs to make his or her own decision, and for now, those decisions necessarily must depend on at least some criteria other than adoption of a standardized mechanism.

We propose an axiomatic approach to selecting a mechanism. The approach is designed to focus effort on information collection and analysis only to the extent needed to make a decision. But the method is fully general and includes cardinalization and selection by maximizing a weighted score as a special case. We have shown that trade-offs among desiderata are typically necessary, and that for the mechanisms currently available, the selection process reaches an outcome quickly. Further, although we have emphasized the selection problem, following our method will guide researchers to further development based on the needs of users who desire particular bundles of characteristics.

ACKNOWLEDGEMENTS

We are very grateful for early suggestions from Nathaniel Borenstein, and for the comments and advice of the UMDL Intellectual Property and Economics Working Group. MacKie-Mason acknowledges support from NSF grant SBR-9230481; MacKie-Mason and White acknowledge support from the NSF/ARPA/NASA Digital Library Initiative grant 1R1-9411287.

REFERENCES

Birmingham, W. (1995, July). *An agent-based architecture for digital libraries. d-Lib* [Online]. Available: http://www.cnri.reston.va.us/home/dlib/July95/07birmingham.html.

Camp, J. (1994). *Privacy and electronic payment systems*. Tech. Rep. Pittsburgh: Carnegie-Mellon University.

Chaum, D. (1987). *Security without identification: Card computers to make big brother obsolete*. Publications by DigiCash [Online]. Available: http://www.digicash.com

Katz, M. & Shapiro, C. (1994). Systems competition and network effects. *Journal of Economic Perspectives, 8*, (2), 93-115.

Manasse, M. S. (1995). *The Millicent protocols for electronic commerce*. Tech. Rep. Systems Research Center, Digital Equipment Corporation [Online]. Available: http://www.research.digital.com/SRC/millicent/papers/mcentney.htm

Mao, W. (1995, May). *Financial transaction models in the electronic world*. Tech. Rep. Hewlett-Packard Laboratories, Bristol [Online]. Available: http://www.hpl.hp.co.uk/projects/ vishnu/main.ps.

Matonis, J. (1995, April). *Digital cash and monetary freedom*. Institute for Monetary Freedom [Online]. Available: http://www.isoc.org/in95prc/HMP/PAPER/136/html/paper.html.

Neuman B. C. (1995, October). Security, payment, and privacy for network commerce. *IEEE Journal on Selected Areas in Communications, 13* (8). [Online]. Available: http://www.research.att.com/jsac/unprot/jsac13.8/jsac13.8.html.

Neuman, B. C. & Medvinsky, G. (1993, November). NetCash: A Design for Practical Electronic Commerce on the Internet. *Proceedings of the First ACM Conference on Computer and Communications Security*. [Online]. Available: ftp://prospero.isi.edu/pub/papers/security/netcash-cccs93.ps.

Neuman, B. C. & Medvinsky, G. (1995, November). Requirements for network payment: The netcheque perspective. *Proceedings of the IEEE Compcon '95*, San Francisco.

Temin, P. (1969). *The Jacksonian Economy*. New York: Norton.

Varian, H. (1992). *Microeconomic analysis*, 3rd ed. New York: Norton.

Wellman, M., Durfee, E. & Birmingham, W. (1996, June). The digital library as a community of information agents. *IEEE Expert*. [Online]. Available: http://ai.eecs.umich.edu/people/wellman/pubs/expert96.html

Williamson, O. (1975). *Markets and hierarchies*. New York: Free Press.

XIWT (Cross-Industry Working Team). (1995). Electronic cash, tokens and payments in the NII [Online]. Available: http://www.cnri.reston.va.us:3000/XIWT/documents/dig_cash_doc/ ElecCashToC.html

APPENDIX 1: CHARACTERISTICS OF PAYMENT PROTOCOLS

Acceptable to users: The payment protocol has (or has the potential to have) a large number of users. Acceptability is a function of: portability, no account required, buyer earns pre-and/or post-transactions float, easy to use, hardware independence, no encryption required, software installation not required, low fixed cost.

Accessible: Users find process to be accessible, easy to effect, and quick. No special expertise required.

Account required: Payment mechanism requires that users maintain an account with a vendor or payment mechanism provider; implies a lack of universality. May limit customer base for vendors accepting only this method of payment.

Anonymous: The identity of one or more parties in a transaction is hidden.

Atomic: A transaction is either completely finished, or it has no effect. That is, interrupted transactions do not have consequences. The transaction is repeatable: If interrupted, it can be redone until successful.

Divisible: Allows for the exchange of multiple low denomination instruments for a single high-denomination instrument.

Easily exchangeable: Easy to exchange as payment for other electronic tokens, paper cash, deposits in bank accounts, etc. Well-accepted and relatively fixed mechanisms exist for converting between various forms of money. Interoperable, fungible. Example: A transit fare card is not exchangeable for a money order; a money order is exchangeable for cash.

Float: Pretransaction, posttransaction. Does the buyer keep the float generated before the transaction occurs? After the transaction occurs? Example: Buyer keeps interest generated in an interest-bearing checking account before and after checks are written. Does not keep interest when using cash because no float is generated.

Hardware independent: For buyers, for sellers. Users do not need specialized hardware.

Immediately respendable: A payee does not have to take an intermediate step after receiving payment to respend it. Example: A check must be deposited in a bank account before the money can be respent; it is not immediately respendable.

Locally scalable: The payment protocol supports many customers simultaneously buying goods; adding users does not create a bottleneck that slows transactions. In addition, the protocol does not place a limit on the total number of a given seller's customers or the number of transactions that can be made with a given seller.

Low financial risk: Is the buyer, seller, or financial intermediary subject to risk? The risk of financial loss to each party involved in the transaction is low/acceptable. The level of risk involved with use of a payment protocol is a function of its security. Loss to buyers can be limited by maximum amount transactions, false positive passwords in case of theft, approval required for large transactions. May involve trade-off with unobtrusiveness.

Low fixed costs: Average cost per transaction of adopting the payment protocol are low. Fixed cost is a function of hardware costs, hardware installation costs, software and software installation costs, account startup costs.

Low transactions cost: *For micro transactions*: The payment protocol is suitable for use in transactions with a value of one cent or less. Such efficiency may involve a trade-off with security; may allow only the use of lightweight encryption or no encryption. *For large transactions*: Fees charged by financial intermediary are low per transaction (as a percentage of the amount of transaction); the payment protocol is secure enough for use in large transactions.

Low transactions delay: The time required to complete a transaction is low.

Monetary value: A payment protocol has monetary value if it represents cash, a bank-authorized credit, or a bank-certified check. Acts as a medium of exchange. Example: A traveler's check has monetary value because its value is guaranteed by the issuing bank. A credit card does not, because its use simply represents a promise to pay sometime in the future.

Nonrefutable: Parties can verify that a transaction took place, the data and/or amount of the transaction. A record may be produced. Nonrefutability may involve a trade-off with privacy.

Offline operation: Use of the payment protocol does not require a network and/or real-time third-party authentication. This characteristic may be important if infrastructure reliability is an issue. A payment protocol requiring on-line operation may be subject to time delays during periods of network congestion; in addition, if the network goes down, the payment mechanism may not be acceptable. Online operation may also limit scalability.

Operational today: Payment mechanism is available for use immediately.

Portable: Security and use of payment mechanism are not dependent on any physical location.

Private: Some or all of the elements of transaction information are hidden from some parties either involved in or observing the transaction. Transaction elements include: amount, date, time, location, product, and identities of the buyer and seller.

Security against unauthorized use: The device is not easily stolen and used. May be secure because of encryption, false positive passwords, and so on.

Storable: Able to be stored and retrieved remotely. Facilitates asynchronous exchange, allows payment mechanism to act as a store of value, adds stability to value of payment mechanism. Retrievable.

Tamper-resistant: Hard to tamper with, copy, forge, double-spend.

Transferable: The payment instrument is not bonded to a particular individual; that is, it can be used by someone other than the original owner. May involve a trade-off with security.

Two-way: Peer-to-peer payments are possible. The payment instrument is transferable to other users without either party being required to attain registered merchant status. Example: A check is two-way; a credit card is not.

Unobtrusive: The buyer does not need to frequently initiate new actions or pay; few steps are required to complete a transaction. May be particularly important in small transactions.

APPENDIX 2: CHARACTERIZATION OF CONVENTIONAL PAYMENT MECHANISMS

	Cash	Check	Credit Card	Debit Card	Money Order	Traveler's Check	Prepaid Card
Easily exchangeable	Yes	Somewhat	No	No	Yes	Yes	No
Locally scalable	Yes	Yes	Yes	Yes	Yes	Yes	Yes
Acceptable to users	Yes	Yes	Yes	Yes	Somewhat	Somewhat	Somewhat
Low transactions delay	Yes	Yes	Yes	Yes	Yes	Yes	Yes
Low transactions cost:							
For small transactions	Yes	No	No	No	No	No	?
For large transactions	No	Yes	Yes	Yes	Yes	Yes	?
Low fixed costs (for seller)	Yes	Yes	Yes	No	Yes	Yes	No
Nonrefutable	No	Yes	Yes	Yes	No	No	No
Transferable	Yes	No	No	No	Yes	No	No
Financial risk:							
Buyer subject to risk?	Yes	No	Up to $50	Limited	Yes	No	Yes
Seller subject to risk?	No	Yes	No	No	No	No	No
Unobtrusive	Yes	Yes	Yes	Yes	Yes	Yes	Yes
Anonymous:							
For buyer	Yes	No	No	No	Yes	No	Yes
For seller	Possible	Possible	No	No	Possible	No	No
Immediately respendable	Yes	No	No	No	Possible	No	No
Security against unauthorized use	No	Some	Some	Yes	No	Some	Possible
Two-way	Yes	Yes	No	No	Yes	Yes	No
Retrievable	Yes	Yes	Yes	Yes	Yes	Yes	Yes
Tamper-resistant	Yes	No	No	Yes	Yes	Yes	Yes
Offline operation	Yes	Yes	No	No	Yes	Yes	No
Divisible	Yes	Yes	Yes	Yes	Yes	Somewhat	Yes
Installation of software required	No	No	No	No	No	No	No
Operational today	Yes	Yes	Yes	Yes	Yes	Yes	Limited
Hardware independent:							
For buyer	Yes	Yes	Yes	Yes	Yes	Yes	Yes
For seller	Yes	Yes	No	No	Yes	Yes	?
Portable	Yes	Yes	No	No	Yes	Yes	No
Accessible	Yes	Yes	Yes	Yes	Somewhat	Somewhat	Yes
Encryption required:							
For buyer	No	No	No	No	No	No	No
For seller	No	No	No	No	No	No	No
Buyer keeps pretransaction float	No	Yes	N/A	Yes	No	No	No
Buyer keeps posttransaction float	No	Yes	Yes	No	No	No	No
Account required	No	Yes	Yes	Yes	No	No	No
Monetary value	Yes	No	No	Yes	Yes	Yes	No

APPENDIX 7.3
Characterization of Digital Payment Mechanisms

	First Virtual	NetBill	Millicent	Ecash	Cyber Cash	Net Cheque	Mondex	NetCash	MicroMint	PayWord
Conventional analog	Credit card	Debit card	Transit fare card	Coins	Credit card	Check/debit	Smart card	Foreign currency	Coins	Credit-based
Status	In use	Trial	Proposed	In use	In use	Trial	Trial	In use	Proposed	Proposed
Easily exchangeable	N/A	Yes[d]	Yes	Yes	N/A	Yes	Yes	No	Yes	N/A
Locally scalable	Yes	Limited	Yes	Yes	Yes	Yes	Yes	Limited	Yes	Yes
Low transactions delay	No	Yes	Yes	Yes	Yes	Yes	Yes	No	Yes	Yes
Low transactions cost:										
For micro transactions	No	No	Yes	No	No	Yes	No	No	Yes	Yes
For large transactions	Yes	Yes	No	Yes	Yes	Yes	Yes	Yes	No	No
Low fixed costs (for seller)	Yes	No	?	No	?	?	No	Yes	?	?
Nonrefutable	No	Yes	No	No[g]	Yes	?	?	No	No	No
Transferable	No	Limited	No	No	No	No	No	Yes	Yes	Yes
Financial risk:										
Buyer risk?	No	Low	Low	Low	Low	Low	Low	Low	Low	Low
Seller risk?	Yes	Low	Low	Low	No	Yes[h]	No	Low	Low	Low
Unobtrusive[a]	No	Yes	No	Yes	Yes	No	No	No	Yes	Yes
Anonymous:										
For buyer	No	No	No	Yes	No	No	No	Weak	No	No
For seller	No	No	No	No	No	No	No	No	No	No
Immediately respendable	No	No	No	Yes	No	No	Yes	Yes	No	No
Two-way	No	No	No	Yes	No	?	Yes	Yes	No	No

	First Virtual	NetBill	Millicent	Ecash	Cyber Cash	Net Cheque	Mondex	NetCash	MicroMint	PayWord
Secure from theft	Yes	Limited	Yes	Yes	Limited	Limited	Somewhat[i]	Variable[j]	Limited	Yes
Secure from eavesdropping	Yes	Yes	Yes	Yes	Yes	Yes	Yes	Variable[j]	Limited	Limited
Divisible	Yes	Yes	Yes	Yes	Yes	Yes	Yes	Limited	Yes	Yes
Atomic transactions	No	Yes	No	No	No	No	No	No	No	No
Suitable for information goods?	Yes	Yes	Yes	Yes	Yes	Yes	Yes	Yes	Yes	Yes
Suitable for tangible goods?	No[c]	No[e]	No	Yes	Yes	Yes	Yes	?	No	No
Privacy	Some	High	Some	High	Some	Some	Some	High	Low	Some
Acceptable to buyers:										
Portable[b]	No	No	No	No	No	No	Yes	No	No	No
Account required	Yes	Yes	Yes	Yes	Yes	Yes	Yes	No	Yes	Yes
Buyer keeps:										
Pretransactions float	Yes	No	Varies	No	Yes	No	No	No	No	Yes
Posttransactions float	Yes	Varies[f]	Varies	No	Yes	No	No	No	No	Yes
Low fixed costs	Yes	Yes	Yes	Yes	Yes	Yes	Yes	Yes	Yes	Yes
Globally scalable	Limited	Yes	Yes	Limited	Limited	Yes	Yes	Limited	Yes	Yes
Version of mechanism analyzed	Debit model	Debit model	Secure without encryption	Mark Twain Bank	0.8	Debit model	Swindon trial	2.0a4	Debit, without variations & extensions	Without variations & extensions
Company	First Virtual Holdings	Carnegie Mellon	DEC	DigiCash	CyberCash	USC-ISI	Mondex International	Software Agents, Inc.	MIT/ Weizmann Institute of Science	

a) The payment mechanism's obtrusiveness is evaluated on the transaction size for which it is optimized.

b) Payment mechanisms are considered not portable even if they could be portable if used on a laptop computer.

c) First Virtual's current implementation is most suitable for information goods; however, in a few cases, tangible goods are being sold using First Virtual.

d) When the user's NetBill account is linked to a credit card account, the characteristic "Easily exchangeable" is not applicable.

e) NetBill is optimized for transactions involving information goods; however, it may be possible to use NetBill for exchanging tangible goods.

f) If the NetBill account is funded through a credit card, it is possible to earn posttransactions float.

g) Ecase is nonrefutable if anonymity is given up.

h) The seller is subject to lower risk if the option of clearing the check in real-time, which may require an extra charge, is exercised.

i) The card can be locked, and if stolen would be useless.

j) The level of security depends on the security of the e-mail program being used.

8

The Social Impact of Electronic Money

Supriya Singh
*Centre for International Research on Communication
and Information Technologies, Melbourne, Australia*

In the discussion of electronic commerce, little is being said about the way the residential consumer uses or will use electronic money to pay for goods and services purchased on-line. There has also been no consideration of the social and cultural meanings of electronic money. In this chapter I address these critical gaps to understand how and why people pay for goods and services the way they do.

The sociological study of electronic money from the users' perspective shifts the focus from the convergence and substitution by digital technologies to examining how these technologies increase the diverse ways customers mix and match forms of money to pay for goods and services. This analysis is framed within the study of money as a social and cultural phenomenon. There are multiple monies in the domestic and market contexts. This is different from the dominant view of money, which sees it as a universal, homogeneous, and impersonal market phenomenon. Multiple monies are earmarked and differ in their source, use, quantity, quality, ownership, and meaning. This sociological approach to money is best illustrated in the work of Viviana Zelizer (1994).

In this analysis of electronic money I draw on two sociological studies of money among middle-income Anglo-Celtic[1] Australians. The first (Singh, 1996b) is part of a broader study of the use of information and communication technologies

---

1. The Anglo-Celtic group covers the dominant ethnic group in Australia, comprising 50.1% of the population. They include those who see their ancestry as English, Irish, Scottish, or other Anglo-Celtic (Australian Bureau of Statistics, 1988, p. 151).

(ICTs) in the home (Singh, Bow, & Wale, 1996). It is based on open-ended interviews with 47 persons from 23 households in Melbourne and its rural hinterland
between March 1995 and February 1996. The sample is over-weighted for middle
and upper income Anglo-Celtic households who own a computer, a modem, and
use personal financial management programs (PFMs), in order to better study the
use of electronic money. I refer to this study in the chapter as the Money Online
study. The second is my doctoral study of money, marriage, and information
(Singh, 1994). This involved open-ended interviews with 37 persons from 21
households in a middle-income, predominantly Anglo-Celtic, Melbourne suburb,
between June 1991 and February 1992. I refer to it as the Marriage Money study.

   In the first part of this chapter, I detail how the users' perspective changes the
questions and the idiom of discussion in electronic commerce. In the second part,
I show how people match the information yielded by forms of money with information required for different kinds of payments. This information varies in range,
timeliness, record, and context. In the third part, I examine the factors that engender trust in the new form of money and online services. In the fourth section, I detail the impact of electronic money on the management and control of money in
marriage. In the concluding section, I show how an understanding of the social and
cultural meanings of electronic money prevents misjudgments and makes for
more effective provision of payments services.

THE USERS' PERSPECTIVE

Providers of electronic goods and services recognize that growth and profitability
are based on an understanding of customers' needs. However, this understanding
is difficult to achieve. This is because the move to the users' perspective involves
a major shift in the questions that are asked. As Dervin (1992) pointed out:

   Almost all our current research applies an observer perspective. We ask users questions which start from our worlds, not theirs: What of the things we
   can do would you like us to do? What of the things we now offer do you
   use?...The difficulty is that the data tell us nothing about humans and what
   is real to them (p. 64).

   The research on information and communication technologies (ICTs) in general
and online services in particular places the technology, service, or application at
the center. The questions, for instance, are: How do you use the plastic card? Will
you use online banking? Online shopping? These questions lead to documenting
the increasing centrality of the new payments instruments and online services in
people's lives. Figure 8.1 illustrates how a focus on plastic cards can easily lead to
the view that we are at the threshold of a cashless society.

   This perspective is important in charting the growth of new technologies. However, if this picture is not complemented by one that places the user and his or her
activities at the center of questioning, it can make for costly misjudgments. Figure

8.2 represents the activity centred approach with the user at the center. It unpacks only some of the activities referred to in Figure 8.1. It makes clear that plastic cards are the first choice only for some activities like gifts and travel, with cash and checks continuing to play an important role in other payments.

The users' perspective alters the framework and direction of questioning. These new frameworks and questions have to be painstakingly discovered by listening to consumers talk of their lives, for much of the public debate is couched in the terms seen important to the providers. In the Money Online study, it led to a focus on *forms of money*, that is a combination of a payments instrument like a plastic card and a transaction mode such as over-the-counter transaction, telephone, fax, or the Internet.

FIGURE 8.1
The Providers' Perspective: The Rise of the Plastic Card

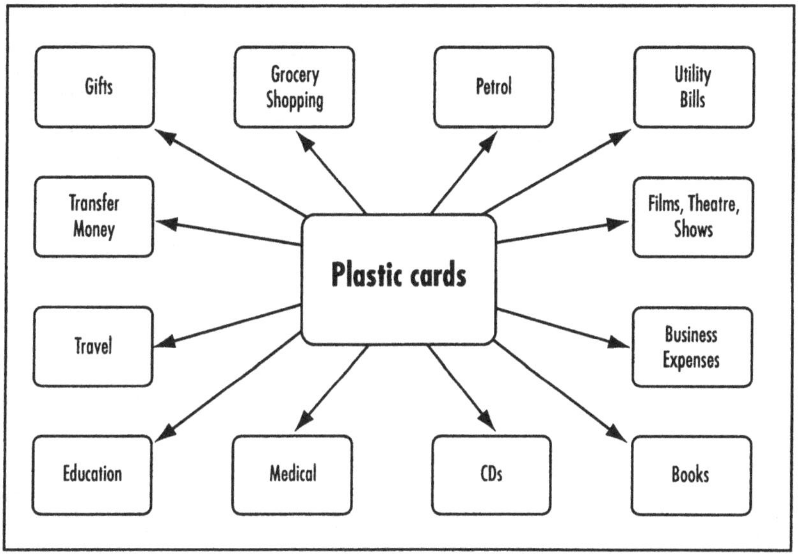

MIXING AND MATCHING FORMS OF MONEY

From the users' perspective, forms of money can be broadly grouped as physical and electronic, depending primarily on the transaction medium used. Physical forms of money include physical payments instruments and physical transaction media. They comprise cash and checks transacted person-to-person and across the bank branch or post office counter. Checks sent by mail also belong here. Electronic forms of money include those that involve physical payments instruments such as plastic cards—credit cards, debit cards, stored value cards, smart cards; direct debit and credit; and the electronic versions of cash and checks. The main transaction modes are electronic but would include plastic cards used across the counter or by mail. This is illustrated in Table 8.1.

Figure 8.2
The Users' Perspective: Diverse Ways of
Mixing and Matching Payments Instruments

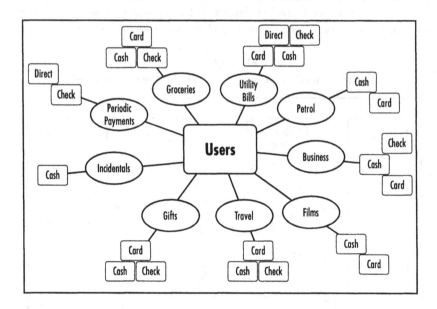

The categories of physical money and electronic money are ideal types. When one speaks of electronic money, there is the assumption that it is virtual money, that is, it is not physical, it is not tangible, and it cannot be held and touched. This is true if one compares cash, that is, currency, and direct debit or credit. It must be remembered, however, that currency is symbolic and representational when compared to barter. Plastic cards transacted physically across the counter or by mail are thus situated along the continuum between the physical and the electronic forms of money. Moreover, the success of particular forms of electronic money such as obtaining cash via electronic funds transfer at point of sale (EFTPOS) and EFTPOS direct debit rests on the fact that there is a physical record and it yields physical cash. Thus one has the situation that physical money like cash may have no record, whereas electronic money obtained from the automated teller machines (ATMs) and EFTPOS is accompanied by a physical record and tangible cash.

Thinking in terms of the various forms of money reminds one that payments have not always been transacted, even in the past, solely within the banking system. Australia Post, for instance, claims to be Australia's "biggest over-the-counter electronic bill paying and agency banking service, handling more than 150 million transactions each year" (Australian Payments System Council, 1996). What is different today is that it is possible to source both payments instruments and transaction modes in the nonbanking sector via the use of stored value cards, e-cash, and the Internet. This leads to the next important question: How do people mix and match these increasingly diverse forms of money?

TABLE 8.1
Forms of Money: Combining Payments Instruments and Transaction Modes

Form of Money	Payment Instrument	Mode of Transaction
Physical forms of money		
"Real" cash	Cash	Person to person, bank branch, post office
"Real" check	Check	Person to person, bank branch, post office, mail
Electronic forms of money		
Direct entry		
Bank direct	Direct debit/credit	Written instruction to bank
Phone direct	Direct debit/credit	Phone
Internet direct	Direct debit/credit	Internet
EFTPOS	Debit card	EFTPOS
Credit direct	Credit card	Written instruction to payee
Electronic cash		
ATM cash	Cash; plastic cards	ATM
EFTPOS cash	Cash; debit cards	EFTPOS
Internet cash	E-cash	Internet e-mail/phone
Electronic wallets	Stored value cards	Person to person, ATM, Internet
Electronic credit		
"Real" plastic	Credit card	Person to person
Mail plastic	Credit card	Mail
Phone/fax plastic	Credit card	Phone/fax
Internet plastic	Credit Card	E-mail/Internet
Electronic check		
ATM check	Check	ATM (deposit)
Internet check	Electronic check	Internet

A focus on the ways Australians pay for goods and services shows that cash, checks, and visiting the bank branch office remain the central payments and transactions mechanisms. The macro picture shows that despite a fast rate of growth in electronic payments and transactions, 22 years after Bankcard was introduced in Australia and posted to bank customers and 16 years after the first bank ATM, physical forms of money continue to dominate.

- Cash remains the most popular and convenient way of paying for everyday transactions of small value in Australia. It is estimated that some 90% of the number of transactions are in cash (Bank for International Settlements, 1994, p. 8).
- The check is the most popular form of noncash payment in Australia. In 1995, its volume (38%) exceeded that of credit cards (10%), EFTPOS (13%), ATM (17%), direct entry credit (18%), and direct entry debit (4%) (Mackrell, 1996, p. 4).
- Although high-value electronic funds transfer is now for the first time higher in value (63%) than checks (35%), checks continue to dominate over retail low-value electronic funds transfer, which remain unchanged between 1991 and 1995 at 2% (Mackrell, 1996, p. 5)
- Payment by cards is increasing, but it still comprised only 0.1% of cashless transactions in value, and 15.4% by volume (Bank for International Settlements, 1994, pp. 46, 48)
- The number of ATMs and EFTPOS outlets is increasing in Australia. At the end of 1993, Australia ranked fifth and seventh in the number of inhabitants per ATM and EFTPOS outlet, respectively, among the 12 major developed countries monitored by the Bank for International Payments (Australian Payments System Council, 1995). However, National Australia Bank data show about 34% of all deposit transactions continue to be done through branches (National Australia Bank, 1996, Table 5.1).

It is difficult to go beyond this picture to detail the use of these different payment and transaction mechanisms. As the Australian Payments System Council noted in its 1994–1995 Annual Report, "There is limited detail on the relative usage by consumers of different payment instruments" (Australian Payments System Council, 1995, p. 41).

Socioeconomic factors like income, literacy, and education are important for drawing the outer limits of access to bank accounts, plastic cards, personal computers and modems. The latest figures on the public record show that:

- An estimated one-tenth of Australian adults have no bank accounts (Singh, 1992).

- Nearly a fifth (18%) of Australian adults have no credit cards (Kavanagh, 1996). The Marriage Money study shows that two-thirds of the low income non-English-speaking background persons with literacy difficulties had no electronic access. Only 17% had a checking account, so the rest were wholly dependent on cash as a payment mechanism.
- Although household ownership of personal computers (PCs) is rising rapidly, only 30% of households have PCs and only 7% have modems (Australian Bureau of Statistics, 1996).
- Poor access to bank branches and ATMs is also a limiting factor in some rural areas. The Money Online study shows this is pushing people more to the use of EFTPOS and checks. Checks cashed at grocery stores or service outlets continue to be an important method of obtaining cash.

These socioeconomic factors are, however, less useful for understanding how people with access use physical and electronic forms of money. The Money Online study shows that even the early adopters of the new technologies continue to mix and match different forms of money. The quality of this mix varies for different households and for persons within the household, despite similarities in income, education, and computer expertise. The mix also differs according to cultural context. Cash, particularly, has different cultural meanings. The opposition between cash and the gift, assumed in the West, is not found in all societies. In Malaysia, crisp new notes in traditional red envelopes are the traditional *ang-pow* gift for the Chinese New Year. In Japan, the liking for clean notes has migrated to the electronic world, where some ATMs clean and deodorize the notes before delivering them.

The Money Online study shows that among middle and higher income Anglo-Celtic households in Australia, the particular mix of forms of money can be partly explained by the strong congruence between forms of money and kinds of payments. It can be noted that:

- Cash obtained from the branch or the ATM and direct debit via EFTPOS are used with grocery payments. Credit cards are not generally used for in middle-income Anglo-Celtic households; there is a strong cultural norm against buying food on credit. Checks are used when there is no cash or EFTPOS.
- Checks and cash across the counter are the most popular way of paying bills.
- Physical cash is most often the only acceptable form of money for both the merchant and the consumer, for incidental expenditure such as parking, photocopying, and buying items of small value. It is also the form of money most associated with gambling in Australia, as

there is a regulatory prohibition against the provision of credit for gambling.

- Direct debit via a standing instruction with a financial institution is the preferred way for paying for periodic payments such as the mortgage.
- Checks and plastic cards across the counter or EFTPOS direct is used for tax-deductible expenditure.
- The credit card, where possible, is the preferred way of paying for large items of discretionary expenditure.
- Internet plastic is at times used for paying for books, CDs, and software ordered over the Internet.

This leads to the next question: What makes one form of money more generally suitable for a particular kind of payment?

INFORMATION, FORMS OF MONEY, AND KINDS OF PAYMENTS

In this section, I argue that one of the important reasons for the congruence between forms and kinds of payments is that forms of money yield information that is required for different payments and income streams. The important dimensions of information are those that relate to time: range, record, and context. The questions behind these information dimensions are: Does it give immediate information or deferred information? Is the information on money spent or also on money still in hand or in the account? Is the immediate record evidential, discretionary, or is there no record at all? Is the transaction context personal or impersonal, physical or virtual? It is these four information dimensions that mark out possible forms of money that are used-for specific kinds of payments and monies. These information dimensions are displayed in Table 8.2.

"Real" cash, that is, cash obtained from branches and paid in a person-to-person transaction, gives immediate information about money spent or received and money in hand or still in the account. It can yield a discretionary record in that one can ask for a receipt, but if one does not want a record, a cash transaction is the most untraceable of all transactions. Cash received via the ATM and electronic funds transfer at point of sale (EFTPOS), that is, ATM cash and EFTPOS (cash out), most often automatically generates a receipt. With ATMs, there is immediate information about money in hand and money still in the account, but the personal transaction element is missing. EFTPOS is similar to getting cash from the branch in that there is a person across the counter, but unlike the ATM it does not give information about money still in the account. This is one of the reasons that people who are uncertain about the sufficiency of funds hesitate to use EFTPOS.

The information yielded by various kinds of cash matches the information required by grocery shopping, incidental purchases, and gambling. With grocery money, the most immediate need is to know how much you have spent, and how much is left. This information is particularly important if the person is operating within a very tight budget or needs to control the flow of money. For most persons, there is no need to account for this money to an outside party. However, the attraction of EFTPOS is that it offers a record of expenditure for budgeting purposes and for monitoring the flow of money from joint marital accounts. Similarly for incidental purchases, where the amount of money involved in each transaction is seen as inconsequential, there is not the same need to keep a record. Gambling money most often does not come from a specific budget, but is seen as part of household shopping or incidental expenses. Hence there is at times a need not to know how much has been spent on gambling (Singh, 1996b).

Checks differ from cash in that with a check you do not immediately get informed by the bank about the money still in the account. But with a check payment, you can prove to the authorities or the recipient, that you did send the check, and if it is cashed, track it down in your statement. The credit card transacted across the counter also has the immediate evidential aspect to the record. However, there is an implicit hierarchy among the forms of money as to which form is seen as more evidential, with the check ranking first, followed by the credit card and then by EFTPOS. It is this greater authority for a check record that makes it popular for paying bills.

Check and credit card payments over the counter add the physical and personal context to the transaction. The plastic card transacted by mail, phone, or fax does not have this physical and personal context, but it does yield a discretionary record such as a copy of the letter or fax, or a receipt number that can be used as a reference point for tracking a transaction. Direct debit or credit via the bank, that is, "bank direct," differs from checks in that it does not give immediate information about the money spent. That is why direct credit is more often used for regular periodic payments, where the amount of money spent is known and certainty of payment is required.

A focus on information dimensions shows the distinctiveness of Internet money, for it is virtual and impersonal in context and often in its record (Singh, 1996c). Internet money is e-cash, electronic checks, and plastic cards transacted over the Internet. It is impersonal as there is no identifiable person at the other end of the transaction as with physical cash, check and plastic, or with EFTPOS direct. It is virtual as it is not associated with a physical payments instrument like cash and checks, nor does it automatically generate an immediate physical record that is evidential in nature.

TABLE 8.2
Information Dimensions of Forms of Money

Forms of Money	Information Dimensions										
	Time		Range		Context				Immediate Record		
	Immediate	Retro-spective	Money spent or received	Money in Hand or Still in Account	Personal	Impersonal	Physical	Virtual	Evidential	Discre-tionary	Untraceable
Real cash	Y		Y	Y	Y		Y			Y	Y
ATM cash	Y		Y	Y		Y	Y		Y		
EFTPOS (cash out)	Y		Y		Y		Y		Y		
E-cash	Y		Y	Y		Y		Y	Still under trial		
Real check	Y		Y		Y		Y		Y		
Plastic check	Y		Y			Y		Y		Y	
Electronic check	Y		Y			Y		Y	Still under trial		
Real plastic	Y		Y		Y		Y		Y		
Mail plastic	Y		Y			Y	Y			Y	
Phone plastic	Y		Y			Y		Y		Y	
Fax plastic	Y		Y			Y		Y		Y	
Net plastic	Y		Y			Y		Y	Still under trial		
Bank direct		Y	Y	Y		Y		Y	Y		
Plastic direct		Y	Y			Y		Y	Y		
Phone direct	Y		Y	Y		Y		Y		Y	

Note: Y, yes.

Internet money is as yet not generally used in households that have Internet access. In the Money Online study, 13 of the 23 households had modems and Internet users. Fourteen persons from 11 households spoke of whether they have bought or would buy online goods and services and whether they would pay for them online. They split neatly in the middle with 7 persons saying they have paid or would pay with Internet money and 7 saying they would not. Of these 7, 5 had already used Internet money for purchasing books, magazine subscriptions, CDs, software, and information services from the United States. Two had used debit cards and three had used credit cards on the Internet to pay for the goods.

All seven who said they would purchase with Internet money were men. The gender dimensions got muddied, however, when one looked at the seven who said they would not purchase on the Internet, as there were three men and four women in this group. Age was not a determining factor in the usage or nonusage of Internet money, as the ages in each group ranged from the 20s to the 50s. All of them were from medium- to high-income households.

Although the lower cost of online payments is at the center of industry discussion, persons do not give this as a reason for going online with their payments. The reasons mentioned by the users are convenience, speed, and the ability to track the different phases of their transaction—for instance, whether the CDs are out of stock and there is going to be a delay, or whether they have been posted or not. The users also mentioned that they are not worried about the lack of security, as physical systems are perhaps more insecure. This contrasts with those who do not use Internet money, for they are primarily worried about the security of their transaction and the privacy of information. These concerns are at times connected to their discomfort with the virtual and impersonal context of Internet transactions, which to a lesser extent also prevents some of them from transacting with a credit card over the phone or fax.

The Money Online study shows that the main difference between those who use and do not use Internet money revolves around issues of trust. This, however, begs another question: What makes for trust in an online environment?

TRUST IN AN ONLINE ENVIRONMENT

It is useful to distinguish between issues of "hard trust" and "soft trust." David Bollier (1996), reporting on the Aspen Institute Roundtable on Information Technology, noted that issues of hard trust "involve authenticity, encryption, and security in transactions" (p. 21), whereas issues of soft trust "involve human psychology, brand loyalty, and user-friendliness" (p. 21). This distinction allows one to emphasise that trust goes beyond technical issues of security and encryption. For a person to use a form of money, he or she has to trust that it is secure. This explains why in Australia, it is estimated that only 1 or 2% of deposits are made through ATMs, whereas in the United States, even with the new generation ATMs, the figure is said to be 5% (Allard, 1996).

Trust may need a variety of "warranting structures." Trust is also encouraged by the speed with which orders are filled, being able to accurately account for the transaction if need be, a willingness by the seller to rectify errors, voice contact at the order-taking stage, and lower prices (Bollier, 1996). These are important supply-side criteria and contribute to the user feeling in control of the transaction. Analysis of the Money Online data indicates that from the users' perspective, it is this ability to control the transaction that makes a person more willing to use a form of money. Trust and control are intertwined, for trust in the system leads to a sense of being in control, and being in control leads to a feeling of trust.

The Money Online study shows that this sense of control comes from the presence of at least one of the following factors:

- A physical payments instrument and/or record of payment,
- A personalized transaction context,
- Ability to track and substantiate a transaction,
- Ensure his or her desired level of privacy,
- Favorable experience of the form of money,
- Knowledge of the service provider and/or the recipient,

The particular mix of these factors for different payments is illustrated in Bob's[2] case.

Bob is an academic in his late 40s, early 50s, with a household income of $A80,000[3] a year. In the last year or two, he has purchased books from Boston, information services from Colorado, and software from the United States. Ryan, 36, is in information technology with a household income of between $A50,000 and $A60,000. Ryan is a regular user of Internet for information and has advertised things for sale on his newsgroup. He has also ordered goods through the newsgroup, but he has not used the Internet for payment or delivery.

Bob says he has "no concern" about using his card on the Net. This comfort is bolstered by his favorable experience with Internet transactions, his personal knowledge of the merchant, and ability to control the transactions. He has previously bought books personally at the bookstore in Boston, and feels comfortable ordering books from the same store online. He has also dealt with it by fax. The most attractive thing about online purchasing for Bob is that it is generally cheaper; it arrives faster and often the goods are not available in Australia. Moreover, he has instantaneous control of the ordering process. He says, "If I see a reference to a book, I can put in my e-mail message, and boom, I've done it within a minute. I don't have to ring anyone up. I've done it. It's happened. And then within a few

---

2. Pseudonyms have been used throughout to refer to the persons interviewed for the Money Online and Marriage Money studies.

3. On 2 September 1996, $A1 was worth US 78.96 cents.

minutes I get a response back…saying they've got my order." He thinks that in the future, payment for online services will become "less intentional."

His case also illustrates that trust does not always hinge on expert knowledge. It sometimes accompanies a lack of knowledge of possible complications. Although liability online remains an unresolved area in Australia, Bob's comfort in online transactions is buttressed by his belief that the risk of using his credit card online is limited to $A50. He is not worried about security online, for he says, "We have no knowledge about the security of the systems which EFTPOS uses…My guess is that it's secure but I've got no knowledge about security."

Bob's story also reminds one that the use of Internet money goes together with physical forms of money. Bob uses EFTPOS and ATMs routinely, but he continues to pay his utility bills by check at the post office or the bank branch. He doesn't pay by phone even when he is paying by credit card. He says, "Check butts are my record. I use them when I do my tax at the end of the year." The measures of control for different kinds of payments are different.

Trust has an important psychological dimension, for other expert users of the Internet do not necessarily feel as comfortable with Internet money even for purchasing books, information services, and software from overseas. Ryan, 36, who is in information technology, with a household income of $A50,000 to $A60,000, is happy to use the Internet for the first stage of purchase. But unlike Bob, Ryan will not put his credit card on the Net because he doesn't think it is "secure at this stage." He has often given his credit card on the phone, but he sees that as more secure than advertising it electronically. However, he does not use telephone or online banking and very rarely pays his bills with the credit card over the telephone. He says, "It's just as easy to go to the bank. I usually have three or four bills like gas, electricity, rates, whatever. I just go to the bank and pay them all off. Sometimes I go to the post office." His need for physical and personal control over a wider range of payment transactions is greater than is the case with Bob. This brings one to the importance of researching the psychological dimensions of trust in online transactions—which is not the focus of this chapter. It is equally important to note that both Bob and Ryan are using electronic money within the social and cultural context of middle-income Anglo-Celtic marriage. This shapes the use of electronic money and also impacts on the way couples manage and control money in marriage.

THE MANAGEMENT AND CONTROL OF ELECTRONIC MONEY

Electronic money is changing the way middle-income Anglo Celtic married couples manage and control money. Traditionally, control of money was ensured by restricting access to money and information about money in the household. In the housekeeping system of money management generally found among middle-income couples, the husband would give the wife a set amount of money for housekeeping. The husband managed the money that was left and had overall control (Edwards, 1984a, 1984b). This was accompanied by a lack of information, for in the past, particularly with the housekeeping system of money management, a woman did not always know how much her husband earned (Komarovsky, 1962; Morris, 1990).

With electronic money, it becomes difficult to restrict access and information about money to both husband and wife. Hence control rests more with the person who monitors the information. Direct crediting of wages, pensions, and benefits to joint accounts and access to the ATM, EFTPOS, and the credit card makes it possible for both husband and wife to withdraw money from the joint account or to have personal credit linked to it. Direct credits give a paper record of money coming in because of wages or other payments, thus detailing income. The ATMs not only generate a record of bank transactions but also are able to give the balance for the account. Credit cards and EFTPOS itemize expenditure. These statements have the added advantage of supplying the answers without the need to ask the spouse about his or her income, transactions, and expenditure.

Much of the social impact of electronic money in Australia is mediated through the marital joint account. Roughly three-fourths of married couples in Australia have a joint account. The joint account symbolizes the jointness of marriage. It also visually marks the marital unit as the most important domestic financial unit, for the joint account is always between husband and wife and not between parents and adult children. A cross-cultural focus on money, particularly in Asia and Africa, suggests that this jointness of money in marriage is culturally distinctive. In Asia and Africa the family or household unit is the more pertinent boundary of domestic money. Moreover, money in marriage is more often separate rather than joint (Singh, 1996a).

The Marriage Money study shows that this first wave of electronic money has been one of the factors that has made for a greater framework of jointness in the management and control of money in marriage. The housekeeping system of management has been replaced by wife management or joint management. Money among couples below the age of 65 years is now more often jointly controlled than before. Where money is separately controlled, husbands continue to control the money more often than wives. This is partly because women continue to be less informed than men about money in the bank, and have less confidence in their own ability to deal with money in the market.

The Money Online study suggests that the second wave of electronic technology represented by personal financial management programs (PFMPs) and the onset of online banking is likely to increase husbands' control of information about money. This is because personal financial management programs concentrate control of information, and at present, they are more often used by men. A survey of Quicken usage—which at present stands at over 100,000 users in Australia—

shows that 81% of its users are male (personal communication, Greg Wilkinson, Managing Director, Intuit Australia).

In the Money Online study, PFMPs are used in 6 of the 23 households for business and personal finances, with 5 using Quicken and 1 using Microsoft Money. In all 6 households, the initiative to go on Quicken came from the man. I examine the change in management and control in the 3 households where Quicken is used mainly for monitoring and controlling money in marriage.

For Jean and John, the move to Quicken came when John took early retirement and there was a need to monitor their retirement investments. John now has taken over all the account keeping for Jean's consultancy business as well as their investment portfolio so as to manage their money in one heap rather than two separate flows. It has meant restructuring their banking and financial structures. Jean says that before this, she kept all the accounts for her business. Now she no longer knows the details of their bank accounts. She sees that information is an important ingredient for the control of money, but losing this information does not worry her. She says, "Figures and all these things, it just doesn't interest me. And John quite likes it. So he can do it." Moreover, she adds, John "monitors it very effectively...So he knows what's getting our best returns and so on."

Goldie, in her 50s and earning between $A60,000 and $A70,000 a year as an academic, is conscious that the way she and her husband manage and control money has changed since her husband Gordon retired. When there were two wages coming in, they jointly managed the money, but Goldie controlled the money, just as her mother and mother-in-law had done before her. Each monitored their own accounts. After retirement, Gordon began to use Quicken to manage the money flows. Goldie's wage is now used for current expenditure, whereas the other income is invested. Now it is Gordon who tells Goldie whether she may write a check or not.

Isabel and Indra's story is interesting, for it shows that PFMPs can also be instrumental for moving from independent control and management to joint control. Both are in their 20s and are in a cohabiting relationship, but have decided to get married in the near future. They have separate accounts but jointly use Quicken to record income and expenditure and as a planning tool. It was Indra's decision to use Quicken so that they could document where their money was going and begin planning to save for a house. For Isabel, the move to Quicken was more difficult. She recognized that they needed to budget and see where their money was going. But for her, it brought up the issue of ceding control. She says that putting her finances on Quicken "was a really big step for me...It was a big emotional thing. I had to really get around it...to actually relinquish that responsibility and control... And eventually I sat down and put all my figures down on Quicken and now we tend to do it together."

In these three households, the women were computer literate. The decision to go with Quicken was triggered by a change in life stage and/or a change of income flows. The stories are varied enough to suggest that the impact of PFMPs depends very much on a person's life stage and on the desire to monitor the flows of money so that they are accountable and calculable. However, the possibility of changed patterns of control and management because of the use of PFMPs needs to be kept in mind when evaluating the use and social impact of electronic money.

CONCLUSION: IMPLICATIONS FOR PROVIDERS OF PAYMENTS SERVICES

Understanding the social and cultural meanings and impact of electronic money from the users' perspective is the key to an effective delivery of payments services. The absence of such an understanding is contributing to banks' declining share in the payments business, particularly in the United States. Daruvala and Stephenson (1996) of McKinsey and Company in New York estimated that the nonbanks already control about one-third of the $127 billion payments business in the United States. There is no corresponding data for the payments business in Australia.

I suggest this loss of payments business in the United States has been partially because banks were asking questions from the providers' perspective rather than the users' perspective. They assumed that electronic forms of money will replace physical forms of money, leading to a cashless society. However, customers continued to mix and match different forms of money. In 1995, electronic transactions still accounted only for 3% of the total number of payments transactions in the United States (Daruvala & Stephenson, 1996). This failure to predict usage cost the banks heavily. Banks found themselves having to invest more in both the physical and electronic delivery networks. In the United States, it has been estimated that the introduction of ATMs in the 1980s added $5 billion in operating expenses, while saving only $200 million from reduced teller positions (Mendonca & Nakache, 1996).

In Australia, the focus continues to be on the lower costs to banks of electronic transactions compared to physical transactions. As the Australia and New Zealand Banking Group (ANZ) submission to the Financial System Inquiry reveals, this is accompanied by an anticipation that ATMs and EFTPOS will make the branch network "obsolete" and that stored value cards will, over the next 5 to 10 years, "largely displace" cash transactions for frequent low-value purchase (ANZ, 1996, pp. 28–29).

Questions from the users' perspectives would direct banks' attention to how they can add value to the various ways of mixing forms of money in different social and cultural contexts. The issues at the center of the payments debate would then be how to increase customers' access to physical and electronic forms of money, and how to give customers the kind of information they desire for the different kinds of payments at the lowest cost. Their efforts would then be directed to giving customers greater control over information about their transactions, so that they are more willing to trust the online environment. As Bowers and Singer (1996) pointed out, bankers should remember "that a widening gap between the information that financial institutions actually provide to their customers and the information that might be delivered was what created the market opportunity" (p. 82) for personal financial management software in the first place.

These questions are also important for regulators who seek to monitor and report developments in the payments system. At present in Australia, the advisory and industry bodies that look after the payments business themselves have no data on how customers use forms of money. As electronic commerce extends its reach, this understanding becomes even more urgent, for electronic transactions will be part of the ways we work, study, play, entertain or inform ourselves, pay for goods and services, or are paid.

ACKNOWLEDGEMENTS

I would like to thank Amanda Bow and Karen Wale who did much of the interviewing, and the external review panel for their valuable comments. The panel comprised Dallas Isaacs, Mouli Ganguly, Patricia Gillard, and Peter White. I am also grateful to Telstra and Nortel for supporting CIRCIT's broad research program *Understanding Demand for Residential Interactive Services*.

REFERENCES

Allard, T. (1996, 11 May). We're withdrawn on ATM deposits. *The Sydney Morning Herald*, p. 3.

Australia New Zealand Banking Group Ltd. (1996). *Submission to the Financial System Inquiry*. Melbourne: Author.

Australian Bureau of Statistics. (1988). *Census 86—Australia in profile: A summary of major findings*. Catalogue No. 2502.0. Canberra: Australian Government Publishing Service.

Australian Bureau of Statistics. (1996, September). *Household use of information technology Australia, February 1996*. Catalogue No. 8128.0. Canberra: Australian Government Publishing Service.

Australian Payments System Council. (1995). *Annual report 1994/95*. Sydney: Author.

Australian Payments System Council. (1996, 16 February). *Giropost information paper*. Unpublished paper tabled at the meeting of the Australian Payments System Council, Sydney.

Bank for International Settlements. (1994). *Payment systems in Australia*. Basle: Author.

Bollier, D. (1996). *The future of electronic commerce: A report of the Fourth Annual Aspen Institute Roundtable on Information Technology*. Aspen, CO: Aspen Institute.

Bowers, T. & Singer, M. (1996). Who will capture value in on-line financial services? *The McKinsey Quarterly, No. 2*, p. 78–83.

Daruvala, T., & Stephenson, J. (1996, October 8–9). *From atoms to bits: Managing an industry in transition*. Presentation at the Bank Administration Institute's National Payments System Symposium, Washington, DC.

Dervin, B. (1992). From the mind's eye of the user: The sense-making qualitative-quantitative methodology. In J. D. Glazier & R. R. Powell (Eds.), *Qualitative research in information management* (pp. 61–84). Englewood, CO: Libraries Unlimited.

Edwards, M. (1984a). The distribution of income within households. In D. Broom (Ed.), *Unfinished business: Social justice for women in Australia* (pp. 120–136). Sydney: Allen & Unwin.

Edwards, M. (1984b). *The income unit in the Australian tax and social security systems*. Melbourne: Institute of Family Studies.

Kavanagh, J. (1996, April 3). Credit cards tossed away. *The Australian*, p. 34.

Komarovsky, M. (1962). *Blue-collar marriage*. New Haven, CT: Yale University Press.

Mackrell, N. (1996, May 16). *The cheque's role in today's payment system*. Unpublished transcript of talk by Neil Mackrell, head of Financial System Department, Reserve Bank of Australia to the AIC Conference on the future of cheques, Sydney.

Mendonca, L., & Nakache, P. (1996). Branch banking is not a dinosaur. *The McKinsey Quarterly, 1*, 136–147.

Morris, L. (1990). *The workings of the household*. Cambridge: Polity Press.

National Australia Bank. (1996). *Submission to Financial System Inquiry*. Melbourne: Author.

Singh, S. (1992). *Banks and migrants: An untapped market*. Melbourne: Consumer Credit Legal Service.

Singh, S. (1994). *Marriage, money and information: Australian consumers' use of banks*. Unpublished doctoral dissertation, Department of Sociology and Anthropology, La Trobe University, Bundoora, Australia.

Singh, S. (1996a). The cultural distinctiveness of money. *Sociological Bulletin, 45*(1), 61–85.

Singh, S. (1996b). *The use of electronic money in the home*. Policy Research Paper No. 41. Melbourne: Centre for International Research on Communication and Information Technologies.

Singh, S. (1996c). The use of Internet money. *Journal of Internet Banking and Commerce, 1*(4) [Online]. Available: http://www.arraydev.com./commerce/JIBC/current.htm

Singh, S., Bow, A., & Wale, K. (1996). *The use of information and communication technologies in the home*. Policy Research Paper No. 40. Melbourne: Centre for International Research on Communication and Information Technologies.

Zelizer, V. (1994). *The social meaning of money.* New York: Basic Books.

Electronic Substitution in the Household-Level Demand for Postal Delivery Services

Frank A. Wolak
Stanford University

MOTIVATION

The past decade has witnessed a dramatic increase in the number of available modes of interpersonal communication and in the range of quality of these modes in terms of speed, reliability, and flexibility. Many of these modes have experienced substantial price reductions over this same time period; for instance, the price of long-distance telephone service has fallen continuously; both the purchase price of fax machines and the cost of using these machines have declined; online information services such as America Online or CompuServe, which charge a zero price for incremental messages sent to other subscribers to their service and to users of the Internet, have experienced explosive growth in the number of subscribers. All of these modes of communication provide attractive alternatives to traditional postal delivery services supplied by the United States Postal Service (USPS).

The relative attractiveness of long-distance telephony, fax communication, and electronic mail is enhanced by the steadily increasing price of postal delivery services over the past decade. From January 1, 1986, to January 1, 1995, the price of a one-ounce first-class letter increased from $0.22 to $0.32, a more than 45% increase in 9 years. This combination of higher prices for postal delivery services and the growing number of viable alternative modes of communication has led to dire predictions about the future demand for postal delivery services. However, aggregate pieces delivered by the USPS have continued to steadily increase up to the present time. From 166.4 billion pieces delivered in 1992, growth in pieces delivered was 2.9% in 1993 and 3.4% in 1994. For 1994, the percentages of the various classes by volume are: first-class, 54%; second-class, 6%; third-class, 39%; and all other classes, 1%. Third-class, principally advertising circulars and mail-order catalogues, has steadily increased its share of pieces delivered. The number of first class pieces delivered has continued to grow, but the bulk of this growth has come from presorted first-class. In fact, from 1993 to 1994, single-piece volume fell by 0.2%, but a 6.8% increase in presorted first-class resulted in a net 2.4% increase in total first class volume (USPS, 1994). The increasing share of pieces delivered going to third class mail and presorted first class, both of which are used primarily by businesses, seems to signal a shift by households away from the consumption of postal delivery services.

The purpose of this chapter is to quantify this shift in the household-level demand for postal delivery services from 1986 to 1994 and determine the extent to which it can be attributed to the appearance of alternative modes of communication versus the concomitant increasing relative price of postal delivery services. Specific questions addressed are: What are the own-price, cross-price, and expenditure elasticities of household-level demand for postal delivery services, and how have these magnitudes changed over time? What is the impact of the increasing penetration of home-computing technology on postal demand? What household characteristics predict differences in household-level postal demand? How has the aggregate demand for postal delivery services by U.S. households changed over this time period?

To perform this analysis, I use the Bureau of Labor Statistics (BLS) Consumer Expenditure Survey (CES), which is a national probability sample of U.S. households generated from the 1980 Census 100% detail file. The BLS administers two distinct surveys to different samples of households: (a) the Quarterly Interview Survey and (b) the Diary Survey. These surveys differ in the number of goods that they cover, the length of time they survey a household, and the kinds of background questions asked about the household. For the Diary Survey, each household is requested to keep two 1-week diaries of all expenditures over consecutive weeks. For the Interview Survey, the household is interviewed every 3 months over a 15-month period; this survey also asks questions about durable goods holdings—cars, housing, and personal computers. Both surveys collect information on household characteristics—hours of work of the head and spouse, occupation of the head and spouse, age and race of the head and spouse, marital status of the head, number of children, dwelling type, and income. This information can be used to link durable goods holdings across the Interview Survey sample and the Diary Survey sample.

The Diary Survey is the source of data on household-level consumption of postal delivery services. The Interview Survey is used to estimate a probit model of the probability of personal computer ownership as a function of household characteristics common to both surveys. Each of the 49,089 households from the Diary Survey for the years 1986–1994 is then assigned an estimated probability of computer ownership using the coefficient estimates from the Interview Survey probit model using the common household demographic characteristics across the two surveys.

Because the purchases of each household in the Diary Survey are only recorded for 2 weeks, a household's purchases of postal delivery services can differ substantially from its consumption of these services. For example, a household can purchase and store these services in the form of blocks or rolls of stamps, using these stamps later to consume postal delivery services. The fact that, for the sample period, approximately 67% of all households do not purchase any postage during the 2-week Diary Survey period attests to the empirical relevance of this difference. The distinction between consumption and purchases within the 2-week sample period creates various complications for the proper recovery of the structure of household-level demand for postal delivery services. The observed purchases of postal delivery services are the combination of actual (but unobserved) consumption and a frequency of purchase process.

A considerable amount of across-household heterogeneity exists in the demand for postal delivery services. The probability of computer ownership affects both the consumption of postal delivery services and the frequency of purchase of postage. A higher household-level probability of computer ownership predicts a decreased amount of consumption of postal delivery services and a reduced probability of postage purchase during the 2-week Diary Survey period by that household. The results from estimating the model of postage expenditures pooling all of the years in the sample imply substitutability between the consumption of telecommunications services and the consumption of postal delivery services.

In order to investigate how the structure of postal demand has changed from 1986 to 1994, I estimated the model allowing for changes over time in the parameters determining the own- and cross-price and expenditure elasticities of household-level demand for postal delivery services. Perhaps the most surprising result to emerge from this analysis is a substantial increase in the absolute value of the own-price elasticity of demand for postal delivery services over time. For example, the own-price elasticity in 1986 is approximately –0.76, but by 1994 this magnitude is –1.3. This trend toward a more price-elastic demand has potentially dire consequences for the ability of the USPS to raise revenues from households through future postage price increases. These results yield an expenditure elasticity of demand decline from 0.36 in 1986 to 0.25 in 1994, indicating that higher expenditure and income households are increasingly substituting away from postal delivery services as their preferred mode for interpersonal communications. There is also an increasing degree of substitutability between postal delivery services and telecommunications services, indicating that, to a greater extent in recent years, declines in the real price of telecommunications services should bring about decreases in the household-level demand for postal delivery services. When I quantify the impact of personal-computing technology and Internet use on the household-level demand for postal delivery services changes over time, I find that, for the early years of the sample, a higher probability of ownership of a personal computer is associated with an increased demand for postal delivery services. However, particularly in 1993 and 1994, when subscribership to online services and access to the Internet in general became widespread, an increased probability of computer ownership predicts a decline in the household-level demand for postal delivery services.

Because I have a weight giving the number of households in the United States represented by each household in the Diary Survey sample each year, we can use these weights to compute estimates of the aggregate household-level demand for postal delivery services as well as the aggregate price elasticities of the demand for postal delivery services. This enables me to assess the likely revenue consequences to the USPS from future postal price increases given the aggregate demand relations I have estimated. We find that for a 10% increase in the price of postage, the most recent year's aggregate own-price elasticity estimate implies a 2.7% reduction in the revenues that the USPS can expect to receive from U.S. households. We can also compute an aggregate elasticity of household-level demand for postal services with respect to the probability of computer ownership. Using the aggregate elasticity estimate for 1994, and assuming a 17% increase in the probability of computer ownership for all households in the sample (a plausible increase given that the estimated penetration of personal computers in the household has more than tripled from 1988 to 1994), yields a 2.7% decline in the aggregate demand for postal delivery services, or a 2.7% reduction in annual household-level revenues, the same percentage decline in revenues brought about by the 10% price increase.

The remainder of the chapter proceeds as follows. The next section discusses the data sources used in my analysis and provides some summary statistics on the changes in the household-level demand for postal delivery services over time. This section also presents estimates of the model for the probability of personal computer ownership as a function of household characteristics common to the Diary Survey and Interview Survey. The next section presents the econometric model for the observed purchases of household-level postal delivery services. This is followed by a discussion of the results from estimating the model and the answers to the questions posed earlier about the structure of household-level demand for postal delivery services that this econometric modeling effort provides. The chapter closes with a summary of results and a discussion of topics for future research.

PREDICTING COMPUTER OWNERSHIP AND TRENDS
IN POSTAGE EXPENDITURES

In this section I first discuss the two major data sources used in the analysis. This is followed by a presentation of the estimates of the probability of computer ownership models estimated using the Interview Survey data. These models are then used to impute a probability of computer ownership for each household in the Diary Survey. The parameter estimates are of some independent interest given the rapid increase in the fraction of households owning personal computers—from 7% in 1988, the first year in which the Interview Survey collected this information, to 25% in 1994, the last year of data currently available. In order to motivate the household-level demand estimation results presented later, summary statistics on household postage and telephone consumption from 1986 to 1994 are given. Finally, I describe the annual changes in the percent of U.S. households owning personal computing technology up to the present time.

As noted earlier, the Diary Survey is the source of data on postage consumption. This survey collects all expenditures for each sampled household for two consecutive 1-week periods. Each sampled household completes a weekly diary document listing every purchase, the good and the amount, made within that 1-week period (except expenditures incurred while away from home, overnight or longer). Every year the Diary Survey sample is redrawn, with each day of the week having an equal probability of being the first day of the reference week for a sampled household. With the exception of the last 6 weeks of the year, when the Diary Survey sample size is doubled to increase the coverage of expenditures unique to the holiday season, the number of Diary Surveys administered is uniformly distributed throughout the year.

For my analysis, total postage expenditures for each household is the sum of all purchases of postage during the 2-week sample interval. Figure 9.1 presents a histogram of the number of purchases of postage during the 2-week diary period. From the figure it is clear that the vast majority of households that purchase postage during their Diary Survey period only make one purchase. For this reason, the subsequent empirical analysis focuses on the decision to purchase within the 2-week period, rather than on the number of purchases made.

The Interview Survey is the source for computer ownership data at the household-level. Beginning with the 1988 survey, households were asked whether they owned a personal computer. Because the Diary Survey and Interview Survey collect the same household characteristics, these variables can be used to match households that share these characteristics across the Interview Survey and Diary Survey samples. The Interview Survey is used to collect information on major expenditure items on a retrospective basis. Each household selected to appear in this sample is interviewed quarterly for five consecutive quarters for these major expenditures and the already mentioned household characteristics in addition to their income. As a result of this sampling scheme, 20% of the Interview Survey sample is rotated each quarter.

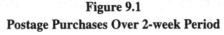

Figure 9.1
Postage Purchases Over 2-week Period

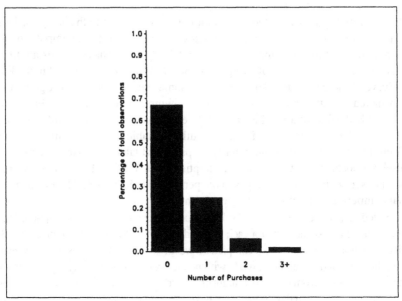

The probability of computer ownership model utilizes all available household characteristics variables that are common to the Interview and Diary Surveys to predict the probability of computer ownership using a probit model. Let X_i denote the vector of household characteristics for the ith household in the Interview Survey. Let y_i^* denote the latent propensity of household i to own a computer. We assume that the household owns a computer if $y_i^* \geq 0$ and does not if $y_i^* < 0$. The event of computer ownership is denoted by the indicator variable y_i, which takes on the value 1 if the household owns a computer and zero otherwise. The propensity to own computers is determined by $y_i^* = X_i'\beta + \varepsilon_i$, where ε_i is an independent identically distributed $N(0,1)$ random variable across households. The log-likelihood function for this model is:

$$L(\beta) = \sum_{i=1}^{N} y_i \ln(\Phi(X_i'\beta)) + (1 - y_i)\ln(1 - \Phi(X_i'\beta))$$

where N is the number of households in the Interview Survey during the year under consideration and $\Phi(t)$ is the standard normal distribution function. Given the maximum likelihood estimate of β, an estimate of the probability of computer ownership as a function of X_i can be computed as $(X_i'\hat{\beta}$, where $\hat{\beta}$ is the is the maximum likelihood estimate of β. Taking the values of X_i for each observation in the Diary Survey for the same year as the Interview Survey, we compute $\Phi(X_i'\hat{\beta})$, the estimated probability of computer ownership for that Diary Survey observation. This estimated probability is used in the model for postage expenditures to measure the extent of electronic substitution in the consumption of postal delivery services.

Appendix 9.1 gives the maximum likelihood estimates of the elements of β for 1988, the first year this information was collected in the Interview Survey, and for 1994, in order to quantify the changes in these probit coefficient estimates over time. This same probit model is estimated separately for each year from 1988 to 1994. To compute the probability of computer ownership for each household during each year of the Diary Survey, we use the probit model parameter estimates from that same year's Interview Survey data. Because computer ownership information was not collected for 1986 or 1987 in the Interview Survey, the parameter estimates from the model estimated for 1988 are used to compute estimated probabilities of computer ownership for all observations from the Diary Surveys in 1986 and 1987.

I now describe the time-series behavior of the sample averages of household-level postage purchases and telephone expenditures from the Diary Survey database. Appendix 9.2 gives the sample average, minimum and maximum household-level purchases of postage, and local and long-distance telephone services during the 2-week interview period for each year from 1986 to 1994 in nominal dollars. This appendix shows an initially increasing average consumption of postal delivery services from 1986 to 1989 and then a steady decline from 1990 to 1994. On the other hand, telephone consumption shows a steady increase throughout the sample period. This appendix also gives these same magnitudes as shares of total nondurable goods expenditure during the 2-week Diary Survey time interval. Viewed relative to total nondurable goods expenditures, the downturn in household-level postage expenditures is even more pronounced. Figure 9.2 plots the monthly Consumer Price Index (CPI) price indexes for postage and the composite of local and long-distance telephone services, nonseasonally adjusted and normalized to have January 1986 equal to 1. This figure shows the large relative price increase in postage versus telephone service over the sample period. These increases in the price of postage later in the sample period, and the accompanying reduction in average household-level purchases, is indicative of a price-elastic demand that brings about reductions in total expenditures in response to a price increase. On the other hand, telephone services expenditures increase over the sample, despite a slight upward trend in the telephone services price index over time.

Figure 9.2
Movements in Postage and Telephone Prices

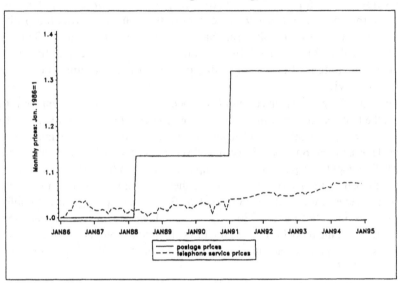

Appendix 9.3 gives the sample average percent of households in the Interview Survey that own computers for each year from 1988 to 1994. This percentage has nearly quadrupled over the 7-year period. This magnitude almost doubled from 1988 to 1989, the same time period in which average household-level expenditures on postal delivery services began to decline. The inverse relation between these two trends suggests the importance of accounting for the relationship between computer ownership and the consumption of postage delivery services in constructing an econometric model of household-level postage expenditures.

ECONOMETRIC MODEL OF POSTAL DELIVERY SERVICES EXPENDITURES

The econometric model of household-level postage expenditures accounts for the infrequency of purchases of postal delivery services within the 2-week sampling interval of the Diary Survey, yet is still consistent with the basic features of consumer demand theory. This model is specified with postage expenditures as a share of total nondurable expenditures within the 2-week Diary Survey period as the dependent variable. The range of real (in January 1986 dollars) total nondurable goods expenditures in the 2-week Diary Survey period for all years in the sample is $25 to $1,700, so there is a considerable amount of variability across households in the share total expenditures going to postal delivery services. This model gives rise to a joint distribution of postal expenditure shares across all the households during the sample period which is then used to estimate the parameters of the econometric model by maximum likelihood techniques.

Before describing the details of the model, we require the following notation. Let y_i* denote the share of total nondurable expenditures going to postage consumption and y_i the share of total nondurable goods expenditures going to postage expenditures, by the ith household for the 2-week interval. Let x_i equal the vector of ratios of the logarithm of prices to total nondurable goods expenditures—$\ln(p_j/M)$, where p_j is the price of the jth good and M is total nondurable goods expenditures—and other demographic variables assumed to shift postage consumption across households. We enter prices and total nondurable goods expenditures as ratios in order to impose the theoretical restriction of homogeneity of degree zero of the resulting demand function in prices and total nondurable goods expenditures. Let w_i denote the indicator random variable that equals 1 if the household purchases postage within the 2-week interval and 0 otherwise. The purchase probability can be written as $\text{pr}(w_i = 1) = \Phi(Z_i'\theta)$, with z_i equal to the vector of household characteristics that shift the propensity to purchase postage; $\Phi(t)$, the standard normal distribution function; and θ, a vector of parameters to be estimated.

Assume that $\log(y_i*) = x_i'\beta + \varepsilon_i$, , where $\varepsilon_i \sim N(0,\sigma^2)$ and $y_i = w_i\, y_i*/\Phi(z_i'\theta)$. The relation $y_i = w_i\, y_i*/\Phi(z_i'\theta)$ implies that if a purchase of postage occurs in the 2-week interval ($w_i = 1$), the household buys the inverse of its postage purchase probability in that 2-week interval $[1/\Phi(z_i'\theta)]$ times its observable demand for postal delivery services for the 2-week interval, y_i*. Consider a simple numerical example to illustrate this aspect of the econometric model. Suppose the household's unobservable 2-week demand for postal delivery services is $10.00 and its probability of purchasing postage within any 2-week interval is 0.5. This implies that when it does purchase postage, it will buy $20.00 = $10.00/0.5 worth to maintain its rate of consumption given its purchase frequency.

Because $\log(y_i*)$ is assumed to be normally distributed, y_i* must take on only positive values. This model assumes that all households consume a nonzero (although it can be extremely small) amount of postal delivery services within a 2-week time period. Let Y equal the vector of all postal expenditure shares, $(y_1, y_2, \ldots, y_N)$, where N is the number of households in the sample. The demand for postal delivery services function and the equation relating consumption to purchases via the purchase probability function yields the log-likelihood function for Y:

$$\log L(Y) = \sum_0 \log(1 - \Phi(z_i'\theta))$$

$$+ \sum_+ y_i[(-\log\sigma + \log\phi[(\log(y_i) + \log\Phi(z_i'\theta) - x_i'\beta)/\sigma] + \Phi(z_i'\theta) - \log(y_i)]$$

Variants of this model were discussed in Blundell and Meghir (1987) and Deaton and Irish (1984). In Wolak (1997) I considered three other forms for this combination of purchase frequency and latent demand for postal delivery services model of postal expenditures that differ in terms of their assumptions about the unobserved demand for postal delivery services. Because all four competing models give rise to joint densities of Y, nonnested hypothesis testing techniques developed by Vuong (1989) can be used to determine which model provides a statistically significantly superior description of the underlying data generation process. The nonnested hypothesis testing results reported in Wolak (1997) find that this *log-infrequency of purchase* model, which implies nonzero consumption of postal delivery services for all households, provides a clearly statistically superior description of the observed household-level expenditure patterns relative to the other three models. Other reasonableness checks of the four models found further support for the superiority of the log-infrequency of purchase of postal delivery services model considered here.

Variables Entering Demand and Purchase Probability Functions

Consumer theory provides a strong guide as to what should enter x_i, the determinants of the demand for postal delivery services. Because $y_i^* = \exp(x_i'\beta + \varepsilon_i)$ is an expenditure share demand function, it follows that the own-price, the prices of all other goods consumed by the household and total nondurable expenditures, should enter x_i. Consequently, we enter the logarithms of the price of postage, price of telephone services, an index of the prices of other nondurable goods besides postage and telephone services, and total nondurable expenditures. We also enter demographic variables describing the characteristics of the household that should alter its consumption of postal delivery services, such as race, number of children, martial status, education, occupation, and age of the head, and the probability of computer ownership.

Economic theory provides less guidance for what variables should enter in z_i, the determinants of the purchase probability. There are a number of reasons why the probability of purchasing postage should differ across households. A major determinant of these differences is the opportunity cost to the household of making a purchase. If it were costless to purchase postage, then all households would purchase only when at least one household member actually consumed postal delivery services. Under these circumstances, the purchase probability within the 2-week sample interval would exactly equal 1 for all households consuming any postal delivery services during this time interval. Consequently, we expect household characteristics that predict the opportunity cost of purchasing postage to be important predictors of this probability— the geographic area in which the household resides, the number of children in the household, the marital status of the head, the education of the spouse and head, the occupation, age, hours of work of the head and spouse, household income, and the probability of computer ownership.

Appendix 9.4 gives estimates of the parameters of this two-equation model for postal expenditures. As noted earlier, we impose homogeneity of degree zero in price and total expenditure on the share equations by requiring that the coefficient on the logarithm of the price index for other nondurable goods equal the sum of coefficients on the prices of postage and telephone services minus the coefficient on the logarithm of total non-durable expenditures. Appendix 9.5 gives a list of the variable definitions used in all of the models. The household demographic variables significantly improve the predictive power of the model, indicating the presence of deterministic differences in postage consumption and frequency of purchase across households based on these observable characteristics. The standard errors estimates in Appendix 9.4 are computed using the misspecification-robust standard error estimates discussed in White (1982). Using these covariance matrix estimates makes the inferences drawn robust to various forms of misspecification of the distributional assumptions used to derive the joint density of Y.

Figure 9.3 plots the smoothed density of household-level expected postal delivery services consumption within the 2-week period for the log-frequency of purchase model.[1] There is a large range in the density of expected 2-week consumption of postage, with a significant positive skew. Figure 9.4 plots the smoothed density of household-level purchase probabilities from this model. Although this density of purchase probabilities is centered at 0.32, many households have estimated purchase frequencies above and below this value. Appendix 9.6 gives the probability derivatives associated with the postage purchase probability for this model. This appendix shows that increases in the probability of personal computer ownership significantly reduces the purchase frequency of postal delivery services.

Appendix 9.7 gives the mean household-level elasticities of postage demand with respect to the prices of postage, telephone services, other nondurable goods and total nondurable expenditures. The structure of the log-frequency of purchase of model implies price and expenditure elasticities which do not vary across households. The last row of Appendix 9.7 gives the sample mean of the elasticity demand with respect to the probability of computer ownership. Because this elasticity varies across households, the sample standard deviation of the household-level elasticities is reported below it. This mean elasticity implies that increases in the probability computer ownership bring about reductions in the demand for postal delivery services at the household level.

---

1. All smoothed density estimates use a kernel density estimator with a Gaussian kernel. The automatic bandwidth selection procedure described in Silverman (1986) is used to determine the amount of smoothing.

Figure 9.3
Density Estimate of Postage Expenditures
(Model: Log-Infrequency)

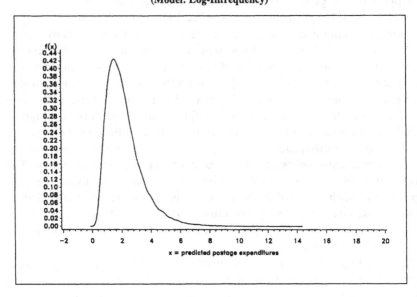

Figure 9.4
Density Estimate of Purchase Probabilities
(Model: Log-Infrequency)

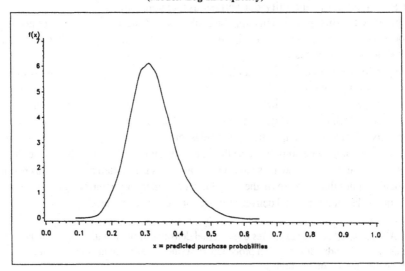

To address the question of the impact of the increasing number of low-cost alternatives to postal delivery services and the increasing price of postal delivery services over the past decade on the demand for postal delivery services, we enlarge the specification of the demand function to allow for time-varying own- and cross-price elasticities, total nondurable expenditure elasticities, and elasticities with respect to the probability of computer ownership. This requires interacting each of these five variables—$\ln(p_{postage})$, $\ln(p_{phone})$, $\ln(p_{other})$, $\ln$(total nondurable expenditures), and the probability of computer ownership—with an annual time trend and including these variables in the vector of regressors, x_i, for the log-consumption share equation. An annual time trend is also interacted with the probability of computer ownership in the purchase probability equation. The remaining parameters in the consumption share equation and the purchase probability equation are assumed to be fixed over time. For these variables, the resulting parameter estimates are very similar to those reported in Appendix 9.4. There are changes in the coefficients on log prices, log total nondurable expenditure, and probability of computer ownership variables that reflect the fact that we have included these variables interacted with a time trend in the model. The full set of parameter estimates is reported in Appendix 9.8.

Appendix 9.9 gives estimates of the sample average price, expenditure and probability of computer ownership elasticities for each year in the sample implied by this model. Recall that the only demand elasticity that varies across households is the one with respect to the probability of computer ownership. Consequently, the last column in this appendix presents the sample means of these elasticity estimates for each year. The own-price elasticity begins at –.0758 in 1986 and ends in 1994 with a value of –1.27. These two price elasticities have very different implications for the ability of the USPS to raise revenues from U.S. households through postage price increases. In general, an $X\%$ price increase of a product with a demand elasticity of ε increases total revenue from the sale of that product by $X(1 + ε)$. Consequently, if the absolute value of ε is less than 1—the product is inelastically demanded—total revenue will increase as a result of this price increase. This is the case for household-level postal delivery services demand for all years up until 1989. If, as is the case for the years following 1989, the elasticity of demand is greater than 1 in absolute value, price increases will bring about total revenue decreases. A natural question to ask is: How much revenue is lost from reductions in the household-level use of postal delivery services by these price increases?

A first step in answering this question is an estimate of annual aggregate household expenditures on postage. Appendix 9.10 gives estimates of the aggregate annual amount of postage expenditures computed using the sample of households for each year and the corresponding weights giving the number of households in the United States represented by each household in the sample. Appendix 9.11 provides summary statistics for the Consumer Expenditure Diary Survey. The documentation to the Diary Survey Public Use Tape describes the procedure we use to estimate annual aggregate household expenditures. The second column of this appendix gives estimates of the average annual expenditures per household on postage, using the procedure to compute this magnitude given in the Diary Survey Public Use Tapes documentation. For the sake of comparison, the third and fourth columns repeat these same two calculations for telephone services. The behavior of average annual household-level postage and telephone services expenditures over time is similar to the behavior of the sample means of the 2-week expenditures on postage and telephone services given in Appendix 9.2.

Using these numbers and the elasticity estimates, I consider the revenue implications of the January 1, 1995, increase in the price of a one-ounce first class letter from $0.29 to $0.32, a little more than a 10% price increase. From Appendix 9.10, the estimated annual aggregate household postage expenditures in 1994 are approximately $5 billion. Using the preceding equation and the own-price elasticity of −1.27 for 1994 yields a 2.7% reduction in annual aggregate expenditures, or approximately a $135 million reduction in annual revenues from sales of postal delivery services to U.S. households. To put these figures into perspective, we should note that, according to these estimates, sales to households is 10% of total USPS mail delivery revenues in 1994. Appendix 9.10 also gives total USPS revenues for each fiscal year, and the last column of the appendix gives the fraction of these revenues that come from expenditures by households. From 1987 to 1994, the share of annual revenues coming from expenditures by households has approximately halved. Given the revenue loss calculation from reduced sales to households due to the recent first-class postage price increase, we can expect further declines in the share of revenues coming from sales to households in the future.

To assess the impact of personal computing technology on the demand for postal delivery services, we perform a similar calculation assuming an equal percentage change in the probability of computer ownership across all U.S. households. Suppose that, as a result of the explosion in services offered via the Internet, the probability of computer ownership increases by 17% for all U.S. households. From Appendix 9.3, in 1994 the fraction of households owning a personal computer is 0.25. A 17% increase in this magnitude would make it $1.17 \times 0.25 = 0.29$, a reasonable increase in the penetration of computers over the course of a single year. There is a substantial amount of heterogeneity in the probability of computer ownership across households, so that the actual final probability of computer ownership (as result of this uniform 17% increase) for each household could be greater or less than this magnitude, although the average probability over all U.S. households would be equal to approximately 0.29. To illustrate this heterogeneity in estimated computer ownership probabilities, Fig. 9.5 plots the smoothed density estimate of the probability of computer ownership for the 1994 Diary Sample. Although the sample mean of the probability of computer ownership is 0.24, there are values substantially above and below this value.

Figure 9.5
Density Estimate of Computer Ownership Probabilities, Year 1994

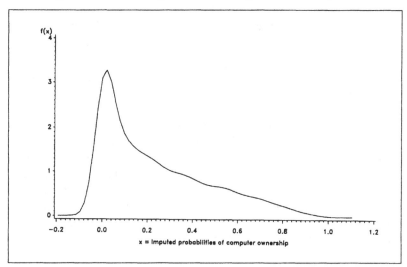

Using the U.S. population average elasticity of demand with respect to the probability of computer ownership for 1994 of –0.158 (computed by taking the average of the estimated household-level elasticities, weighted by the Diary Survey weights, over all of the households in the 1994 Diary Survey) implies a 2.7% decrease in the demand for postal delivery services. Assuming no accompanying change in the price of postal delivery services, this nationwide increase in probability of personal computer ownership implies a 2.7% reduction in revenues from household-level postage expenditures, which is exactly the same reduction in revenues brought about by the 10% increase in the price of postal delivery services discussed earlier. Consequently, for the aggregate demand for postal delivery services by U.S. households, the increasing penetration of personal computing technology and the accompanying more widespread access to the Internet should have significant adverse impacts on the revenues generated from households that rival those from substantial postage price increases.

The other elasticities in Appendix 9.9 show plausible trends over the sample period. For example, there appears to be an increasing degree of substitutability between postal delivery services and all other nondurable goods from 1986 to 1994. The expenditure elasticity shows a downward trend from 0.362 in 1986 to 0.251 in 1994. Another interesting result to emerge from this appendix is the initial slight complementarity between telephone services and postal delivery services, which shifts towards substitutability between these two goods from 1991 onward.

CONCLUSIONS AND DIRECTION FOR FUTURE RESEARCH

The two major results to emerge from this analysis are: (1) Postal price increases of the magnitude recently enacted on January 1, 1995, should lead to significant reductions in aggregate household-level expenditures on postal delivery services, and (2) annual increases in the pervasiveness of personal computer technology at historical rates should lead to reductions in aggregate household-level expenditures on postal delivery services that are at least as large as those that would result from a postal price increase on the order of 10%.

Clearly, there are many caveats associated with these results. One obvious direction for future research is to investigate alternative functional forms for both the demand for postal delivery services and the frequency of purchase model. An additional extension would be to investigate models that explicitly utilize the number purchases made within the 2-week Diary Survey period. These kinds of models can allow own-price and other prices to impact both the decision to purchase and the amount to consume. The relatively small number of multiple purchases observed may imply that multiple purchases in the 2-week period occur primarily for noneconomic reasons. This appears to be a fruitful direction for future research, given the potential large payoff in terms of a richer model for expenditures on postal delivery services.

ACKNOWLEDGEMENTS

I thank Matt Shum for outstanding research assistance. Financial support for this research was provided by the Markle Foundation and the National Science Foundation.

REFERENCES

Blundell, R., & Meghir, C. (1987). Bivariate alternatives to the tobit model. *Journal of Econometrics, 34*, 179–200.

Deaton, A., & Irish, M. (1984). Statistical models for zero expenditures in household budgets. *Journal of Econometrics, 23*, 59–80.

Silverman, B. W. (1986). *Density estimation for statistics and data analysis*. London: Chapman and Hall.

United States Postal Service. (1994). *The 1994 annual report of the Postmaster General*. Washington, D.C: United States Postal Service. Author.

Vuong, Q. (1989). Likelihood ratio tests for model selection and non-nested hypotheses. *Econometrica, 57*, 307–334.

White, H. (1982). Maximum likelihood estimation of misspecified nodels. *Econometrica, 50*, 1–26.

Wolak, F. A. (1997). Changes in the household-level demand for postal delivery services from 1986 to 1994. In M. A. Crew & P. Kleindorfer (Eds.), *Managing change in the postal and delivery industries*. Boston: Kluwer Academic. p. 162–191.

APPENDIX 9.1
Selected Results from Computer Ownership Probits

	1988 Estimates		1994 Estimates	
	Estimate	Std Error	Estimate	Std Error
Constant	–3.03e+00	4.62e-01	-2.14e+00	3.14e-01
February	–5.79e-01	1.80e-01	-8.05e-02	5.16e-02
March	–1.95e-01	1.40e-01	-1.14e-01	5.10e-02
April	1.18e+00	1.19e-01	4.82e-02	7.48e-02
May	1.17e+00	1.18e-01	1.84e-01	7.50e-02
June	1.33e+00	1.16e-01	1.03e-02	7.42e-02
July	1.20e+00	1.17e-01	6.51e-02	7.65e-02
August	1.23e+00	1.16e-01	5.23e-02	7.37e-02
September	1.31e+00	1.13e-01	4.64e-03	7.40e-02
October	1.31e+00	1.13e-01	8.30e-02	7.20e-02
November	1.35e+00	1.14e-01	5.00e-02	7.35e-02
December	1.46e+00	1.13e-01	1.26e-02	7.56e-02
Northwest	2.27e-01	1.26e-01	-7.50e-02	8.32e-02
Midwest	1.54e-01	1.27e-01	-7.84e-02	8.30e-02
South	1.38e-01	1.27e-01	-7.33e-02	8.31e-02
West	2.34e-01	1.25e-01	1.19e-01	8.26e-02
SMSA	–7.28e-02	8.83e-02	-2.02e-01	6.27e-02
Homeowner	5.90e-02	2.55e-01	2.35e-01	1.21e-01
Renter	–5.75e-02	2.55e-01	9.18e-02	1.21e-01
Dorm resident	2.37e-01	3.18e-01	1.19e+00	1.66e-01
Family size	–8.15e-02	5.38e-02	-7.62e-02	3.48e-02
Pers. < 18	1.63e-01	5.68e-02	1.42e-01	3.76e-02
Pers. > 64	5.03e-04	9.04e-02	-4.19e-02	5.79e-02
Number of earners	5.85e-02	4.95e-02	1.28e-01	3.26e-02
Number of vehicles	5.97e-02	1.59e-02	1.28e-01	3.26e-02
White	–2.93e-01	2.16e-01	5.99e-02	1.09e-02
Black	–4.23e-01	2.32e-01	-6.42e-02	2.00e-01
Male	1.16e-01	6.22e-02	-2.91e-01	2.06e-01
Married	–2.91e-01	1.90e-01	6.07e-02	3.76e-02
HS grad	2.29e-01	8.89e-02	-2.31e-01	1.19e-01
>HS, < College	5.66e-01	8.84e-02	3.81e-01	5.86e-02
College grad	8.14e-01	9.18e-02	8.61e-01	5.85e-02
Age	1.05e+00	1.24e+00	1.15e+00	6.03e-02
Age^2	–2.26e+00	1.43e+00	2.46e+00	7.20e-01

	1988 Estimates		1994 Estimates	
	Estimate	Std Error	Estimate	Std Error
Spouse's age	1.62e+00	1.00e+00	1.84e+00	6.11e-01
Spouse's age^2	−1.43e+00	1.27e+00	-1.78e+00	7.50e-01
Prof. occupation	1.98e-01	6.67e-02	1.01e-01	5.13e-02
Tech. occupation	1.98e-01	6.673-02	-9.97e-02	7.88e-02
Self-employed	1.51e-01	9.64e-02	-1.06e-01	1.11e-01
Retired	−4.99e-03	1.44e-01	-2.57e-02	6.42e-02
Hours of work	−2.28e-03	1.78e-03	-2.38e-04	1.08e-03
Spouse's hours of work	1.11e-03	1.87e-03	1.37e-03	1.28e-03
Positive income	3.59e-02	1.03e-02	5.10e-02	6.28e-03
Negative income (dummy)	1.38e-01	8.05e-02	7.02e-02	5.57e-02
Log-likelihood function value	−1,881.159		−4,554.942	
N	10,122		9,967	

APPENDIX 9.2
Average Expenditures and Shares by Year

		Mean	Std Deviation	Minimum	Maximum
Year = 1986	Postal share	0.01	0.02	0.00	0.62
$N = 5,839$	Postal expn. ($)	2.34	6.67	0.00	200.58
	Telephone share	0.04	0.08	0.00	0.82
	Telephone expn. ($)	16.00	32.76	0.00	400.00
Year = 1987	Postal share	0.01	0.03	0.00	0.07
$N = 6,024$	Postal expn. ($)	2.61	9.51	0.00	399.98
	Telephone share	0.05	0.09	0.00	0.84
	Telephone expn. ($)	17.62	34.96	0.00	401.94
Year = 1988	Postal share	0.01	0.03	0.00	0.75
$N = 5,264$	Postal expn. ($)	2.80	8.33	0.00	357.78
	Telephone share	0.05	0.09	0.00	0.82
	Telephone expn. ($)	18.21	35.53	0.00	429.65
Year = 1989	Postal share	0.01	0.03	0.00	0.89
$N = 5,317$	Postal expn. ($)	3.02	8.20	0.00	214.00
	Telephone share	0.05	0.08	0.00	0.93
	Telephone expn. ($)	19.36	38.26	0.00	584.94

		Mean	Std Deviation	Minimum	Maximum
Year = 1990	Postal share	0.01	0.02	0.00	0.39
N = 5,446	Postal expn. ($)	2.90	6.62	0.00	87.00
	Telephone share	0.05	0.09	0.00	0.78
	Telephone expn. ($)	20.29	41.07	0.00	551.00
Year = 1991	Postal share	0.01	0.02	0.00	0.46
N = 5,550	Postal expn. ($)	2.51	7.08	0.00	120.55
	Telephone share	0.05	0.08	0.00	0.78
	Telephone expn. ($)	20.79	41.87	0.00	518.72
Year = 1992	Postal share	0.01	0.02	0.00	0.75
N = 5,436	Postal expn. ($)	2.56	7.57	0.00	222.78
	Telephone share	0.05	0.09	0.00	1.00
	Telephone expn. ($)	23.66	45.78	0.00	729.16
Year = 1993	Postal share	0.01	0.02	0.00	0.31
N = 5,299	Postal expn. ($)	2.23	6.85	0.00	122.37
	Telephone share	0.05	0.09	0.00	0.86
	Telephone expn. ($)	23.74	47.63	0.00	742.00
Year = 1994	Postal share	0.01	0.02	0.00	0.72
N = 4,914	Postal expn. ($)	2.13	6.76	0.00	157.32
	Telephone share	0.05	0.09	0.00	0.90
	Telephone expn. ($)	25.55	48.83	0.00	520.00

APPENDIX 9.3
Annual Sample Percentages of Personal Computer Ownership for Interview Survey Sample

Year	Number of Households	Percent Owning Personal Computer
1988	10,122	7.0
1989	9,907	14.5
1990	10,015	16.2
1991	9,878	17.4
1992	10,028	19.1
1993	10,097	22.3
1994	10,011	24.9

APPENDIX 9.4
Log Infrequency of Purchase Model
(Log Likelihood Function Value: 19,070.6)

	Share Equation		Purchase Probability	
	Estimate	Std Error	Estimate	Std Error
Constant	−6.35e+00	1.76e-01	−6.12e-01	9.78e-02
log price$_{post}$	−5.59e-02	1.93e-01		
log price$_{tel}$	8.95e-02	1.08e-01		
Nondurable expn.	6.82e-01	1.29e-02		
Northwest	2.66e-02	5.33e-02	−3.33e-02	3.05e-02
Midwest	−5.13e-02	5.27e-02	−4.08e-02	3.01-02
South	−8.43e-02	5.30e-02	−8.29e-02	3.02e-02
West	2.89e-02	5.44e-02	−5.29e-02	3.10e-02
SMSA	−1.93e-01	4.12e-02	−5.21e-02	2.33e-02
Family size	−1.90e-02	1.70e-02		
Pers. < 18	−3.82e-02	1.89e-02		
Pers. > 64	4.85e-02	2.02e-02		
Number of earners	−1.35e-02	1.45e-02		
White	−3.38e-03	1.06e-01	3.66e-03	5.91-02
Black	−3.64e-01	1.13e-01	−1.43e-01	6.24e-02
Male	−1.85e-01	2.60e-02	−5.42e-02	1.49-02
Married	2.46e-01	4.89e-02		
HS grad	3.19e-01	3.13e-02	1.65e-01	1.77e-02
> HS, < college	4.52e-01	3.55e-01	2.73e-01	2.04e-02
College grad	6.48e-01	4.05e-02	3.89e-01	2.32e-02
Age	1.22e+00	9.28e-02	1.64e-01	4.65e-02
Spouse's age	3.40e-01	9.67e-02	5.12e-01	3.19e-02
Prof. occupation	1.95e-01	3.19e-02	1.38e-01	1.91e-02
Tech. occupation	1.06e-01	3.09e-02	9.40e-02	1.80e-02
Self-employed	5.29e-02	4.65e-02	−1.28e-02	2.59e-02
Retired	1.43e-01	3.78e-02	7.03-02	2.29e-02
Hours of work			−2.02e-04	3.58e-04
Spouse's hours of work			3.63e-04	3.22e-04
Positive income			1.03e-02	2.65e-03
Negative income (dummy)			−1.52e-01	1.73e-02
Computer ownership prob.	−4.15e-01	1.08e-01	−7.90e-01	6.15e-02
December	3.94e-01	2.85e-02	5.67e-02	1.73e-02
σ	9.80e-01	5.32e-03		

Note: Standard errors are heteroscedasticity-consistent in sense of White (1982).

APPENDIX 9.5
Log Infrequency of Purchase Model
(Log Likelihood Function Value: 19,070.6)

Variable Name	Definition
Northwest	1 if household resides in Northwest Census region
Midwest	1 if household resides in Midwest Census region
South	1 if household resides in Southern Census region
West	1 if household resides in Western Census region
	(omitted category is rural residents in all regions)
SMSA	0 if household resides in a Census SMSA; 1 if not
Family size	Number of members in household
Pers. < 18	Number of persons <18 years of age in household
Pers. > 64	Number of persons >64 years of age in household
Number of earners	Number of earners in household
White	1 if household head is white
Black	1 if household head is black
	(omitted category are Native Americans and other ethnic groups)
Married	1 if household head is married
Male	1 if household head is male
HS grad.	1 if household head is high school graduate
>HS, <college	1 if household head has some college education
College grad.	1 if household head is college graduate
	(omitted category is high school noncompleters)
Age	Age of household head
Spouse's age	Age of spouse (if applicable)
Prof. occupation	1 if household head is a professional
Tech. occupation	1 if household head is in a technical occupation
Self-employed	1 if household head is self-employed
Retired	1 if household head is retired
Hours of work	Weekly hours of work for household head
Spouse's hours of work	Weekly hours of work for spouse (if applicable)
Positive income	Household's income, if >0
Negative income (dummy)	1 if household income is negative
Computer ownership prob.	Imputed computer ownership probability (details in text), December
	1 if survey month is December

APPENDIX 9.6
Purchase Probability Derivatives for Log Infrequency Model

	Mean dP/dX	Std Error
Constant	−2.34e-01	2.02e-02
Northwest	−1.18e-02	1.01e-03
Midwest	−1.72e-02	1.48e-03
South	−2.93e-02	2.52e-03
West	−2.09e-02	1.80e-03
SMSA	−1.84e-02	1.58e-03
White	1.29e-03	1.11e-04
Black	−5.05e-02	4.35e-03
Male	−1.92e-02	1.65e-03
HS grad	5.83e-02	5.02e-03
> HS, < college	9.64e-02	8.30e-03
College grad	1.37e-01	1.18e-02
Age	5.80e-02	4.99e-03
Spouse's age	1.81e-01	1.55e-02
Prof. occupation	4.88e-02	4.20e-03
Tech. occupation	3.32e-02	2.86e-03
Self-employed	−4.51e-03	3.88e-04
Retired	2.48e-02	2.14e-03
Hours of work	−7.12e-05	6.13e-06
Spouse's hours of work	1.28e-04	1.10e-05
Positive income	3.64e-03	3.13e-04
Negative income (dummy)	−5.37e-02	4.62e-03
Computer ownership prob.	−2.79e-01	2.40e-02
December	2.00e-02	1.72e-03

APPENDIX 9.7
Sample Mean Household-Level Elasticities from the Log-Infrequency Model

		Log-Infrequency
ε_{pp}		−1.06
ε_{pt}		−7.64e-02
ε_{po}		6.48e-01
ε_{pm}		3.18e-01
$\varepsilon_{p,comp}$	Mean	−6.87e-01
	deviation	6.55e-02

APPENDIX 9.8
Log Infrequency Model with Time Trends
(Log Likelihood Function Value: 19,195.2)

	Share Equation		Purchase Probability	
	Estimate	Std Error	Estimate	Std Error
Constant	−6.50e+00	1.76e-01	−7.28e-01	9.85e-02
log price$_{post}$	2.42e-01	5.33e-01		
(yearly trend) * log price$_{post}$	−6.36e-02	1.32e-01		
log price$_{tel}$	−1.22e-01	2.21e-01		
(yearly trend) * log price$_{tel}$	2.88e-02	4.67e-02		
log nondurable expn.	6.15e-01	2.09e-02		
(yearly trend) * log expn.	8.70e-03	4.80e-03		
Northwest	6.54e-03	5.32e-02	−4.97e-02	3.05e-02
Midwest	−6.17e-02	5.26e-02	−5.78e-02	3.01e-02
South	−9.52e-02	5.29e-02	−9.16e-02	3.02e-02
West	−3.96e-03	5.44e-02	−8.67e-02	3.11e-02
SMSA	−1.87e-01	4.12e-02	−4.95e-02	2.33e-02
Family size	−1.42e-02	1.70e-02		
Pers. <18	−4.19e-02	1.89e-02		
Pers. >64	5.19e-02	2.02e-02		
Number of earners	−2.28e-02	1.45e-02		
White	1.05e-01	1.07e-01	9.47e-02	5.99e-02
Black	−2.41e-01	1.13e-01	−4.23e-02	6.32e-02
Male	−2.04e-01	2.60e-02	−7.35e-02	1.50e-02
Married	2.24e-01	4.89e-02		
HS grad.	3.12e-01	3.13e-02	1.60e-01	1.77e-02
>HS, <college	4.07e-01	3.59e-02	2.35e-01	2.06e-02
College grad.	5.56e-01	4.17e-02	3.11-01	2.39e-02
Age	1.31e-+00	9.31e-02	2.50e-01	4.70e-02
Spouse's age	3.22e-01	9.66e-02	4.83e-01	3.19e-02
Prof. occupation	1.45e-01	3.23e-02	9.55e-02	1.94e-02
Tech. occupation	8.85e-02	3.09e-02	7.83e-02	1.80e-02
Self-employed	3.34e-02	4.65e-02	−3.18e-02	2.60e-02
Retired	1.36e-02	3.80e-02	6.44e-02	2.28e-02
Hours of work			−3.16e-04	3.58e-04
Spouse's hours of work			−9.02e-05	3.24e-04
Positive income			7.79e-03	2.66e-03
Negative income (dummy)			−1.60e-01	1.74e-02
Computer ownership prob.	1.09e+00	1.76e-01	7.22e-01	1.02e-01
yeartrend * comp. ownership	−2.16e-01	2.40e-02	−1.81e-01	1.19e-02
December	3.55e-01	2.87e-02	2.42e-02	1.75e-02
σ	9.79e-01	5.32e-03		

Note: Standard errors are heteroscedasticity-consistent in sense of White (1982).

APPENDIX 9.9
Sample Mean Elasticity Estimates: Log-Infrequency Model with Time Trend

	ε_{pp}	ε_{pt}	ε_{po}	ε_{pm}	$\varepsilon_{p,comp}$
1986	−7.58e-01	−1.22e-01	5.18e-01	3.62e-01	1.18e-01
1987	−8.21e-01	−9.29e-02	5.66e-01	3.48e-01	9.19e-02
1988	−8.85e-01	−6.41e-02	6.15e-01	3.34e-01	7.06e-02
1989	−9.49e-01	−3.53e-02	6.63e-01	3.20e-01	6.64e-02
1990	−1.01e+00	−6.51e-03	7.12e-01	3.07e-01	3.81e-02
1991	−1.08e+00	2.23e-02	7.61e-01	2.93e-01	1.79e-03
1992	−1.14e+00	5.11e-02	8.09e-01	2.79e-01	−4.09e-02
1993	−1.20e+00	7.99e-02	8.58e-01	2.65e-01	−9.94e-02
1994	−1.27e+00	1.09e-01	9.06e-01	2.51e-01	−1.68e-01

APPENDIX 9.10
Postal Revenue and Estimated Annual Household Expenditures

	Estimated Aggregate Postage Expn. ($, billion)	Estimated Average Postage Expn. ($)	Estimated Aggregate Telephone Expn. ($, billion)	Estimated Average Telephone Expn. ($)	Total USPS Annual Revenue ($, billion)	Revenue Share of Households (%)
1986	5.07	58.19	36.70	421.67	29.12	17.39
1987	5.83	66.89	41.32	474.22	30.50	19.11
1988	5.85	65.87	42.64	480.38	33.92	17.23
1989	6.69	74.26	45.63	506.51	36.67	18.24
1990	6.38	69.88	50.11	549.16	37.89	16.83
1991	5.80	62.71	52.16	563.45	41.92	13.85
1992	6.06	64.20	59.79	633.04	44.72	13.56
1993	5.90	62.75	60.82	647.07	45.91	12.85
1994	5.05	54.35	64.19	690.65	47.75	10.58

APPENDIX 9.11
Consumer Expenditure Diary Survey Summary Statistics
(Number of Observations: 49,089)

	Mean	Std Deviation	Minimum	Maximum
Postage purchase indicator	0.33	0.47	0.00	1.00
Nondurable expenditures	393.26	289.79	20.00	1,698.08
Postage expenditures	2.57	7.60	0.00	399.98
Telephone expenditures	20.46	41.00	0.00	742.00
Number of postage purchases (in 2 weeks)	0.45	0.78	0.00	11.00

	Mean	Std Deviation	Minimum	Maximum
Postage expenditure share	0.01	0.02	0.00	0.89
Telephone expenditure share	0.05	0.09	0.00	1.00
Nondurables prices (January 1986 = 1)	1.15	0.11	0.96	1.30
Telephone price (January 1986 = 1)	1.04	0.02	1.00	1.08
Postage price (January 1986 = 1)	1.18	0.13	1.00	1.32
Computer ownership probability	0.16	0.16	0.00	0.98
Northwest	0.19	0.39	0.00	1.00
Midwest	0.23	0.42	0.00	1.00
South	0.26	0.44	0.00	1.00
West	0.21	0.41	0.00	1.00
SMSA	0.19	0.39	0.00	1.00
Homeowner	0.64	0.48	0.00	1.00
Renter	0.33	0.47	0.00	1.00
Dorm resident	0.01	0.10	0.00	1.00
Family size	2.61	1.50	1.00	16.00
Pers. <18	0.74	1.13	0.00	12.00
Pers. >64	0.29	0.60	0.00	5.00
Number of earners	1.44	1.00	0.00	9.00
Number of vehicles	1.65	1.13	0.00	21.00
White	0.89	0.31	0.00	1.00
Black	0.10	0.30	0.00	1.00
Male	0.65	0.48	0.00	1.00
Married	0.58	0.49	0.00	1.00
HS grad.	0.31	0.46	0.00	1.00
>HS, <college	0.23	0.42	0.00	1.00
College grad.	0.25	0.43	0.00	1.00
Age	0.47	0.17	0.15	0.90
Spouse's age	0.25	0.25	0.00	0.90
Prof. occupation	0.22	0.42	0.00	1.00
Tech. occupation	0.18	0.39	0.00	1.00
Self-employed	0.07	0.25	0.00	1.00
Retired	0.15	0.35	0.00	1.00
Hours of work	31.69	20.54	0.00	90.00
Spouse's hours of work	14.66	19.59	0.00	90.00
Positive income	2.66	2.65	0.00	48.64
Negative income (dummy)	0.12	0.33	0.00	1.00
December (dummy)	0.14	0.34	0.00	1.00

III

INTERNET REGULATION
AND CONTROL

Can and Should the FCC Regulate Internet Telephony?

Robert M. Frieden
Pennsylvania State University

On March 4, 1996, America's Carriers Telecommunication Association (ACTA), a trade group representing primarily medium and small long-distance telephone companies, filed with the Federal Communications Commission (FCC) a controversial Petition for Declaratory Ruling, Special Relief and Institution of a Rulemaking.[1] ACTA alleged that providers of Internet telephony software operate as uncertified and unregulated common carriers[2] in contravention of Commission Rules and Regulations. The trade association suggested the need for regulatory parity: FCC jurisdiction and common carrier regulation of companies selling Internet telephony, because conventional telephone service providers face such government oversight. ACTA also warned that increasing use of Internet resources for

---

1. *"Provision of interstate and international interexchange telecommunications services via the "Internet" by non-tariffed, uncertified entities; petition for declaratory ruling. Special relief, and institution of rulemaking."* Petition of America's Carriers Telecommunication Association, RM No. 8775 (filed March 4, 1996) Available [Online]: http://www.fcc.gov/Bureaus/Common_Carrier/Other/acatapet.html (hereafter cited as ACTA Petition).

2. Legislative, regulatory and judicial action have eliminated a bright line dichotomy between common carriers and private, non-common carriers. See E. M. Noam, Beyond liberalization II: The impending doom of common carriage, *Telecommunications Policy 18*, No. 6 435-452 (June 1994); R. Frieden, Contamination of the Common Carrier Concept in Telecommunications, *Telecommunications Policy 19* No. 9, 685–697 (December 1995).

telephony "could result in a significant reduction of the Internet's ability to handle the customary types of Internet traffic."[3]

Much of the extensive opposition to the ACTA Petition predictably objected to FCC regulation of software enterprises never before subject to Commission oversight. Such opponents argued that the FCC cannot impose common carrier regulation, because Internet telephony constitutes enhanced services under the Commission's Computer Inquiries[4] and information services[5] as defined by the Telecommunications Act of 1996.[6] Others[7] alleged that definitions contained in the revised Communications Act—access software,[8] access software providers[9] and interactive computer service[10]—cover the functionality involved in Internet telephony and expressly qualify for exclusion from the common carrier classification.[11]

Beyond the immediate question of jurisdiction and regulatory scope, the ACTA Petition raises several broader issues largely ignored by the commenting parties:

- What, if any, steps should the FCC undertake to eliminate regulatory asymmetry, that is, different and inconsistent regulatory treatment of competing enterprises and services,[12] where doing so would obligate it to assert jurisdiction and to regulate a previously unregulated industry, as opposed to the typical situation where fostering regulatory symmetry would result in fewer or eliminated regulations?

---

3. ACTA Petition at 7.

4. For an analysis of how the FCC attempted to create a "bright line" separation between unregulated enhanced service functions and regulated basic transport capacity see R. M. Frieden, The Computer Inquiries: Mapping the communications/information processing terrain, *Federal Communications Law Journal 33*, 55–115 (1981); R. M. Frieden, The third computer inquiry: A deregulatory dilemma, *Federal Communications Law Journal 38*, 383-410 (1987).

5. Section 3 (20) of the Telecommunications Act of 1996, P.L. 104-104, 110 Stat. 56, signed into law February 8, 1996, codified at 47 U.S.C. § 153(20)(1996), defines information services as "the offering of a capability for generating. acquiring, storing, transforming, processing, retrieving, utilizing or making available information via telecommunications."

6. See, e.g., Joint Opposition of Netscape Communications Corporation, Voxware, Inc. and Insoft, Inc., (May 8, 1996). [Online]: http: www.technologylaw.com/techlaw/acta-comm.html#n53

7. See, e.g., *Opposition of the Information Technology Industry Council* (May 8, 1996). Available [Online]: http://www.itic.org/fccacta.htm

8. The Communications Act, in a section addressing obscenity or harassing telephone calls, defines assess software as: "software (including client or server software) or enabling tools that do not create or provide the content of communications but that allows a user to do any one or more of the following: (A) filter, screen, allow or disallow content; (B) pick, choose, analyze, or digest content; or (C) transmit, receive, display, forward, cache, search, subset, organize, reorganize, or translate content" [47 U.S.C. § 223(h)(3)].

- When should the FCC maintain regulatory asymmetry, despite some legal and economic arguments favoring a single, consistent regulatory regime?
- To what extent does Internet telephony support or frustrate longstanding efforts to foster universal service, and how should the FCC balance the longstanding objective of achieving ubiquitous access to plain old telephone service (POTS) with its new and broader mandate to "encourage the provision of new technologies and services to the public"[13] including a "high-speed, switched, broadband telecommunications capability that enables users to originate and receive high-quality voice, data, graphics, and video telecommunications using any technology"?[14]

This chapter examines the consequences of continuing to regulate incumbent common carriers, despite market entry by ventures operating free of regulation. Although incumbents demand that a "level competitive playing field" necessitates either their deregulation, or the regulation of newcomers, the chapter concludes that regulatory asymmetry promotes competition, serves objectives articulated in the Telecommunications Act of 1996, promotes innovation, stimulates downward pressure on rates, and helps achieve universal service goals. Although Internet telephony may fit within the loose, new definition of telecommunications contained in the Telecommunications Act of 1996,[15] the enabling software does not. This means that the FCC can avoid having to confront key policy and regulatory issues

---

9. The Communications Act defines assess software providers as: "providers of software (including client or server software) or enabling tools that do any one or more of the following: (A) filter, screen, allow or disallow content; (B) pick, choose, analyze, or digest content; or (C) transmit, receive, display, forward, cache, search, subset, organize, reorganize, or translate content" [47 U.S.C. § 223(e)(4)].

10. The Communications Act defines interactive computer service as "any information service, system, or access software provider that provides or enables computer access by multiple users to a computer server, including specifically a service or system that provides access to the Internet and such systems operated or services offered by libraries or educational institutions" [47 U.S.C. § 223(e)(2)].

11. "Nothing in this section shall be construed to treat interactive computer services as common carriers or telecommunications carriers" [47 U.S. C. § 223(e)(6)]. The clear meaning of this section may help explain why ACTA did not seek common carrier regulation of Internet service providers who lease the telecommunications lines necessary to provide access to the Internet, including the necessary transmission of traffic between users and various network nodes and servers.

12. The FCC appears inclined to eliminate regulatory asymmetry at least under conditions where it can deregulate a particular carrier or service to foster a "level" competitive playing field. See *Motion of AT&T corp. To be reclassified as a non-dominant carrier,* Order, 11 FCC Rcd 3271 (1995); *Motion of AT&T corporation to be declared non-dominant for international service,* Order, FCC 96-209 (released May 14, 1996).

13. Communications Act of 1934, as amended, 47 U.S.C. § 157.

raised by the ACTA petition in much the same manner as it can reject a petition to regulate Internet radio services as broadcasting.

However, future growth in Internet telephony may challenge the wisdom of regulatory forbearance, making it likely that the Commission may have to consider the impact of such services on universal service objectives. Likewise, the potential for migration of incumbent carrier traffic and revenues will grow substantially as new Internet telephony options enter the marketplace, including user-friendly access via widely used Internet browsers like Netscape Navigator,[16] service packages offered by industry leaders,[17] and even telephones in lieu of costly personal computer configurations.[18]

INTERNET TELEPHONY TECHNOLOGY—THE BASICS

Internet telephony uses the digital, packet-switched nature of the Internet along with its routing and addressing standards to provide real time audio conferencing.[19] Internet switching and routing technology manages the transmission and processing of text, graphics, data, audio, or video. The Internet's TCP/IP protocol[20] provides a standard vehicle for subdividing content, such as, a voice conversation, into a stream of packets that are routed via any available path between the sender and intended call recipient. Each packet has space reserved for destination

14. Telecommunications Act of 1996, Sec. 706(c)(1), Advanced Telecommunications Capability (requiring the FCC and each State regulatory commission with jurisdiction over telecommunication services to encourage the timely deployment of advanced telecommunications capability to all Americans).

15. The Telecommunications Act of 1996 defines telecommunications as "the offering of telecommunications for a fee directly to the public, or to such classes of users as to be effectively available directly to the public, regardless of the facilities used," in Sec. 3(46), codified at 47 U.S.C. § 153(46).

16. For example, "Netscape has incorporated...real-time audio and video technology into the latest release of its Internet software." Joint Opposition of Netscape Communications Corporation, Voxware, Inc. and Insoft, Inc., to the America's Carriers Telecommunication Assoc. Petition for Declaratory Ruling, Special Relief, and Institution of Rulemaking at note 9. Available [Online]: http://www.technologylaw.com/techlaw/acta_comm.html#n53 (filed May 8, 1996, hereafter cited as Netscape Opposition).

17. See, e.g., Lucent jumps into Internet telephony, *Telecommunications Reports* 62, No. 39, at 21 (September 30, 1996, reporting on Lucent's plans to offer an Internet server capable of routing nonessential facsimile transmissions, voice mail and Internet telephone calls and deeming such product marketing the final touch on its separation from AT&T).

18. "Unlike other Internet telephony products that require a computer with special software to complete calls, AlphaNet's service will work over normal telephone lines and fax machines 'at a considerable price advantage over traditional voice networks.'" In AlphaNet unveils "Mondial" Internet telephony service, *Telecommunications Reports* 62, No. 39 at 37 (September 30, 1996).

information so that intermediary routing facilities have information necessary to determine how and where to send the packets onward toward their intended destination. This header information includes a sequence of digits that correspond to an Internet address, much like the numbering sequence in direct distance dialing via telephone.

Packet switching efficiently uses available switching and routing capacity. Likewise it can operate despite outages, blockages, and busy conditions, because the Internet Protocol (IP) addressing scheme makes it possible for multiple efforts to route traffic onward in the event that initial efforts fail. Resending misdelivered or unreceived packets and routing them via different and possibly circuitous links requires software processing to reassemble the packets in proper order. For traffic and services that do not require immediate, real-time delivery, such as, electronic mail, possible delays and reassembly present little problem. However, Internet telephony requires immediate, "real-time" delivery of the packets in their proper order. Any delay, loss, or improper sequencing of packets will resort in distortion, or the temporary loss of the audio stream.

Heretofore, Internet telephony has lacked the quality, reliability, and security to be considered comparable to conventional telephone services. On the other hand, the cost of such service substantially undercuts the current per-minute rates paid for conventional services. Unlike conventional telephone service, Internet-mediated telephony requires a significant initial capital outlay of about $2,000 for a personal computer, modem, sound card, speakers, microphone, software, and Internet access. Conventional telephone services use an inexpensive, "dumb" terminal, the telephone handset, but users incur per-minute charges that can exceed $1.00 a minute for many international destinations. Internet telephony requires an "intelligent" terminal, but users incur little, if any, charges above the initial investment in hardware and software, and the monthly fees for local telephone or cable television service, which provide a link from users to Internet access providers.

WHO'S THERE?

A key problem in determining the permissible and prudent scope of FCC regulation lies in identifying who provides Internet telephony service as opposed to enterprises involved in the design, marketing, and software integration with

19. For a helpful nontechnical introduction to Internet telephony see the Internet Telephony Consortium, *Frequently asked questions: How can I use the Internet as a telephone?* Available [Online]: http://itel.mit.edu/voice_faq.html.

20. "The common denominator for e-mail communications is the use of a standard programming protocol, TCP/IPMTransmission Control Protocol/Internet ProtocolMupon which inter-computer communications are based." R. A. Horning, Has Hal signed a contract: the statute of frauds in cyberspace, *Santa Clara Computer & High Tech Law Journal 12*, 253, 258 (August 1996).

hardware needed by consumers to make an Internet-mediated telephone call. The ACTA petition focused on the software providers, but it could have targeted:

- Internet service providers, like Netcom, PSI, and America Online, which link individuals with the networking capabilities of the Internet and own or lease the transmission facilities needed to transport packets to and from subscribers' terminals;
- Internet browser developers, like Microsoft and Netscape, which provide the user-friendly link to Internet resources, and which bundle Internet telephony software typically as a point-and-click icon on the browser;
- Internet telephony hardware manufacturers, like Lucent, IBM, and Intel, that develop the computer chips, servers, and other devices that make it possible to route telephony packets via the Internet;
- Most telecommunications ventures, whether regulated or not, which consider the Internet a potential new profit center even if some of the services cannibalize existing revenue streams, including local and long-distance telephony; or
- Any venture that uses Internet telephony in conjunction with the sale or marketing of a good or service.

Internet telephony can occur only through the integration of products and services representative of different market segments in telecommunications, information processing, computer hardware, and software. Although constituting an integral part of an Internet telephony service, software alone cannot deliver any service. Accordingly, companies like Netscape Communications Corporation can characterize themselves as "leaders in the new market for Internet voice and video services by creating the open standards, protocols, compression technologies and software products necessary to enable transmission of interactive voice and video information over the Internet."[21] Promoting user-friendly access to Internet telephony, which includes a telecommunications transport function, does not make Netscape a common carrier, nor should the FCC use regulation to hamper the development of technologies simply on the prospect that they might in the future compete for customers and revenues now served by incumbent, regulated carriers.

ASYMMETRIC REGULATION IN TELECOMMUNICATIONS

Despite its current predisposition toward regulatory parity, in many instances the FCC has established a different regulatory regime for competing telecommunication services and carriers. For example, the price, but not necessarily the cost, of a minute of telecommunication use depends on such factors as:

- The perceived value of the service;[22]
- Which regulatory agency has jurisdiction over cost allocation and tariffing;[23]

---

21. Netscape Opposition at paragraph 10.

- Whether the service is domestic or international;[24]
- Whether another carrier or end-user seeks facilities interconnection;[25]
- The type of carrier[26] or enterprise[27] providing service;[28] and
- The type of line or facility providing service[29] and whether the service can access the public switched telephone network (PSTN).[30]

With increasing frequency the FCC has expressed its intention to rely on facilities-based and resale competition rather than use heavy-handed "command and control" regulatory oversight. Economic analysis has driven this migration and provided the methodology for both predicting and measuring enhanced consumer welfare. Yet despite such reliance on marketplace self-regulation and the scientific discipline of economic analysis, the FCC and its state regulatory counterparts persist in market-distorting activities. Such intervention continues by design rather than as a product of "deregulatory lag" resulting from a backlog of procompetitive initiatives.

The FCC has maintained regulatory asymmetry at various times and for particular circumstances on justifiable public policy reasons, political expediency, the belief that it must temporarily insulate operators from competition,[31] or the perceived need to incubate competition.[32] In addition to jurisdictional and legal limitations, which would prohibit the FCC from regulating Internet telephony software designers, the preceding reasons likewise support leaving

---

22. Both the FCC and state regulatory commission have allowed carriers to price some services on the perceived value consumers accrue. For example, some local exchange telephone service rates have increased as a when the number of accessible subscribers reaches a benchmark. "In most states, the Bell Operating Companies and larger independents charge higher rates in metropolitan areas than in rural areas—a pricing practice that dates back to the turn of the century and is traditionally justified in the belief that the value of the service provided is higher for subscribers with larger local calling areas." Federal Communications Commission, *FCC releases semiannual study on telephone trends*, 1991 FCC LEXIS 4305 at *10 (August 7, 1991).

23. Typically an intrastate long-distance minute of use significantly exceeds the price of an interstate long-distance minute of use. Ironically, an intrastate state call originated via a cellular telephone may be significantly cheaper than the corresponding rate for a call originated over wireline facilities. The rate differential results, in part, from rate-making policies, which may include cross-subsidies to local exchange service, as opposed to actual cost of service differences.

24. International message telephone service substantially exceeds domestic rates on a per-minute and mileage-band basis, primarily because international carriers have negotiated toll revenue division agreements that have failed to drop commensurately with cost reductions. For a discussion of these international accounting rates see R. Frieden, International toll revenue division: Tackling the inequities and inefficiencies, *Telecommunications Policy 17*, No. 3, 221–233 (April 1993); R. Frieden, Accounting rates: The business of international telecommunications and the incentive to cheat, *Federal Communications Law Journal 43*, 111–139 (April 1991).

Internet telephony unregulated. Despite the invitation to foster an immediate level competitive playing field, a simple cost/benefit analysis would result in the FCC finding that the cost to individual companies, like the members of ACTA and their stakeholders, are greatly offset by the benefits accruing to other companies and to the public in general if the Commission refrained from regulating Internet telephony altogether.

IS INTERNET TELEPHONY TELECOMMUNICATIONS?

The FCC would not have to make a possibly controversial cost/benefit analysis if it could conclude that Internet telephony does not constitute telecommunications as defined by the Communications Act of 1934. However, newly legislated definitions of telecommunications, telecommunications carrier, and telecommunications service arguably are expansive enough to include Internet telephony as a new common carrier service. The revised Communications Act now defines telecommunications as "the transmission, between or among points specified by the user, of information of the user's choosing, without change in the form or content of the information as sent and received."[33] The purveyors of Internet telephony software do not necessarily provide the needed Internet access and transmission function. However, the FCC might consider the provider of such access as also providing a

---

25. The Telecommunications Act of 1996 and preexisting FCC regulations differentiate the terms and conditions for interconnection between carriers as opposed to customer-carrier interconnection. The Telecommunications Act orders favorable and potentially zero-cost interconnection between certain types of carriers. For example, Section 251 requires all local exchange carriers "to establish reciprocal compensation arrangements for the transport and termination of telecommunications" [47 U.S.C. § 251(b)(5)]. End-users and interexchange ("long distance") carriers must pay higher "access charges."

26. During a time when interexchange carrier competitors of AT&T received inferior access to the PSTN, the Commission authorized discounted access charges. However, the Commission never stated that the discounts were cost-based as opposed to a rough justice solution designed to reflect both inferior access and the Commission's desire that carriers like MCI acquire market share. See, e.g., *Exchange network facilities for interstate access (ENFIA)*, CC Docket No. 78-371, Report and Order, 71 FCC 2d 440 (1979); on recon., 93 FCC 2d 739 (1983), aff'd in part and remanded in part sub nom., MCI Telecomm. Corp. v. FCC, 712 F.2d 517 (D.C. Cir. 1983). The FCC recently decided that, on an interim basis, wireline and wireless service providers should compensate each other on a reciprocal basis for terminating calls. See *First Report and Order and Implementation of the Local Competition Provisions in the Telecommunications Act of 1996, Interconnection between Local Exchange Carriers and Commercial Mobile Radio Service Providers*, CC Docket No. 96-98, FCC 96-325, 1996 WL 452885 (F.C.C.)(released August 8, 1996).

27. "Captive" long-distance callers from hotel rooms and callers not familiar with "dial around" options for avoiding price gouging for pay-phone service recognize the vast price differences for long-distance telephone service.

transmission function if the Commission were to couple Internet access with the transmission functionality derived from the owned or leased transmission facilities needed for any Internet service, including telephony.

The second and third prongs of the definition are met, because users choose what to transmit, such as voice messages, and the intended recipient receives the same voice message.[34]

The definition of telecommunication appears not to exempt functionality that involves intermediary processing that temporarily renders content incomprehensible to users. However, under the policies already established in the FCC's *Computer Inquiries,* processing Internet packets using the TCP/IP protocol might qualify as non-common-carriage, enhanced services even if the final output constituted a voice message.[35]

The Communications Act now defines telecommunications service as "the offering of telecommunications for a fee directly to the public, or to such classes of users as to be effectively available directly to the public, regardless of the facilities used."[36] Again a software seller does not provide telecommunication services, but the for-pay services of Internet access providers approach the required level of ubiquity given the number of deep-pocketed providers entering the market and their use of 800 numbers to provide access to subscribers even in remote areas. Internet access may be "effectively available" to the public, despite the fact that a relatively small percentage of the public is effectively able currently to use the service offered. Because the definition contains a reference that the nature of the fa-

---

28. Certain types of services have qualified for exemption from regulatory burdens that impose extra costs. For example, enhanced services qualify for non-common-carrier status and their users are exempt from having to pay an access charge payment otherwise applicable to basic service subscribers. A 1987 FCC initiative to eliminate the exemption generated substantial opposition by users who claimed the Commission had proposed to impose a "modem tax." "In 1983 we adopted a comprehensive 'access charge' plan for the recovery by local exchange carriers (LECs) of the costs associated with the origination and termination of interstate calls. [citing MTS and WATS Market Structure, Memorandum Opinion and Order, 97 FCC 2d 682 (1983)]. At that time, we concluded that the immediate application of this plan to certain providers of interstate services might unduly burden their operations and cause disruptions in provision of service to the public. Therefore, we granted temporary exemptions from payment of access charges to certain classes of exchange access users, including enhanced service providers." In *Matter of amendments of Part 69 of the Commission's rules relating to enhanced service providers,* CC Docket No. 87-215, Notice of Proposed Rulemaking, 2 FCC Rcd 4305 (1987)(proposing to imposed access charges on enhanced service lines), terminated, 3 FCC Rcd 2631(1988)(proposal abandoned on ground that despite the apparent discrimination in charges "a period of change and uncertainty" besetting the enhanced services industry justified ongoing exemption from access charge payments). Currently the FCC requires users of ISDN services to pay only one subscriber line charge, an access payment, despite the fact that ISDN circuits can derive more than one voice-grade equivalent channel.

cility used does not matter, the Commission might infer that telecommunication services subject to the Communications Act may include access methods other than that provided by incumbent operators. Arguably some of the services provided by Internet access providers might constitute telecommunication services, despite the definitions contained in the Telecommunications Act of 1996 that exempt the kinds of "customary" services to which ACTA refers and Congress assumed, such as electronic mail, the World Wide Web and the transmission of data packets that correspond to letters, graphics, and audio and video works.[37]

When the Internet provides real-time delivery of voice and data services, the Commission may infer that such services are not customary, to use ACTA's characterization, or that the potential for interference with other telecommunication policy objectives like universal service bears further scrutiny. Notwithstanding its desire to refrain from defining the scope and array of services available via the Internet, the Commission might have to consider whether certain Internet services like telephony can be provided on an unregulated basis, despite their potential for adversely impacting incumbent interexchange and local carriers. Bear in mind that some local exchange carriers have qualified for access to universal service subsidies and some degree of regulatory safeguards, because of their essential mission as providers of first and last mile access.

The new definition of telecommunications carrier creates the possibility that providers of access to the Internet, as well as service providers, might fit into the

---

29. The FCC's access charge regime established a different pricing structure for switched and special access. The former includes regular dial-up services and requires end-users to pay a monthly flat-rated subscriber line charge, currently $3.50 for residential and small business users and $6.00 for other business users. The latter includes leased, private line users, who certify that the line does not "leak" into the PSTN through the use, for example, of an on-premises switch like a private branch exchange, which could couple the private line with trunks that access the PSTN provided by Local Exchange Carriers ostensibly for local switched services. See *MTS/WATS Market Structure (Phase I)*, 93 FCC2d 241 (1983), modified on recon., 97 FCC2d 682, further modification on recon., 97 FCC2d 834, partially aff'd and partially remanded sub nom., Nat'l Ass'n Regl. Util. Comm'rs v. FCC, 737 F.2d 1095 (1984), cert. den., 105 S.Ct. 1224; further modification, 99 FCC2d 708 (1984), 100 FCC2d 1222, further recon. den., 102 FCC 2d 899 (1985). See also *Investigation of access and divestiture related tariffs*, 101 FCC2d 911(1985) recon. denied, 102 FCC2d 503 (1985), and *Investigation of access and divestiture related tariffs*, 101 FCC2d 935 (1985).

30. International private line services, which do not access the PSTN, are exempt from the accounting rate regime. Their per-minute costs are significantly lower than switched services. Undetected private line leakage has become commonplace, making it possible for resellers to provide a service functionally equivalent to international message telephone service at a fraction of the cost. See R. Frieden, (1996). "The Impact of boomerang boxes and callback services on the accounting rate regime," In D. Wedemeyer and R. Nickelson, (Eds.) *Proceedings of the Pacific Telecommunications Council Eighteenth Annual Conference.* (pp. 781–790). Honolulu: Pacific Telecommunications Council.

telecommunications carrier classification with the application of software that makes the Internet a conduit for telephony. It may seem difficult to envision "involuntary" common carriers, because Internet access and service providers surely do not seek common carrier status, nor do they as yet offer an essential service, or hold themselves out indiscriminately to the public.[38] However, the FCC's public interest mandate and concerns about the potential for adverse financial consequences on incumbent telephony providers may trigger the need for the Commission to classify Internet service elements as telecommunications.

The Communications Act now defines telecommunications carrier as "any provider of telecommunications services, except [for]...aggregators [which provide pay telephone services to the transient population]."[39] This provision classifies a telecommunications carrier as a common carrier "to the extent that it is engaged in providing telecommunication services,"[40] but a separate provision in the Communications Act now accords the Commission flexibility to refrain from applying almost any provision that it determines is no longer needed to ensure just and reasonable rates, consumer protection, and service in the public interest.[41]

ARE INTERNET TELEPHONY MINUTES OF USE DIFFERENT FROM REGULAR MESSAGE TELEPHONE SERVICE?

Even if the FCC concludes that Internet telephony constitutes telecommunications, the Commission need not apply the same regulatory regime, particularly if Internet telephony does not constitute the functional equivalent of common carrier telecommunications like message telephone service (MTS). One way to make

---

31. For example, the FCC required the formation of a single consortium comprising all individual mobile satellite service applicants. The Commission then granted the American Mobile Satellite Corporation a geostationary mobile satellite service monopoly on the belief that the market could not support multiple operators and because of its perception that the United States would have the most success in coordinating satellite orbital slots with other nations and the Inmarsat cooperative if a fully licensed operator existed. See *Amendment of Parts 2, 22 and 25 of the Commission's rules and policies to allocate spectrum for and establish other rules and policies pertaining to the use of radio frequencies in a land mobile satellite service for the provision of various communication services*, Gen. Docket No. 84-1234, Report and Order, 2 FCC Rcd. 1825 (1986), 2d Report and Order, 2 FCC Rcd. 485 (1987), recon. den. 2 FCC Rcd. 6830 (1987), on further recon., 4 FCC Rcd. 6029 (1989), partially reversed and remanded sub nom., Aeronautical Radio, Inc. v. FCC, 928 F.2d 428 (D.C. Cir. 1991), tentative decision on remand, 7 FCC Rcd. 266 (1992), aff'd sub nom., Aeronautical Radio, Inc. v. FCC. 983 F.2d 275 (D.C. Cir. 1993).

32. *Establishment of domestic communication satellite facilities by nongovernmental entities*, First Report And Order, 22 FCC2d 86 (1969), Second Report and Order, 35 FCC2d 844 (1972), on recon., 38 FCC2d 665 (1972)(imposing a moratorium on accepting applications from AT&T to construct, launch, and operate a domestic satellite).

33. 47 U.S.C. § 153(43)(1996).

such an equivalency evaluation is to apply antitrust case law[42] and economic analysis of users' cross-elasticities. The FCC has employed an interdisciplinary "functional equivalency" analysis which in this instance would involve its assessment of the "likeness" between Internet telephony and common carrier services.[43]

Heretofore, the Commission has used a likeness comparison to determine whether a carrier has engaged in unlawful discrimination. Presumably if two services are not like, then they need not have the same price, or be subject to the same regulatory regime.

ASSUMING INTERNET TELEPHONY FALLS WITHIN THE DEFINITION OF TELECOMMUNICATIONS, SHOULD THE FCC REGULATE IT?

The ACTA petition stated that continuing regulatory asymmetry will "distort the economic and public interest environment in which ACTA carrier members and nonmembers must operate."[44] Although providing no projections on traffic migration and revenue diversion, the ACTA petition alleged that "Continuing to allow [unregulated Internet telephony]...threaten[s] the continued viability of ACTA's members and their ability to serve the public and acquit their public interest obligations under federal and state laws."[45] In addition to arguing that definitions in

---

34. Some parties filing oppositions to the ACTA petition claim that although packet switching by itself constitutes a basic service under the Commission's *Computer Inquiries*, interconnections among the various networks comprising the Internet result in data and protocol conversions that constitute enhanced services. "Not only do all Internet applications employ computer processing, but the Internet itself is a network of interconnected clients, hosts, routers and gateways that request, store, direct, transport, retrieve and utilize data to deliver to Internet users information different from a subscriber's transmissions" (Netscape Opposition to ACTA Petition at 9).

35. Enhanced services "employ computer processing applications which act on the format, content, code, protocol or similar aspects of the subscriber's transmitted information; provide the subscriber additional, different, or restructured information; or involve subscriber interaction with stored information" [47 C.F.R. § 64.702 (1995)].

36. Telecommunications Act of 1996, Sec. 3(46), codified at 47 U.S.C. § 153(46).

37. The FCC has concluded that its definition of enhanced services would not include the provision of telecommunication services. See *Implementation of the Local Competition Provisions in the Telecommunications Act of 1996*, CC Docket No. 96-98, CC Docket No. 95-185, First Report and Order, FCC 96-325, 1996 West Law 452885 (F.C.C.), n. 1416 (released Aug. 8, 1996).

38. See National Ass'n. of Regl. Util Comm'nrs v. FCC, 525 F.2d 630 (D.C. Cir. 1976), cert. denied, 425 U.S. 992 (1976).

39. See case cited in note 38, Sec. 3(44), codified at 47 U.S.C. § 153(44).

40. See case cited in note 38.

41. See case cited in note 38, Sec. 10(a) Regulatory Flexibility, codified at 47 U.S.C. § 160 (a).

the Telecommunications Act directly cover Internet telephony and subject it to provisions of the Communications Act, the ACTA petition urged the FCC to assert indirect jurisdiction in much the same way as the Commission found it necessary to regulate cable television. The Commission asserted "ancillary" jurisdiction over cable television, because that service had the potential for adversely impacting broadcast television.[46]

The ACTA petition appears to extrapolate from the cable television regulatory model to assume that the trade association need only allege the potential for adverse financial impact and the FCC would have to assert ancillary jurisdiction. Like cable television, Internet telephony has the potential to "siphon" traffic and divert revenues. But unlike cable television the potential migration of traffic and revenues does not threaten the continuing viability of an essential industry, local and long-distance telephone service, or the ability of these industries to meet their public interest mandates including subsidization of universal service. Neither the FCC nor the Supreme Court favors foreclosing competition merely on grounds that a competitor might lose revenues, or even go out of business.[47] With the enactment of the Communications Act of 1934 as amplified by the Telecommunications Act of 1996, Congress intended to foster competition,[48] and not to insulate from competition a particular operator, or class of operator, absent compelling justifications.

In calling for parity in regulatory treatment, the ACTA petition seeks a contrarian outcome: In a time when nondominant carriers, including AT&T, have few regulatory responsibilities, the FCC properly has a bias against expanding its regulatory wingspan. The Commission has no legal right to regulate software. It

---

42. See, e.g., United States v. E.I. DuPont de Nemours and Co., 351 U.S. 377 (1956) ("reasonably interchangeable" products constitute the relevant product market in antitrust analysis).

43. MCI Telecommunications Corp. v. FCC, 842 F.2d 129, 1303 (D.C. Cir. 1988) established a three part test for "likeness": (1) whether the services are "like"; (2) if they are "like," whether there is a price difference; and (3) if there is a difference, whether it is reasonable.

44. ACTA Petition at 4.

45. ACTA Petition at 4.

46. ACTA Petition at 9 citing United States v. Southwestern Cable Co., 392 U.S. 57 (1968).

47. See Federal Communications Commission v. Sanders Brothers Radio Station, 309 U.S. 470 (1939)(economic harm to an incumbent radio station by market entry is not grounds to foreclose the authorization of an additional radio station).

48. "It shall be the policy of the United States to encourage the provision of new technologies and services to the public. Any person or party (other than the Commission) who opposes a new technology or service proposed to be permitted under this Act shall have the burden to demonstrate that such proposal is inconsistent with the public interest." Communications Act of 1934, as amended, 47 U.S.C. § 157 (a).

does not automatically impose the same regulatory classifications and burdens on operators, including some noncommon carriers that can secure access to the public switched telephone network (PSTN). If the FCC can analogize Internet telephony to software or enhanced services, and if it can explain how Internet telephony constitutes an interactive computer service and not telecommunications provided by telecommunications carriers, then the Commission cannot impose common carrier regulatory requirements on Internet telephony providers.

PUBLIC POLICY AND PRACTICAL REASONS FOR NOT REGULATING INTERNET TELEPHONY IN ANY EVENT

Internet telephony constitutes a key means for individuals to exploit technological innovation for private benefit without public harm. It constitutes a kind of "self-help" for individuals who collectively can provide the same impetus for competition, downward rate pressure, innovation, and other public benefits as can occur when Congress, the FCC and the courts make it possible for facilities-based or resale competition.

Unregulated Internet telephony reminds decision makers that changes in user strategies at the micro level are important, because they reflect actual buyer behavior instead of theory. The ACTA petition appears to assume broadsweeping traffic and revenue migration without providing evidence that users in large numbers, as opposed to hobbyists, will tolerate the initial expense, inconvenience, and relatively poor quality of service. Regardless of the rate of growth in demand, Internet telephony follows a variety of self-help strategies that collectively have forced carriers and regulators to make business and policy accommodations. Even if only a relatively small percentage of users embrace such "grey market" strategies as international callback private branch exchanges,[49] leaky private branch exchanges (PBXs)[50] and other incumbent carrier or service bypass techniques, companies and regulators may assume the potential for vast changes to the status quo. So even if Internet telephony does not reach the market penetration levels the ACTA petition assumes will occur, the potential is real for triggering much-needed changes in business and regulatory policies.

---

49. See G. Retske. (1995). *The international callback book—An insider's view.* New York: Flaitiron.

50. 'Leaky PBX' is a term used to describe a PBX or other similar device through which a private line subscriber can 'patch' an interstate call to off-network destinations in the local exchange. WATS-Related and Other Amendments of Part 69 of the Commission's Rules CC Docket No. 86-1, Second Report and Order, FCC 86-377, 1986 WL 290930 (F.C.C.) n. 46 (released August 26, 1986) citing MTS and WATS Market Structure, Second Supplemental Notice of Inquiry and Proposed Rule Making, 77 FCC2d 224, 241 (1980).

SELF-HELP STRATEGIES AND TECHNIQUES PROMOTE RATIONAL AND NON-DISCRIMINATORY PRICING

Real or perceived opportunities to bypass incumbent networks, services, and pricing regimes can trigger changes well before actual traffic and revenue figure mandate an accommodation. The doctrine of market contestability encourages regulators to change policies in anticipation of market entry. Perhaps a corollary lies in the predisposition not to impose regulation until such time as market entry is proven to have harmed the public interest. Put another way: Unless and until such self-help causes such severe traffic or revenue migration as to handicap universal service accomplishments and goals, the FCC should remain steadfastly opposed to expanding its regulatory wingspan.

THE EXAMPLE OF INTERNATIONAL CALL-BACK SERVICES

High international accounting rates[51] and commensurately high end-user charges in many nations present a lucrative market for entrepreneurs providing call-back services to evade or interpret liberally regulations or laws that prohibit, limit, or condition market entry. "Code-calling" via callback services providers enables callers in high-cost locations to place a call to the United States or another country with low outbound international calling rate, hang up, and soon receive a second dial tone with outbound calling capability, or the intended call recipient on the line.[52]

Rather than reject call-back as a violation of International Telecommunication Union (ITU) Rules and Recommendations or at least the "spirit" of such provisions, the U.S. Department of State officially noted that such tactics do not violate any treaty commitment of the United States, and the FCC has refused to revoke authorizations granted to international carriers providing callback services. In a letter to FCC Chairman Reed Hundt on March 22, 1995, Ambassador Vonya B. McCann stated the view that no treaty or general concept of law obligates the United States to "require that §214 authorizations for call-back configurations be denied or licenses revoked upon assertion by for-

---

51. For background on international accounting rates and the settlement process see chapters cited in note 24.

52. In its most simple form, code calling involves the assignment of a unique telephone number or code to each subscriber. When the subscriber wants to make an international call, he or she dials a local or international telephone number and hangs up after two or more rings. A switch or private branch exchange also known as a "boomerang box," receives the call and is able to consult a database of subscriber identities and telephone numbers as a function of having received a call to a particular number. After determining who seeks calling access, the boomerang box delivers long-distance dial tone to the calling party and generates a billing record.

eign carriers that call-back operators operating in the United States are violat-
ing their countries' laws."[53] The letter also stated that the United States had
made no commitment in any ITU forum to prohibit call-back services:

> In sum, we did not undertake any obligation in ratifying the Melbourne
> Agreement [which modified the ITU's International Regulations to account
> for developments in telecommunication services] that would obligate the
> United States (or any other Member) to consider the call-back configuration
> an "international telecommunication service" regulated by the Melbourne
> Agreement or to enforce foreign laws regarding call-back.[54]

Although agreeing not to permit call-back services to users in nations expressly
deeming the tactic illegal, the FCC has refused to invalidate call-back service operator authorizations granted under Section 214 of the Communications Act.

In the absence of actual facilities-based competition, which may or may not result despite assumptions made by Congress when enacting the Telecommunications Act of 1996, self-help tactics like call-back and Internet telephony provide
immediate "virtual competition." Although the actual numbers of users may be
low and the revenue diversion small, such techniques evidence how vulnerable
markets are, particularly ones that have generated supracompetitive prices, or have
been so saddled with expenses and subsidy burdens that users actively seek ways
to lower their total service costs.

International call-back services will force incumbent carriers to reduce their negotiated accounting rates as well as their user charges. So too can Internet telephony force facilities-based carriers to establish rational and cost-based prices,
particularly the cost of accessing the PSTN. Users may pursue Internet telephony
simply to avoid having to pay outlandish international rates, and domestic rates
still artificially boosted by above cost local access charges.[55]

INTERNET TELEPHONY CAN FORCE A REAL DEBATE ON HOW TO ACHIEVE UNIVERSAL SERVICE GOALS

Internet telephony has the potential to promote achievement of universal service
goals recently broadened by Congress.[56] However, it also may reduce the flow of
money accruing to the current funding mechanism[57] used to promote the universal

53. Letter from Ambassador Vonya B. McCann, United States Coordinator International
Communications and Information Policy, United States Department of State to Chairman
Reed Hundt, Chairman, Federal Communications Commission at p. 4 (March 22, 1995).

54. Letter cited in note 53 at 3 and at note 10, p. 9, discussing the extent to which ITU
regulations require extraterritorial application of domestic law and stating that "the United
States and like-minded countries refused to bind themselves in Art. 1.7 or elsewhere in the
Melbourne Agreement to enforcing the domestic laws of other ITU Members."

55. See G. W. Brock. (1995, April). *The economics of interconnection.* (Three studies
commissioned by Teleport Communications Group.)

service mission. On one hand, the Internet has stimulated demand for second tele-phone lines into homes, resulting in additional subscriber line charge payments. On the other hand, flat-rated local exchange rates, based on historical voice calling patterns, make it possible for users to occupy lines for hours without incurring ad-ditional local exchange carrier charges. The current access charge regime applies a per-minute Universal Service Fund (USF) charge on interexchange carriers whose customer rates presumably reflect that expense. To the extent that Internet telephony can substitute for telephone calls subject to the per-minute USF charge, the aggregate sum of funds collected to subsidize universal service will fall.

However, the potential shortfall in universal service funding resulting from traf-fic migration to the Internet does not warrant regulation of Internet telephony. It does evidence the need to consider changes in how to fund universal service goals,[58] and perhaps as well whether usage-insensitive pricing can be sustained even for local exchange services.

AVOIDING THE FALSE PRIVATE/COMMON CARRIER DICHOTOMY

On numerous occasions the FCC has attempted to justify regulatory asymmetry by ascribing the private carrier classification to nonessential services that the Commission believes would stimulate competitive benefits. Such a strategy cre-

---

56. A Federal-State Joint Board on Universal Service to recommend changes to any uni-versal service policy (47 U.S.C. 254, 1996). This section sets out several guiding princi-ples: (a) access to quality services at just, reasonable and affordable rates; (b) access to advanced services throughout the nation now defined to include low-income consumers, and those in rural, insular, and high-cost areas as well as advanced telecommunications ser-vices access for schools, health care providers and libraries; (c) equitable and nondiscrim-inatory contributions by all providers of interstate telecommunications services to universal service funding; and (d) specific and predictable support mechanisms. See 47 U.S.C. § 254(b); see also *Federal-State Joint Board on Universal Service, Notice of proposed rule-making and order establishing joint board*, FCC 96-93 (released March 8, 1996); reprinted in 61 Fed. Reg., 10, 499 (March 14, 1996).

57. The FCC adopted the current Universal Service Rules in 1984. Amendment of Part 67 of the Commission's Rules and Establishment of a Joint Board, Decision and Order, 96 FCC 2d 781 (1984). The rules provide for subsidies from interstate interexchange carriers to local exchange carriers ("LECs"), whose average cost per loop is substantially higher than the national average cost per loop. LECs with an average cost per loop above 115% of the national average cost per loop can allocate a specified percentage of these costs to the interstate jurisdiction. In 1993, the FCC initiated a rulemaking to examine and reevaluate the Part 36 jurisdictional separations rules governing USF assistance. Amendment of Part 36 of the Commission's Rules and Establishment of a Joint Board, Notice of Proposed Rulemaking, 8 FCC Rcd. 7114 (1993). With enactment of the Telecommunications Act of 1996, the FCC and the Federal-State Joint Board faced a rigorous timetable for substantial-ly overhauling the universal service subsidization process.

ates a regulatory "safe harbor" even for enterprises that increasingly offer "like," functionally equivalent services that access the PSTN and provide alternatives to customers previously "captive" to incumbent common carriers. The use of a private/common carrier dichotomy motivates market entrants to craft services that qualify for the private carrier classification, such as, enhanced services under the Computer Inquiries and International Value Added Network or permissive private line resale and shared use under Recommendations shaped at conferences of the International Telecommunication Union.[59] Securing the private carrier classification and the regulatory exemption it provides has become a market-distorting goal, because ventures seek to characterize services as enhanced and value-adding regardless of whether a discrete market exists and whether consumers perceive a difference between what the private carrier offers vis-à-vis what regulated, incumbent common carriers offer.

Congressional concern about regulatory asymmetry and market distortion has resulted in two amendments to the Communications Act requiring the FCC to apply a single common carrier classification. In 1993 Congress revised Section 332 of the Communications Act[60] to require the FCC to deem as common carriage the services of all enterprises providing commercial mobile radio services, irrespective of prior FCC determinations that had classified some services as private carriage.[61] The Telecommunications Act of 1996 required the FCC to classify as common carriage the provision of any telecommunications service.[62] However,

---

58. Section 254 of the Telecommunications Act of 1996, 47 U.S.C. § 254 (1996), expands the concept of universal service to include insular areas, such as Pacific Island territories, low-income consumers, health care providers for rural areas, elementary and secondary school classrooms, and libraries. Rates for rural health care services shall be "reasonably comparable" to charges for similar service in urban locales, and service provided to meet an educational purpose must be discounted with the difference offsetting the carrier's universal service payments, or qualifying it for reimbursement from the universal service fund.

59. See *International Telecommunication Union, International Telegraph and Telephone Consultative Committee Blue Book*, Vol. II, Fascicle II.1, Recommendation D.1, Sec. 7.1.1, General Tariff Principles, Charging and Accounting in International Telecommunications Services. This recommendation addresses the conditions under which facilities-based carriers should allow the shared use and resale of private lines that historically had been provisioned for the exclusive use of only one large volume customer. The Recommendation suggests that administrations can condition, consult, and agree to the scope of access to public networks provided to users of international private leased circuits.

60. In 1982, Congress amended the Communications Act by adding Section 3(gg) and Section 332(c). See United States Congress, H.R. Rep. No. 97-765, 97th Cong., 2d Sess. (1982).The purposes of adding these provisions were: (a) to define private land mobile service; (b) to distinguish between private and common carrier land mobile services; and (c) to specify the appropriate authorities empowered to regulate these same services. See the same report at 54.

both amendments to the Communications Act of 1934 distinguish between the baseline classification of common carriage and the extent of regulatory burdens that the FCC should apply to any particular carrier.

Rather than allow the FCC to apply an increasingly suspect private/common carrier dichotomy, the Congress accorded the Commission discretion over how much of conventional common carrier regulatory burdens a particular type of common carrier should bear. Section 332 authorizes the FCC to forbear from applying to commercial mobile radio service providers most regulatory burdens established in Title II of the Communications Act.[63] Section 401 expands the regulatory forbearance option to any common carrier upon a determination by the FCC that such deregulation would serve the public interest.[64] The Telecommunications Act of 1996 creates an opportunity for the FCC to apply a single regulatory classification, common carriage, without having to impose symmetrical regulatory burdens. Rather than perpetuate a dichotomy in carrier classifications, the Commission should take advantage of the opportunity lawfully to apply unequal regulatory burdens to a now larger set of common carriers.

The FCC should refrain from attempting to finesse a private carriage classification for Internet telephony. It should classify Internet telephony service providers as common carriers and then decide affirmatively to forbear, for the time being, from applying any common carrier regulatory requirement. This action would require the FCC to define who within the large set of software, hardware, and telecommunications service providers warrant the common carrier classification as opposed to simply being collaborators, integrators, and packagers. Applying the Telecommunications Act of 1996 definition of telecommunications service provider, the Commission could exempt Internet browser and other software developers, hardware manufacturers, and even ventures that use Internet telephony in

---

61. The revised Section 332 created two new categories of mobile services: commercial mobile radio service (CMRS) and private mobile radio service (PMRS). CMRS is defined as "any mobile service (as defined in section 3(n)) that is provided for profit and makes interconnected service available (A) to the public or (B) to such classes of eligible users as to be effectively available to a substantial portion of the public" [Omnibus Budget Reconciliation Act of 1993, Pub.L. No. 103-66, Title VI, § 6002(b)(2)(A), 6002(b)(2)(B), 107 Stat. 312, 392 (1993), amending the Communications Act of 1934, § 332(d)(1)]. 47 U.S.C. § 332(d)(1). The Telecommunication Act of 1996 explicitly exempts CMRS providers from the definition of common carrier, local exchange carrier "except to the extent that the...[FCC] finds that such service should be included in the definition of such term" [47 U.S.C. § 153 (26)(1996)].

62. Section 251 of the Communications Act of 1934 as amended requires all telecommunications carriers to provide, upon request, direct or indirect interconnection with other telecommunications carriers [47 U.S.C. § 251 (a)(1)]. It also requires all such carriers to be treated as common carriers.

63. See 47 U.S.C. § 210 et seq.

conjunction with the sale or marketing of a good or service. On the other hand, the FCC could classify as telecommunication service providers all ventures that lease or own packet transport facilities used to link end users with Internet-mediated telephony, regardless of whether they might previously have qualified as enhanced services providers, or another type of private carrier.

Applying a single common carrier classification provides the FCC with an opportunity to oversee, but not necessary to regulate, the growing set of service providers involved in telecommunications and telecommunications services as defined by the Telecommunications Act of 1996. Such oversight does not expand the FCC's regulatory wingspan unless and until the Commission can articulate a compelling case for reregulation, or the application of traditional common carrier regulatory burdens to new ventures. The Commission may have to take such action if newcomers adversely affect the universal service mission, or if the recently empaneled federal state Joint Board on universal service recommends that all telecommunication service providers, presumably including Internet telephony providers, contribute to a revamped mechanism for funding universal service programs.

CONCLUSION

The FCC has considered the private carrier classification a regulatory safe harbor for enterprises it believes can better stimulate competitive benefits if unregulated. However, countervailing concerns about regulatory asymmetry and a level competitive playing field make it less likely that the Commission can continue to erode the dichotomy between common and private carriers particularly in the face of a presumption of common carriage contained in the Telecommunications Act of 1996. Amendments to the Communications Act have created a preference for a single common carrier classification with the proviso that the FCC can selectively reduce or eliminate common carrier regulatory requirements like the tariff filing requirement, a duty the Commission had sought previously and unsuccessfully to eliminate for certain types of common carriers.[65]

By creating a presumption that telecommunication service providers operate as common carriers, the Telecommunications Act of 1996 appears to require the FCC

---

64. 47 U.S.C. 160, Sec. 10 (1996). The FCC may forbear from applying any provision of the Communications Act and existing Commission regulations if it determines that such oversight is unnecessary to ensure just and reasonable rates, consumer protection, and service in the public interest. Quite soon after enactment of the Telecommunications Act of 1996 the FCC initiated a proceeding designed to eliminate the tariff filing requirement for long distance common carriers providing interstate services. See *Policy and rules concerning the interstate interexchange marketplace implementation of Sec. 254(g) of the Communications Act of 1934, as amended*, CC Docket No, 96-61, Notice of Proposed Rulemaking, FCC 96-123, 1996 FCC Lexis 1472 (rel. March 21, 1996).

to determine that Internet telephony constitutes a common carrier service. However, this legislation also authorizes the FCC to conclude that compelling public interest factors militate against applying common carrier regulatory requirements to a particular category of carrier. As a result of this two-step process the difference between common and private carriage grows even more murky, because the FCC may now lawfully free common carriers of traditional regulatory burdens. Such ambiguity fosters regulatory flexibility, but it also creates an uncertain environment where stakeholders perceive the incentive and opportunity to seek competitive advantages by qualifying for private carrier status. Carriers unable to so qualify in turn claim that a level competitive playing field requires regulatory parity, including the option of seeking exemptions from common carrier regulatory requirements.

Even in the absence of legal justifications for different regulatory responses, the FCC should not have to apply consistent, symmetrical regulations in instances where a difference makes sense. For example, the Commission does not auction all frequency spectrum, despite the fact that operators who had to bid competitively surely enter the market with a substantial financial handicap incumbents and licensees of nonauctioned spectrum did not have to bear. Likewise the FCC does not have to apply the full panoply of regulatory requirements on all common carriers.

The ACTA petition approaches the problem of asymmetrical regulation with a contrarian, proregulation strategy, at a time when deregulation predominates. Congress now expects the FCC to support deployment of new technologies and requires opponents to prove that a new technology will not serve the public interest. For several years the FCC consistently has sought to reduce regulatory requirements, recently exemplified by its decision to reclassify AT&T as a nondominant carrier and the Commission's proposal to forbear from requiring most carriers to file tariffs. AT&T realized that its best regulatory strategy would be to seek deregulatory parity, rather than to encourage the FCC to impose new or additional regulatory burdens.

Carriers having to compete with Internet telephony services should focus their efforts on key regulatory adjustments, rather than grand-scale regulatory realignments. The ACTA petition primarily seeks a financial remedy: eliminating Internet service providers' exemptions from access charge and universal service subsidy

65. See *Policy and rules concerning rates for competitive common carrier services and facilities therefore*, Notice of Inquiry and Proposed Rulemaking, 77 FCC2d 308 (1979); First Report and Order, 85 FCC2d 1 (1980); Second Report and Order, 91 FCC2d 59 (1982), recon. den., 93 FCC2d 54 (1983); Third Report and Order, 48 Fed. Reg. 46,791 (1983); Fourth Report and Order, 95 FCC2d 554 (1983) rev'd and remanded sub nom., AT&T v. FCC, 978 F.2d 727 (D.C. Cir. 1992); Fifth Report and Order, 98 FCC2d 1191 (1984); Six Report and Order, 99 FCC2d 1020 (1985), rev'd and remanded sub nom. MCI Telecommunications Corp. v. FCC, 765 F.2d 1186 (D.C. Cir. 1985), aff'd sub nom. MCI Telecom. Corp. v. Am. Tel. & Tel. Co., 114 S.Ct. 2223 (1994).

payments. By convincing the FCC that a larger set of Internet telephony and enhanced service providers should contribute to universal service funding obligations, already fettered carriers could achieve the proper sharing of a financial
burden.[66]

This outcome simply requires the FCC to conclude that Internet telephony constitutes a telecommunication service. It does not require the Commission to impose the entire range of common carrier regulatory burdens. Rather than expect
the FCC to apply uniform common carrier regulatory burdens, ACTA would have
fared better in seeking the broader sharing of access charge and universal service
subsidy obligations.

---

66. Section 254(d) of the Communications Act of 1934, as amended, requires "[e]very
telecommunications carrier that provides interstate telecommunications services...[to]
contribute, on an equitable and nondiscriminatory basis, to the specific, predictable, and
sufficient mechanisms established by the Commission to preserve and advance universal
service." 47 U.S.C. § 254(d)(1996).

Bedrooms, Barrooms, and Boardrooms on the Internet[1]

L. Jean Camp
Harvard University

Donna M. Riley
Carnegie Mellon University

Determining the First Amendment rights, institutional responsibility, and institutional perspective using the model of four media types is not feasible in the new electronic media. Media classifications are failing in many open networks. For example, cellular technology is undeniably broadcast technology, yet it fits (appropriately) under the regulatory rubric of wired telephony. Defining First Amendment rights in electronic forums is critical to preserving them (Pool, 1983).

There has been debate over the question of whether network services fall under the rubric of broadcast, publisher, distributor, common carrier, or some yet undetermined media type (Beall, 1987; Becker, 1989; Berman, 1995; Di Lello, 1992; Krattenmaker & Powe, 1995; Sassan 1992). Every media classification has an argument for and against it. The efforts to find a single media classification, or invent one, are doomed to failure because the provision of an electronic bulletin board (bboard) can be done as a publisher, a distributor, a broadcaster, or a common carrier. The differences between bboards and newsgroups are much too subtle to be contained in a four-way or even five-way media classification. The differences are suffi-

1. Although Dr. Camp is now at Harvard University, this work was completed at Carnegie Mellon University while funded by the United States Postal Service. The opinions presented in this work represent only the opinions of the authors. This work does not represent the opinions of the Department of Engineering and Public Policy, Carnegie Mellon University, the United States Postal Service, or Harvard University.

ciently subtle that rather than a single media type or single space (Naughton, 1992), the only appropriate model for cyberspace is that of multiple varied physical spaces. There is no such thing as cyberspace, but there are many cyberspaces.

Electronic spaces are as varied as physical spaces. Previously the responsibilities of the producers and distributors of speech were determined by the medium of transmission: printed paper, over-the-air broadcast, or closed private circuit (as in telephony). However, the classification of speech by media type is inadequate on the Internet. Instead, system configuration, tradition, applications in use, and purpose determine the types of spaces that are created on a common medium.

We show in this chapter how the media model is ill-suited to the Internet by illustrating how different actions on Usenet news and the Web are broadcast, distributing, publishing and common carriage. We further describe how the bboards at Carnegie Mellon University all serve different purposes, which are analogous to different spaces: bedrooms, classrooms, club meetings, departmental conference rooms, and public spaces. Finally we consider three cases where the space analogy may have worked, but the media analogy did not.

MEDIA TYPES

The differences in traditional media types are based on the inherent technological capabilities to control content and distribution. With Internet services, the classification depends on the configuration and use of the service, not the inherent technological capabilities. In this section we show how Usenet news and Web services do and do not fit into traditional media types.

Usenet News and Web Pages as Broadcasts

Broadcasters are under particularly stringent regulation for two reasons: the natural scarcity of bandwidth, and the lack of filtering mechanisms for viewers.

Broadcasters have an inherent capability to control what they broadcast. The facilities for distributing broadcast information are concentrated, and the person or institution that controls those facilities can exercise absolute control over the distribution. Because of the nature of broadcasting, the decision to broadcast one stream of information prohibits anyone else in the service area from using the same spectrum to broadcast any information. Thus broadcasters both select what they will broadcast and prevent others from broadcasting different information.

With broadcast transmissions, there is no way to prevent someone in the broadcast area from receiving the information. Thus broadcasters have no control over distribution.

Usenet news posts and html documents are like broadcast in that they enable the transmission of information to large numbers of people. Once on Usenet or on the

Web, the information may be downloaded by any of the millions of subscribers, thereby becoming inaccessible even to cancelbots. Because of the ease of secondary distribution of electronic information, once information has been sent over the network or placed in a publicly accessible forum it is nearly impossible to prevent widespread distribution.

Usenet news posts and html documents are like broadcast because the information comes directly into the home, and it may be impossible to retrieve or destroy the information after it has been broadcast. According to one advocate of the Communications Decency Act, it was exactly these broadcast characteristics that enabled the regulation to pass. Dee Jepsen of the advocacy group Enough is Enough described the need for the Communications Decency Act as follows: "hard-core, child pornography and 'indecent' material, which is harmful to minors, are being transmitted over the Internet directly into our homes" (Jepsen, 1995).

Like broadcast, accessing much of the information in Usenet groups and most information on the Web requires little technical expertise. On the Web one need only know how to use a mouse to find questionable material.

Providers of Web and Usenet services are unlike broadcast systems in that they are not constrained by a natural scarcity of bandwidth. Maintaining or providing access to an information service does not prevent others from doing so. Also unlike broadcasters, there is a level of technical sophistication required to obtain some types of information (such as downloading images) available on Usenet.

Providers of Usenet services are unlike broadcasters in that the information is organized and can be searched on a high level. Newsgroups are in named hierarchies, which users can choose to explore, or not. Thus the sudden appearance of explicit images is extremely unlikely. With information on a newsgroup users can choose to look into the newsgroup or not. The Usenet hierarchy is completely unenforceable, but is self-sustaining through the practices of publicly ignoring and privately flaming inappropriate posters, as well as the judicious use of cancelbots.

With the Web, badly named links and misdirection can cause a user of information services to be confronted with unwanted information, or may allow a user to stumble unintentionally onto offensive imagery.

Usenet service providers may arguably provide one broadcasting capability: the creation of groups. There is some information seen by every user of Usenet: the names of the groups. The creation of Usenet discussion groups is initiated by a create group command. Organizations where users can issue create group commands at will serve as broadcasters because the information goes into every feed, and the user need not actively choose to investigate to encounter the name. However, it is notable that the names of Usenet groups are necessarily text, which makes the use of group creation abilities to broadcast images arguably impossible. Thus the sudden appearance of explicit images is not enabled by the creation of groups.

One example of a group name is "alt.Barney.die.die.die." Certainly there is nothing wrong with poking fun at a children's show. However the simple fact that the knowledge of this group was extremely common, although few would search for "Barney" or "die," illustrates the broadcast characteristics of newsgroup titles. Note that groups have respected the desires of others in keeping names modest; for example, the discussion group for gay men is soc.motss (meaning members of the same sex), and virulent racists tend to misc.activism.militia.

The attempt to regulate the Internet as a broadcast medium through the Communications Decency Act has failed. The regulation of the Internet as a broadcast medium has been found to be unconstitutional, and thus is not an option (*American Library Association, Inc., et al.,. v. United States, et al., 1996*; Stets, 1996).

Usenet News and Web Pages as Publishing

Publishers are assumed to have full knowledge of the information they present— as newspapers have editorial control. Publishers are given unique freedom to expose private facts to public scrutiny for the public good (Ginger, 1975; *New York Times Co. v. United States, 1971*).

That a provider of information services can be found to be a publisher is illustrated in the case of Prodigy. Prodigy is a publisher as illustrated by the well known court decisions (*Stratton Oakmont, Inc., v. Prodigy Servs. Co.* 1991). However, Prodigy was found to be a publisher not of Usenet but of its proprietary newsgroups and possibly the posts of its subscribers. Prodigy filtered messages and rejected those found to be unacceptable for the family environment, which Prodigy targets as its market.

Providers of information services are like publishers in that the primary goal is to give readers, electronic or physical, access to selected information.

All system operators are like publishers in that they cannot reasonably be held liable when information they produce is incorrect and leads to bad decision making by the receiver (*Daniel v. Dow Jones, 1987*). Providers of Usenet services are also similar to publishers in that they have no control over secondary dissemination of the materials they provide.

Providers of Usenet and Web servers are like publishers in that they can control the initial distribution of information. Operators of private subscriber systems can choose to restrict their customers to employees or those that can provide proof of age; firewalls, Internet Protocol (IP) address, and password protection is available for Web pages.

Treating Internet service providers (ISPs) as publishers would result in prohibition of vast amounts of information, as suggested by *Stratton Oakmont, Inc., v. Prodigy Servs. Co.* (1991). Such a regulatory regime would require that Internet service providers search all posts and approve all pages. This has implications not only for speech rights but also for privacy rights.

Usenet News and Web Pages as Distributors

Providers of information services are like distributors in that they have the ability to filter at the highest level of content. That is, information services providers can select Usenet groups or domains where access is allowed, while prohibiting others. Like publishers, system operators have been found not to be liable when information they produce is incorrect and leads to bad decision making by the recipient (*Daniel v. Dow Jones, 1987*).

Providers of Internet services are like distributors in that they cannot have knowledge of all information that is provided through their service. Only high-level filtering is possible for many providers of Internet services. Filtering such as provided by Prodigy requires large investments both in filtering software and in labor, because each flagged message must be evaluated by a human.

Like distributors, providers of Internet services may have reason to know some content. For example, any ISP would know the content of any group named *.pedophilia would be unacceptable, and in fact requests to create such groups are denied.

Providers of Internet services are unlike distributors in that a limited ability to distinguish users means a limited ability to control access. The sheer volume of information that flows through a Usenet feed and the amount of information requested by users' clients from various Web servers make detailed filtering extremely difficult.

CompuServe is a distributor because it provides access to information "published" or written by the subscribers, whereas CompuServe itself does not provide content (*Cubby v. CompuServe, Inc., 1991*). CompuServe doesn't edit the content in specific posts and forums, but rather chooses which forums are allowed. This is analogous to a bookstore, which selects specific books but cannot reasonably know what is in every book.

However, regulating Usenet and Web services as secondary publishers creates difficulties. High-level decisions on content mean broad prohibitions. Prohibited content may have a particularly valuable character for reasons of health. For example, broad prohibition of obscene words forced breast cancer survivors on America Online (AOL) to create a chat group for "hooter" cancer survivors, and prevented women from listing themselves as breast cancer survivors in their .profiles (roughly equivalent to .plan files). Arguably, electronic discussions of sex, as takes place on alt.sex.wizards and alt.sexual.abuse.recovery, are particularly valuable on the Internet because there is no threat of violence, and there is potential for anonymous or pseudonymous discussion.

Usenet and the Web as Common Carriage

Providers of Usenet and Web servers also arguably have characteristics of common carriers. Providers of Internet services provide routing for others' packets with nondiscrimination. Providers of Usenet feeds provide the entire feed, with the recipient responsible for filtering.

Usenet news posts and html documents are like telephone conversations in that they are so numerous that knowledge of all content is not possible.

However, Internet provision of Usenet and Web services is not common carriage in that there is no requirement to provide nondiscriminatory services to individual users. ISPs can remove users at will, unlike telephone companies. Internet providers can choose whether or not to act when a subscriber is sending or receiving harassing or offensive e-mail, and an ISP can choose to block certain domains or sites.

The difficulty with regulating the Internet and Internet service providers as though Usenet news and Web services were common carriage is that there is no need or right to act. If Internet services are common carriage then an ISP could not remove Internet access privileges for abuse of other individuals, or other behavior commonly defined as abuse of the Internet itself, such as spamming (see Glossary).

INTERNET COMMON LAW

The law is not coming to the Internet—it has always been there. It has never been the case that communicating a death threat to a federal official has been legal if it was communicated through e-mail. That the President is only recently on the Internet means that this law is only recently applied—it does not mean that death threats were ever legal.

Recent arrivals to the Internet, or newbies, often see the Internet as a place of chaos—actually it is quite orderly. There are rules of the road on the information highway. The fundamental common law is, "You spam, we flame." Spams and cancelpoodles (see Glossary) always create responses. Complaints to employers and ISPs of offenders is common, as is simple flaming.

Where an item is posted or how it is linked defines the acceptability of the information on the Internet. Hate posts, which may seem unacceptable anywhere, are sometimes flamed for being in the wrong place. This is substantially different from objecting to one's right to speech—it is an objection that the reader's right to ignore such speech in daily life is being violated.

There are vastly different levels of civility, and vastly different signal-to-noise ratios in different areas of the Internet. Cross-posting does create confusion and blurs such lines. However, some things are certain. For example, alt.peeves will remain forever less civil than rec.crafts.textiles.needlework.

This range of acceptability suggests a recognition: that there is no cyberspace, but many cyberspaces. The attempt to fit all ISPs, or all Usenet groups, or all Web pages into a single category will fail just as attempts to determine universal laws for speech will fail. In the next section we consider physical spaces and their electronic analogs.

PHYSICAL SPACES

One reason denizens of the Internet call it cyberspace is that the various attributes applicable to defining different physical spaces can also be used to define different virtual spaces. Note that the physical space of a speech has been a defining characteristic of its acceptability; you can neither falsely yell "Fire!" in a crowded theater nor expound on philosophies of racial superiority at work. Yet there are places for both outrageous exclamations and offensive arguments in the physical world. We argue that the same is true in the virtual world.

The organization and language of users on the Internet suggest an option: parallels of physical space. Consider how the description of files and services currently reflects an awareness of how different services offer different types of space: cyberspace, home page, chat room, mailbox, dungeon, and home directory, to name a few.

The argument that the same constraints apply to electronic newsgroups as to physical bulletin boards reflects a recognition that there are different spaces; what governs the appropriate use of a physical bulletin board is in no small degree its location. The bulletin board in the grocery store differs significantly from the bulletin board in the office.

Why are cyberspaces more like spaces than media? First, media type assumes that the technology determines content control. Multiple spaces are set up using the same fundamental construction technologies and principles. Similarly electronic spaces with very different characteristics can be set up using the same electronic tools.

Second, media types assume control exists at a single location. For telephony the location is at the end user, and for broadcasters it is at the originator. With electronic services there can be multiple levels of control. An institution may limit access to an internal Web page, or allow it to be open. In either case the author of the Web page may choose to provide access to a select group, or to many.

Third, there is limited assurance of speech protection in various media types. Obscenity can be outlawed by the states, for example, yet raiding a home or private poetry reading would be extreme. Because the equivalent spaces on the Internet are not as closed, such a raid or prohibition may be likely to happen. Authors have advocated protecting cyberspace as a whole as an electronic soapbox (Berman, 1995). The Internet offers a unique opportunity for speech—particularly the speech of the despised or threatened. Individuals can interact with those that support violence against their groups without fear of personal reprisal. For example, few of the liberals who argue on the militia group would go to a militia meeting to confront the armed self-proclaimed patriots physically. Physical threat is far less effective at silencing speech when the space is virtual and the participants dispersed. Yet it is precisely those dialogues that would be limited by a threatening environment that are most likely to be subject to limitation—areas for hate speech and sexual speech provide forums for debates that would otherwise be unlikely to occur. Recognizing the value in these debates implies protecting them from closure, and assuring providers and participants free reign to advocate the most outrageous stands without fear of liability. It also requires protection of speech from other users who would silence opponents through cancels.

Fourth, the space analogy offers an opportunity to declare that there are special cases where control must be asserted without requiring complete control. A case where control may be required would be when an ISP is being used to stage a malicious attack on other systems, or is sending out a collection of perfectly legal but speech-destroying cancelbots. With media types, control in an extreme situation shows evidence of the ability to control and therefore results in broad liability.

TABLE 11.1
Fits and Failures of Media Types at Carnegie Mellon University

Media Type	Space Equivalent	Media Characteristics: Fits	Media Characterstsics: Failures
Announcement groups (official.cmu)	Broadcaster	Administrative offices	Complete control Complete availablility	May offer information unsuitable for children
Department and class groups (org.epp)	Publisher	Department offices	Limited knowledge Possibly limited availability	Implies administrative control over departmental debates
Class (academic.epp.19-101)	Distributor	Classroom	Limited knowledge of content Possibly limited availability	Administration control over content is highly constrained— liability for obscenity inappropriate
Club (assocs.*)	Common carrier	Club meeting	No knowledge of content High traffic	Controversial groups may be unable to maintain virtual presence No universal access requirement
Graffiti (graffiti.*)	Common carrier	Public forum	No knowledge of content High traffic	No universal access requirement No action against cancels

Finally the space analogy has the advantage that its understanding is intuitive. The social contract that may appear to fail in a workspace may be failing simply because it is not understood to be a workplace by all parties. Such an understanding would bring to the electronic forum far better internalized understanding among users than the application of media types.

Table 11.1 shows how different discussion groups at Carnegie Mellon University can correspond to different spaces; if forced into the media type paradigm, these would need to be in different media types.

Bedrooms

In the bedroom an individual has the right to speak and act in most offensive ways that would not be acceptable outside the home. Should a group gather in a bedroom there would be no action against speech absent criminal intent.

Law enforcement must have reason to even enter a person's bedroom. There must be some evidence of a crime, or a clear and present danger. Lewd, obscene, and pornographic speech finds protection in the home.

Private distribution lists created for the consensual discussion of adult topics may be an example of a bedroom operating on the Internet. Dial-up pornography electronic bulletin board systems (BBSs) may be a second example. A user's personal e-mail can serve as a bedroom.

Not only do citizens have the right to speak freely in our bedrooms, but we have the right to bar ideas from our bedrooms. Just as technical mechanisms have developed to filter material from our desktops and mailboxes (Spertus, 1996), legal mechanisms have been developed to protect our bedrooms. Speech that has only political implications on the street corner, such as the advocacy of violence against a select group, has much different implications in terms of politics and present danger when reiterated in the bedroom of a member of the targeted group.

Bedrooms can provide more freedom of speech than telephones, particularly sexual speech. As gender issues are part of today's most hotly contested political debates, determination to prohibit passionate speech as obscenity can strictly limit these debates. Network owners have been allowed to prohibit constitutionally protected speech and have refused to enter into contracts with those that present undesirable viewpoints. Declaring that some areas are legitimately the proper space for undesirable opinions and obscene speech, and that retributions for such speech made in the appropriate forums is prohibited, has the potential to provide greater freedom of speech than current regulation of telephony network owners (Barron, 1993).

Workspaces

In the workspace, the spirit of respect for others is an important aspect of maintaining a professional environment. Depending on the work being done there, other values of how colleagues should treat one another will prevail; in a corporate environment, there may be expectations for more formal speech. In a university, the spirit of inquiry and openness may dominate.

In a university setting, workspaces may include departmental newsgroups, and some student groups that maintain a professional environment or that students consider to be preparation for their careers, such as a newspaper or business organization.

Speech is controlled in a workspace. Speech that would be dangerous because of implications of physical violence in a bedroom can create the danger of economic deprivation if repeated in workspaces, as recognized by the Equal Employment Opportunity Commission (EEOC).

Classrooms

In the classroom, there should be a spirit of inquiry that enables the broaching of any subject and the expression of any viewpoint. There should also be a spirit of respect for others and a code of behavior that prohibits personal attacks and intimidation. An environment that is free from harassment and intimidation yet that accords individuals due respect is most conducive to the uninhibited sharing of ideas.

Course discussion newsgroups at universities may function as a classroom space. Clarification of homework, queries, and comments between students, professors, and teaching assistants sometimes result in terse or angry debates. The advocacy of ethnic cleansing, racist remarks, or personal insults remain as inappropriate or appropriate as in the classroom—depending on the topic at hand. In the classroom the instructor and other students work together to maintain the appropriate learning atmosphere by responding to inappropriate behavior.

Just as one would be unlikely to come to class in pajamas, some speech appropriate in the bedroom is inappropriate in a classroom. Conversely, as one never need listen to offensive speech in the bedroom, one must listen to ideas in the classroom. The purpose of being in a classroom is to encounter ideas. Unlike workplaces, the exchange of passionate ideas is an ideal in the classroom.

The Town Square

In the town square, free expression is paramount. It is a space designed for the exchange of ideas political and personal. Personal attacks, outrageous arguments, hate speech, and the like are welcome here. Direct criminal threats and slander are the bounds of speech here; propriety is not at issue.

On the Internet, town squares may include groups that wish to create dialogue on various issues, such as political organizations. The entire alt hierarchy is a town n-cube, far surpassing the space and time limitations imposed on a physical square.

It is clear that a given entity on a college campus could serve more than one purpose. For example, an African-American students' organization could be involved with political dialogue, with offering support to African-American students on campus, and with the actual business of running the organization. It might choose to have virtual analogs that imitate business meetings, discussion groups, and open political forums.

Note that discussion in a town square can actually be more limited than the discussion in a bedroom. Just as public nudity is prohibited, some arguments and actions protected in the bedroom can be prohibited in the town square. Thus the ability to create spaces in addition to the information agora implies the ability to provide a level of freedom of debate and discussion unmatched in any single medium.

THE MEDIA FAILURE

We next consider three examples in which the existence of media analogies as the dominant paradigm for regulating electronic communications created incentives (in the form of potential liability) for poor policy decisions in the university environment. Others have argued that the current regulatory regime allows limitations on speech, but they have focused on case law rather than the effects of the shadows of liability (Hammond, 1995). We choose examples in the academic environment not only because it is this environment with which we are most familiar, but also because universities have the creation and sustenance of dialogue as a primary purpose. Examples from business or government domains would be more complicated by questions of misuse of resources.

In the first case, Carnegie Mellon University (CMU) implemented a broad policy censoring all Usenet newsgroups in the alt.sex.* hierarchy (Camp & Riley, 1995; Steinberg, 1995). Although the stated purpose of the policy was to bring Carnegie Mellon in compliance with Pennsylvania obscenity law, what resulted was the denial of access to all conversations about sex and sexuality, and the allowance of some legally obscene material (e.g., child pornography) to remain available to the CMU community.

In the second case, two female students sued Santa Rosa Junior College after having been the subject of sexually explicit rants posted to an all-male class discussion board (Dorgan, 1994). The entire situation could have been prevented had the professor, students, and university been clear from the start about the purpose of the discussion board. Instead, what resulted was a temporary job termination, involvement of the Department of Education's Office of Civil Rights, and an embarrassing settlement for the college.

Third is the much-hyped Jake Baker case, in which a student at the University of Michigan posted an abduction, rape, and murder fantasy involving a female student in one of his classes. He then sent a second e-mail saying, "It's not enough anymore to think about it, I have to do it" (MacKinnon, 1995). Although most of the debate surrounds the federal case in which Baker was arrested for transmitting a threat across state lines, we concern ourselves with the response from the University of Michigan itself.

Carnegie Mellon University

In November 1994, Carnegie Mellon University (CMU) announced that it would remove all newsgroups in the alt.sex hierarchy, as well as selected groups in the alt.binares.pictures.erotica hierarchy, and a few others (Camp & Riley, 1995). The newsgroups were selected for removal by title, not by substance. This action was intended to make clear that CMU was acting as a distributor, not as a publisher, because it avoided knowledge of the content of Usenet newsgroups. Hence, groups such as alt.sex.fat, which is a support and discussion group for issues of sexuality and being overweight, and alt.sex.NOT, which is a discussion group about abstinence, were banned. Similarly, valuable information about sexual health was banned through the removal of groups like alt.sex.safe and alt.sex.wizards.

The policy's stated goal was to comply with state and federal law. CMU determined that it was a distributor of Usenet information and thus had to act where the university could not deny knowledge of obscenity. Thus CMU cut a broad swath through the Usenet feeds. Yet CMU let remain groups that regularly contain child pornography, such as alt.binaries.pictures.girls. By trying to minimize perceived liability the administrators at Carnegie Mellon both limited important sexual discourse and failed to remove material that is uniquely prohibited.

However, had CMU selected for content and removed only the child pornography postings, it would be in a position of monitoring. Monitoring is expensive technically—it involves processor time and coding costs. It is also hazardous in that such monitoring would have created the potential for CMU to be a publisher; like Prodigy, the university would have created an infrastructure theoretically capable of monitoring every post. As long as the media types framework applies, institutions like CMU cannot take the necessary steps to comply with child pornography laws without making themselves vulnerable to a publisher's standard of libel.

What is the correct analogy to the physical world? What are Usenet newsgroups?

If Usenet is the information agora, as others have suggested, then the University would have a responsibility to tolerate all speech (again with the exception of child pornography). If Usenet and local newsgroups (commonly called bboards at CMU to differentiate from Usenet groups) are a collection of cyberspaces, not one cyberspace, then the variations as shown in Table 11.1 are legitimate, and the university can act accordingly.

Consider a more obscure event at CMU. The Women's Center was bombarded by hate messages in the spring of 1993. The flame war that ensued was so vigorous that readers who subscribed to the board to receive announcements of business meetings found it too cumbersome to wade through the flamefest to the business messages. Flamers on both sides thought the Women's Center newsgroup should be a place for unfettered debate. Others wanted a haven, a supportive environment in a competitive and aggressive institution.

The situation was resolved by a traditional feminist "both-and" solution: The single space was split in two. Both a discussion board and an announcement board were created, so people could find passionate, contentious, even offensive debate, or read announcements and supportive messages. In this way, women maintained a safe space and protected free speech at Carnegie Mellon.

This solution does not fit into the media type rubric. If CMU as an institution is a publisher or distributor of its bboards, it would be liable for the content of all newsgroups it carries. All spaces would become "safe"—as CMU chooses to define it—and the conversation on the discussion groups would be restricted. A student objecting to the offensiveness of a statement would have a case in the university disciplinary process, regardless of where the statement was posted. Either the discussion bboard would go unused or its content would come to resemble the official Women's Center bboard. This is not how other speech is now regulated; different standards apply to speech in classrooms, dorms, workspaces, and public areas.

If CMU were a common carrier, any newsgroup dispute would be outside of official jurisdiction. In such a case, all boards run the risk of reading like the discussion bboard, regardless of purpose. Bboards used to conduct student meetings or departmental business could be disrupted with countless hostile messages. There would be no recourse for the bombarded organization. Important messages would not reach members who lack the patience or thick skin to locate the nonvitriolic messages. Underrepresented or unpopular groups could be unable to maintain any electronic presence.

There was not a single standard applied to both newsgroups. In technical terms, the university had the same level of control in both the assocs.womens-center and assocs.womens-center.discussion groups. These two bboards are of the same media type. Yet one serves as a protected space for organization and mutual support by an underrepresented group, and the other serves as a forum for even the most offensive speech.

In the second example, CMU asked the question: what kind of space is the Women's Center? The answer was, a space controlled by the Women's Center Collective. The collective decided that the Women's Center is both a space for passionate dialogue and a space for safe discussions. This distinction is enabled by time in the physical world, as support groups do not meet during protests. As cyberspaces cannot separated along such a dimension, a virtual separation was created. This illustrates that although there may not be a clear one-to-one mapping from virtual to physical spaces, consideration of solutions in physical spaces can lead to solutions in virtual spaces. The application of media types to all bboards at Carnegie Mellon would prevent such subtle distinctions.

Santa Rosa Junior College

When Roger Karraker taught a journalism class at Santa Rosa Junior College (SRJC), he maintained an electronic conferencing system. The professor used the electronic conferencing system to motivate his students to have wide-ranging discussions unlimited by classroom constraints, and to expand the learning experience by familiarizing his students with the electronic medium. At the students' request, he established single-sex spaces for the continuation of classroom discussions.

When the student newspaper published a swimwear ad featuring a woman in a thong bikini, one female student picketed the paper to protest the objectification of women. After another woman defended her, they both became targets of sexually derogatory comments on the all-male bulletin board. One of the men told the women about the comments, and they complained to Karraker, who shut down both single-sex boards. However, one student was not satisfied with this response and complained to the administration, which put Karraker on paid leave. The other students retaliated against the loss of the professor by attacking the woman who had made the original complaint. She dropped out of college, and threatened a sex discrimination suit (Dorgan, 1994; Lewin, 1994).

The Office of Civil Rights (OCR) in the Department of Education demanded that the college have an across-the-board policy on all its electronic services against sexually explicit comments that could create a hostile educational environment. With respect to Usenet, this would have resulted in the termination of the alt.sex hierarchy in addition to the removal of much of the soc and talk hierarchies. In order to meet this condition, the college would have had to eliminate many distribution lists as well. Arguably, the OCR treated SRJC as a publisher because this particular case originated in the journalism class, which published the school newspaper. Certainly the journalism class did have newsgroups that were published by the class as journalistic endeavors. Because one newsgroup at Santa Rosa was a publication, all newsgroups must be publications under the media rubric.

Consider the answer to the more subtle question: What is the correct analogy to the physical world? Was it a locker room? A group therapy session? Surely in these contexts the speech is to be protected. The group therapy analogy is probably most appropriate, considering that students were required to abide by a confidentiality rule regarding the contents of the discussions. Indeed, in such a situation, personal attacks are to be expected and the privacy of the participants is paramount.

If one considers the forum to be an extension of the classroom, as Karraker first characterized it, professional limitations may be placed on speech. Vulgar personal attacks are prohibited in a classroom, and Karraker as the course instructor would have a personal and possibly legal responsibility to intervene.

Regardless of the classification, choosing an analogy for that particular electronic space would not have defined all electronic spaces associated with the journalism class. The attempt of the OCR to fit the entire college computing system into that of a publisher was a misguided solution that ignored the nuances of cyberspace, and thus resulted in threats to online speech.

The professor's choice to create a specified therapy session in a classroom created an environment that resulted in the professor's suspension, loss of access to some online speech for journalism students, threats to many unrelated electronic forums on campus, and one young woman's abandonment of academic goals because she perceived this space as an extension of the classroom. The lack of a clear conceptual model created great harm when there was no intention to do so. Because all the speech was electronically assisted, communication did not make all the spaces equivalent. Here again, the lack of consideration of the difference in cyberspaces resulted in the loss of access to online free speech at Santa Rosa Junior College.

University of Michigan

Jake Baker posted a set of detailed steps to kidnap, rape, torture, and murder another student. Jake Baker named that student and identified her living area in alt.sex.stories. What is the university's responsibility under different media types?

First, if the University of Michigan is a common carrier, its responsibility is essentially to do nothing. By definition it does not have any knowledge of the information posted to alt.sex.stories and thus has no obligation to intervene. Michigan could have maintained its contention of common carriage by refusing to act.

If Michigan is a publisher or a distributor, it must choose to act. In this case the options included deleting the message, or investigating Jake Baker's intent, or both. However, acting on that message implies broad future control of information and future liability in cases of libel, copyright, and other infringements that may not pose the same dangers implied by this particular case.

Consider the question: What kind of space is alt.sex.stories? What is Michigan's responsibility if the issue is one of cyberspaces? An appropriate analogy might be an erotica reading night at a campus cafe. People who attend the readings do so knowingly and expect to hear erotic stories of an explicit nature. The university can choose as part of its mission to provide students with the opportunity to explore erotic literature, and to support its students' writing experiments by providing such a forum. However, upon hearing a specific plan of violence directed at a specific individual, it is reasonable that the administration would act to prevent any clear danger. In choosing to act in a single case to prevent physical harm, the university would not commit to prohibiting erotic story hours, or be considered a distributor of all such papers written on campus.

In short, the media type classification encourages irresponsibility. When an institution acts in a dangerous case, this institution risks creating huge liability by changing its media classification. One action in a particular case can be then applied to all cases using the standards of the media type to determine liability. Consider the implications of this standard if code, including hazardous code, is speech.

CONCLUSION AND IMPLICATIONS

The classification of the Internet into a media type, or the creation of a new media type, is tempting for many reasons. First, the media model for electronically assisted communication has served well for decades. The model of four media has remained solid through the early days of the wireless revolution, and several court cases have seen fit to apply a specific media type to a specific provider.

Space as the determinant has advantages and disadvantages. The advantage is that control over speech may be exercised in some forums, without hindering the free flow of information on the Internet as a whole. Users can define and declare the limits of their own spaces. It would enable a broad and consistent policy where control results in liability, without blanket restrictions directed at all providers. Treating cyberspace as spaces also assists in users' right to address and access a public forum, as well as the users' right to be left alone in their own electronic spaces.

The disadvantage is that there is no graceful unified theory as now exists with media types. There is no overarching theory of speech in spaces, but rather a quilt of court decisions and regulations. Like all frameworks, regulation of electronic spaces as space could be manipulated.

As in physical spaces there exist common carriers, publishers, distributors, and broadcasters in cyberspace. A single person can use all of these services—by sending e-mail through a simple gateway, by creating a Web page, by setting up a reflector, and by spamming.

Cyberspace has more modes and models than can be contained in the four media types. The regulation of cyberspace as a multidimensional world of spaces is far more complex than the search for the appropriate media type. Yet the virtual world requires such complexity.

This chapter is meant to suggest further research. Certainly the debate over the regulation of speech in physical spaces has not ended. The debate over speech in cyberspaces may prove as difficult. Labeling virtual spaces as spaces can offer an alternative to the multivariant high-tech but sometimes incomprehensible rating systems being suggested. Or is it best as an interface to simplify self-rating using these protocols? When considering time, place, and manner on the Internet, what time is it? What implications does the concept of cyberspaces have to privacy? Using cyberspaces can an employee have a (private) desk drawer as well as a desktop?

The clearest question for further research is, if cyberspaces are spaces, how then can they be protected as appropriate? We hope by this chapter to inspire debate within the legal community. We do not seek here to provide a complete answer to this question, for we understand that the computer science adage, "Of course it's complicated, that's why they call it code," has multiple applications.

GLOSSARY

.plan file—On UNIX systems, a user's plan file displays personal information that the user chooses to make available to other users who access the file.

bboard—Electronic bulletin board; in this piece refers in general to topic groups in a BBS or a Usenet newsgroup, or specifically to CMU's campus-wide newsgroups.

BBS—Electronic bulletin board system, a message database where people can log in and leave messages for others in topic groups—usually requires a subscription unique to the bboard, not available through Usenet.

cancelbot—A program that removes a spam (see definition) because it was widely posted, identifies the message by its unique identifier and removes it from all locations, removes posts on the basis that spamming is unacceptable behavior.

cancelpoodle—A program that removes a single message from the few newsgroups to which the post is relevant because of the content of the post, technically the same as a cancelbot but directed for a different purpose and at a different target.

chat room—Bboard on America Online.

dungeon—Short for multiuser dungeon or MUD, set of real-time chat rooms that have the feel and structure of an adventure game.

firewall—Widely used security product that limits the flow of information between a specific domain and the wider network; may block information based on its location, file type, content, and so on.

flame—An e-mail or post message intended to insult and provoke, differentiated from a troll in that it is directed with hostility towards a specific individual(s).

home page—One's personal presence on the World Wide Web; often includes a photo, a listing of links to favorite Web sites, personal and professional information, and so on.

IP address—Internet Protocol address, provides electronic location information.

post—A message that is publicly distributed via a BBS, bboard, or newsgroup.

spam—The sending of many identical messages separately to a large number of newsgroups.

troll—A post that is intended to insult or provoke.

Usenet newsgroups—One of the 10,000 topic areas on Usenet, a distributed bulletin board system.

REFERENCES

American Library Association, Inc., et. al. v. United States, et al. (1996). Civil, No. 96-1458, 3rd Circuit U.S. Court of Appeals.

Beall, R. (1987). Developing a coherent approach to regulating bboards. *Computer Law Journal, 7*, 499–516.

Barron, J. A. (1993). The telco, the common carrier model and the First Amendment—The Dial-A-Porn precedent. *Rutgers Computer and Technology Law Journal, 19*, 371, 385–391.

Becker, L.E. (1989). Liability of computer bulletin board operators. *Connecticut Law Review, 22*, 203–238.

Berman, J. (1995). Abundance and user control: renewing the democratic heart of the First Amendment in the age of interactive media. *Yale Law Journal, 104*, 1619–1637.

Camp, L. J., & Riley, D. (1995, March). Women, children, animals and the like: Protecting an unwilling electronic populace. *Proceedings of the Fifth Conference on Computers, Freedom and Privacy* (pp. 120–139). Burlingame, CA.

Cubby v. Compuserve, Inc. (1991). 776 F. Supp. 135.

Daniel v. Dow Jones. (1987). 520 NYS2d 334, 338.

Di Lello, E.V. (1992). Functional equivalency and its application to freedom of speech on computer bulletin boards. *Columbia Journal of Law and Social Problems, 26*, 199–247.

Dorgan, M. (1994, September 17). Free speech in cyberspace: Santa Rosa case tests government limits on computer bulletin boards. *San Jose Mercury News*. p. A-1

Ginger, A. F. (1975). *Pentagon Papers case collection: Annotated procedural guide and index*, Dobbs Ferry, NY: Oceana.

Hammond, A. (1995). Regulating the multi-media chimera: Electronic speech rights in the United States. *Rutgers Computer and Technology Law Journal, 21*, 1–87.

Jepsen, D. (1995). *Testimony for Senate Judiciary Hearing on Cyberporn*, Washington, DC, July 24.

Krattenmaker, T. G. & Powe, L. A., Jr. (1995). Converging First Amendment principles for converging communications media. *Yale Law Journal, 104*, 1719–1744.

Lewin, T. (1994, September 22). Dispute over computer messages: Free speech or sex harassment? *New York Times*. p. A1:1

MacKinnon, K. (1995, March 9). *Policing the Internet: Jake Baker and beyond*, Transcript of Michigan Telecommunications and Technology Law Review Conference. Available [Online]: http://www.umich.edu/~mttlr/archives/bakerconf/

Naughton, E.J. (1992). Is cyberspace a public forum? Computer bulletin boards, free speech and state action. *Georgetown Law Journal, 81*, 409–441.

New York Times Co. v. United States. (1971). 403 US 713.

Pool, I. S. (1983). *Technologies of freedom.* Cambridge, MA: Harvard University Press.

Sassan, A. J. (1992). Comparing apples to oranges: The need for a new media classification. *Software Law Journal, 5*, 821–844.

Spertus, E. (1996, April). *Social and technical means for fighting on-line harassment.* Presented at Virtue and virtuality: Gender, law, and cyberspace. MIT, Cambridge, MA.

Steinberg, E. (1995, January). *Living on the slippery slope.* Presented at Law in the misinformation age: The First Amendment, privacy and electronic networks. Duke Law School, Durham, NC.

Stets, D. (1996, June 13). Judge rejects the Internet obscenity law. *Philadelphia Inquirer*, p. 1:1.

Stratton Oakmont, Inc., v. Prodigy Servs. Co. (1971). N.Y. Sup. Ct. May 25.

12

Rating the Net

Jonathan Weinberg
Wayne State University Law School, Detroit, Michigan

Internet filtering software is hot. Plantiffs in *ACLU v. Reno*[1] relied heavily on the existence and capabilities of filtering software (also known as blocking software) in arguing that the Communications Decency Act was unconstitutional.[2] President Clinton has pledged to "vigorously support" the development and widespread availability of filtering software.[3] Free speech activists see that software as providing the answer to the dilemma of indecency regulation, making it possible "to reconcile free expression of ideas and appropriate protection for kids."[4] Indeed, some of the strongest supporters of blocking software are First Amendment activists who sharply oppose direct government censorship of the Net.

---

1. 929 F.Supp. 824 (E.D. Pa. 1996) (striking down the Communications Decency Act).

2. See *ALA Plaintiffs' Post-Hearing Brief, American Civil Liberties Union v. Reno*, 929 F.Supp. 824 (E.D. Pa. 1996) (No. 96-963), at Parts I.C.1 & I.C.2. Available [Online]: http://www.eff.org/pub/Legal/Cases/EFF_ACLU_v_DoJ /960429_ala_post-hearing.brief.

3. Statement by the President (June 12, 1996). Available [Online]: http://www.eff.org/pub/Legal/Cases/EFF_ACLU_v_DoJ 960612_clinton_cda_decision.statement

4. Daniel Weitzner, Deputy Director, Center for Democracy and Technology, quoted in P. H. Lewis (1996, March 1). Microsoft backs ratings system for Internet, *New York Times*, p. D5; see J. Berman & D. J. Weitzner (1995), Abundance and user control: Renewing the democratic heart of the First Amendment in the age of interactive media, *Yale Law Journal, 104*, 1619, 1634-1635.

Internet filtering software, further, is here. As of this writing, the Platform for Internet Content Selection (PICS) working group has developed a common language for Internet rating systems, making it much easier to create and market such ratings.[5] Two heavily promoted ratings systems (SafeSurf and RSACi) allow content providers to rate their own World Wide Web sites in a sophisticated manner. Microsoft's World Wide Web browser incorporates a feature called Content Advisor that will block Web sites in accordance with the rules of any PICS-compliant ratings system, including SafeSurf and RSACi.[6] Stand-alone blocking software—marketed under such trademarks as SurfWatch, Cyber Patrol, CYBERSitter, KinderGuard, Net Nanny, and Parental Guidance—is gaining increasing sophistication and popularity.

It is easy to understand the acclaim for filtering software. That software can do an impressive job at blocking access, from a given computer, to sexually explicit material that a parent does not wish a child to see. The PICS standard for describing ratings systems is an important technical achievement, allowing the development and easy use of a variety of sophisticated ratings schemes.

In the midst of the general enthusiasm, though, it is worth trying to locate the technology's limitations and drawbacks. Blocking software is a huge step forward in solving the dilemma of sexually explicit speech on the Net, but it does come at a cost. People whose image of the Net is mediated through blocking software will miss out on worthwhile speech—through deliberate exclusion, through inaccuracies in labeling inherent to the filtering process, and through the restriction of unrated sites.

---

5. See P. Resnick & J. Miller (1996), *PICS: Internet access controls without censorship.* Available [Online]: http://www.w3.org/pub/WWW/PICS/iacwcv2.htm. PICS was developed by the World Wide Web Consortium, the body responsible for developing common protocols and reference codes for the evolution of the Web, with the participation of industry members and onlookers including Apple, America Online, AT&T, the Center for Democracy and Technology, Compuserve, DEC, IBM, MCI, the MIT Laboratory for Computer Science, Microsoft, Netscape, Prodigy, the Recreational Software Advisory Council, SafeSurf, SurfWatch, and Time Warner Pathfinder.

6. Microsoft calls its World Wide Web browser Internet Explorer; Content Advisor first appeared in Internet Explorer's version 3.0. Content Advisor makes it easiest to use RSACi ratings. RSACi is an Internet ratings system established by the Recreational Software Advisory Council (RSAC), which was created by the Software Publishers Association in 1994 to create a rating system for computer games. RSAC formed a working party in late 1995, including representatives from Time Warner Pathfinder, AT&T, PICS and Microsoft, to develop RSACi. See infra note 12 and accompanying text. For Content Advisor to use ratings from a PICS-compliant ratings service other than RSACi, the user must copy that service's .RAT file into his Windows System folder, and then click on "Add A New Ratings Service" in the IE Options menu. See *Safesurf March News.* Available [Online]: http://www.safesurf.com/nletter/summ96.htm

In the first part of this chapter I offer some general background on blocking software. In the second part I consider the extent to which inaccuracy is inevitable in rating the Net. It is easy to find anecdotes about sites inappropriately blocked by filtering software, and complaints that ratings systems are insufficiently fine-tuned to label particular sites accurately. Do these bad results reflect a limitation inherent in the nature of ratings systems? In the third part I consider the treatment of unrated sites. Relatively few Internet content sources today carry ratings. What portion of the Net universe can we expect to carry ratings once the technology is mature? To the extent that ratings systems' reach is less than complete, what implications does that have for Net speech as a whole? In the fourth part I examine the extent to which adults' access to the Net is likely, in the near future, to be filtered through blocking software.

BACKGROUND

I assume in this chapter that the reader is familiar with the Internet; for those who are not, the court's findings of fact in *ACLU v. Reno* provide an excellent guide.[7] I focus here on software blocking access to the World Wide Web, and leave the question of screening access to Usenet news for another day.[8]

Blocking access to sexually explicit material on the World Wide Web is difficult. There are millions of individual pages on the Web, and the number is increasing every day. An astonishingly small fraction of those pages contain sexually explicit material.[9] Every Web page (indeed, every document accessible over the Internet) has a unique address, or URL,[10] and the URLs of some Web pages do contain

---

7. American Civil Liberties Union v. Reno, 929 F.Supp. 824, 830-49 (E.D. Pa. 1996).

8. It is fairly easy for software to limit access to Usenet news. Since each newsgroup has a name describing its particular topic, software writers can do a reasonably effective job of blocking access to sexually explicit material simply by blocking access to those newsgroups (such as alt.sex.stories) whose names indicate that they include sexually explicit material. The question of identifying the newsgroups to be censored, however, can nonetheless be quite delicate. See *Illegal material on the Internet*, http://dtiinfo1.dti.gov.uk/safety-net/r3.htm; *Turnpike news ratings*, http://www.turnpike.com/ratings.

9. "While it is difficult to ascertain with any certainty how many sexually explicit sites are accessible through the Internet, the president of a manufacturer of software designed to block access to sites containing sexually explicit material testified in the Philadelphia litigation that there are approximately 5,000 to 8,000 such sites, with the higher estimate reflecting the inclusion of multiple pages (each with a unique URL) attached to a single site. The record also suggests that there are at least 37 million unique URLs. Accordingly, even if there were twice as many unique pages on the Internet containing sexually explicit materials as this undisputed testimony suggests, the percentage of Internet addresses providing sexually explicit content would be well less than one-tenth of one percent of such addresses." From *Shea v. Reno*, 930 F. Supp. 916, 931 (S.D.N.Y. 1996) (citations omitted).

clues as to their subject matter. Because nothing in the structure or syntax of the Web requires Web pages to include labels advertising their content, though, reliably identifying pages with sexually explicit material is not an easy task.

First-generation blocking software compiled lists of off-limits Web pages through two methods. First, the rating services hired raters to work through individual Web pages by hand, following links to sexually explicit sites, and compiling lists of URLs that were to be deemed off-limits to children. Second, they used string-recognition software to automatically proscribe any Web page that contained a forbidden word (such as "sex" or "xxx") in its URL. The software was not appreciably configurable by home users; once a parent installed the software on a home computer, the question of which sites would be blocked was answered entirely by the ratings service.

The PICS specifications contemplate that a ratings system can be more sophisticated. A ratings service may rate a document along multiple dimensions: that is, instead of merely rating the document as "adult" or "child-safe," it can give it separate ratings for (say) violence, sex, nudity, and adult language. Further, along any given dimension, the ratings service may choose from any number of values. Instead of simply rating a site "block" or "no-block" for violence, a ratings service can assign it (say) a rating of 1 through 10 for increasing amounts of violent content. These features are important because they make possible the creation of filtering software that is customizable by parents. A parent subscribing to such a ratings service, for example, might seek to block only sites rated over 3 for violence and 8 for sex. Finally, the PICS documents note that ratings need not be assigned by the authors of filtering software. They can be assigned by the content creators themselves, or by third parties.[11]

Most rating services today follow the PICS specifications. Their particular approaches, however, differ. The Recreational Software Advisory Council (RSAC) has developed an Internet rating system called RSACi.[12] Participating content providers rate their own sites along a scale of 0 through 4 on four dimensions: violence, nudity, sex, and language. RSAC does not itself market blocking software; instead, it licenses its service to software developers. Another system in which content providers rate their own speech is called Safesurf; in that system, content

---

10. The term is an acronym for Uniform Resource Locator. See *Internet engineering task force, Uniform Resource Locators* [RFC-1738] (1994). Available [Online]: ftp://ds.internic.net/rfc/rfc1738.txt

11. See Resnick & Miller, cited in note 5.

12. The heavy hitters behind RSACi have been RSAC, established in 1994 by the Software Publishers Association, and such industry players as Microsoft, AT&T and Time Warner Pathfinder. See note 6. As of April 15, 1996, content providers have been able to rate their own sites using the RSACi system by completing a questionnaire at the RSAC web site, http://www.rsac.org.

providers choose from nine values in each of nine categories, from "profanity" through "gambling."[13]

Rating services associated with individual manufacturers of blocking software include Cyber Patrol, Specs for Kids, and CYBERSitter. Cyber Patrol rates sites along 15 dimensions, from "violence/profanity" to "alcohol & tobacco," but assigns only two values within each of those categories: CyberNOT and CyberYES.[14] Specs for Kids rates documents along 11 dimensions, including "advertising," "alternative lifestyles," "politics," and "religion," and assigns up to five values (including "no rating") in each of those categories.[15] CYBERSitter, by contrast, maintains a single list of objectionable sites; it affords users no opportunity to block only portions of the list.[16]

ACCURACY

Since blocking software first came on the market, individual content providers have complained about the ratings given their sites. Not all of those complaints relate to problems inherent to filtering software. For example, some programs tend to block entire directories of Web pages simply because they contain a single "adult" file. That means that large numbers of innocuous Web pages are blocked merely because they are located near some other page with adult content.[17] Indeed, it appears that some programs block entire domains, including all of the sites hosted by particular Internet service providers.[18] This is highly annoying to affected content providers. It may be a temporary glitch, though; over time, it is plausible that the most successful rating services will—properly—label each document separately.[19]

---

13. The categories are: profanity, heterosexual themes, homosexual themes, nudity, violence, intolerance, glorifying drug use, other adult themes, and gambling. See *SafeSurf Rating System*: http://www.safesurf.com/classify/index.html

14. See *CyberNOT list criteria*, http://www.microsys.com/cyber/cp_list.htm

15. See *Specs glossary*, http://www.newview.com/cust/ss_sg_lvl3a_cf_fcs.html

16. See *CYBERSitter product information*: http://solidoak.com/cysitter.htm

17. For the most part, Cyber Patrol drops all but the first three characters of the filename in the URL, thus blocking some innocuous pages. See B. N. Meeks & D. B. McCullagh, (1996, July 3), Jacking in from the "Keys to the Kingdom" port, *CyberWire Dispatch*. [Online], Available: http://www.eff.org/pub/Publications/Declan_McCullagh/cwd.keys.to.the. kingdom.0796.article. In some instances, Cyber Patrol blocks at a higher level, so that (for example) it excluded all of Jewish.com because personal ads were not stored in a separate subdirectory. E. Berlin & A. Kantor, Who will watch the watchmen? Available: http:// www.iw.com/current/feature3.html. In at least one case, Cyber Patrol has blocked entire Internet service providers. See note 18. Surfwatch has suggested to content providers that they accommodate its decision to block on the directory level by segregating their adult material in separate directories. See *Censorship sucks*. Available [Online]: http://cocacola.whoi.edu/~scott/surf.html; e-mail from Chris Kryzan for wide distribution, c. June 15, 1995 (on file with author); see also surfwatching, http://www.links.net/dox/surfwatch.html.

Other problems arise from the wacky antics of string-recognition software.[20] America Online's software, for example, ever alert for four-letter words embedded in text, refused to let users register from the British town of "Scunthorpe." (The online service solved the problem, to its own satisfaction, by advising its customers from that city to pretend they were from "Sconthorpe" instead.)[21]

Controversies over sites actually rated by humans are less amenable to technological solution. One typical dispute arose when Cyber Patrol blocked animal-rights web pages because of images of animal abuse, including syphillis-infected monkeys; Cyber Patrol classed those as "gross depiction" CyberNOTs.[22]

Sites discussing gay issues are commonly blocked, even if they contain no references to sex. Surfwatch, in its initial distribution, blocked a variety of gay sites including the Queer Resources Directory, an invaluable archive of material on homosexuality in America,[23] and the International Association of Gay Square Dance Clubs. It responded to protests by unblocking most of the contested sites.[24] Other blocking programs, on the other hand, still exclude them: Cyber Patrol blocks a mirror of the Queer Resources Directory, along with Usenet newsgroups including clari.news.gays (which carries AP and Reuters dispatches) and alt.journalism.gay-press.[25] CYBERSitter is perhaps the most likely to block any reference to sexual orientation, forbidding such newsgroups as alt.politics.homosexual. In the words of a CYBERSitter representative: "I wouldn't even care to debate the issues if gay and

---

18. According to an article in the electronic version of *Internet World*, CYBERSitter blocks all sites hosted by cris.com. A CYBERSitter representative suggests that Internet service providers "are responsible for their content," and that the owners of cris.com are to blame because they "will not monitor their [customers'] sites." According to another recent report, CYBERSitter now blocks all sites at the WELL, the pioneering California electronic community and Internet access provider (e-mail from Bennett Hazelton to the fight-censorship mailing list, October 27, 1996; on file with author). It has threatened to block at least one other large service provider. See note 30 and accompanying text. Cyber Patrol blocks all pages hosted by crl.com (including a real estate agency's Web pages). A Cyber Patrol representative says that the company is reviewing its policy of blocking entire domains (Berlin & Kantor, see note 17).

19. The RSACi and Safesurf systems allow content providers to label each page individually, and that should also be true for future self-rating systems. At a recent PICS developers' workshop, the prevailing view was that filtering software should expect to find PICS-compliant labels only in the individual documents, not in the directories or elsewhere in the site. See *PICS Developers' Workshop Summary*. Available [Online]: http://www.w3.org/pub/WWW/PICS/picsdev-wkshp1.html. This makes sense only in the context of software that makes the blocking decision for each page individually. The more difficult question is whether the problem will persist in connection with third-party rating services. At least one third-party rating service (Specs for Kids) does rate on the document level, and such a service should be more attractive to many users; the question is whether the increased sales will outweigh the added expense of that granularity. The recent CyberSitter/Peacefire controversy (see note 30 and accompanying text) suggests that my optimism may be unfounded.

lesbian issues are suitable for teenagers...We filter anything that has to do with sex. Sexual orientation [is about sex] by virtue of the fact that it has sex in the name."[26]

The list of blocked sites is sometimes both surprising and alarming. Cyber Patrol blocks Usenet newsgroups including alt.feminism, soc.feminism, clari.news.women, soc.support.pregnancy.loss, and alt.support.fat-acceptance.[27] It blocks gun and Second Amendment Web pages (including one belonging to the NRA Members' Council of Silicon Valley). It blocks the Web site of the League for Programming Freedom (a group opposing software patents). It blocks the Electronic Frontier Foundation's censorship archive.[28] CYBERSitter blocks the National Organization of Women web site.[29] It blocks the Web site of Peacefire, a teen-run cyber-rights group, although the site contains no questionable material other than criticism of CYBERSitter.[30]

You might think that a better answer lies in rating systems, such as RSACi and SafeSurf, in which content providers evaluate their own sites. Surely, you might figure, an author could hardly disagree with a rating he or she selected. The matter, though, is not so clear. When an author evaluates his or her site in order to gain a rating from any PICS-compliant rating service, the author must follow the algorithms and rules of that service. Jonathan Wallace, thus, in an article called "Why I Will Not Rate My Site,"[31] asks how he is to rate "An Auschwitz Alphabet,"[32] his powerful and deeply chilling work of reportage on the Holocaust. The work contains descriptions of violence done to camp inmates' sexual organs. A self-rating

20. For sheer wackiness, though, nothing can match a CYBERSitter feature that causes Web browers to white out selected words but display the rest of the page (so that the sentence "President Clinton opposes homosexual marriage" would be rendered "President Clinton opposes marriage" instead). See e-mail from Solid Oak Software Technical Support to Bob Stock, October 24, 1996 (on file with author).

21. See *Risks Digest*, *18* (7) (April 25, 1996), http://catless.ncl.ac.uk/Risks/18.07.html#subj3.1. On string-recognition software, see note 44 and accompanying text.

22. See Meeks & McCullagh, cited in note 17: http://www.mit.edu/activities/safe/labeling/cp-bans-animal-rights

23. http://www.qrd.org.

24. See *Surfwatch censorship against lesbigay WWW pages*. Available [Online]: http://www.utopia.com/mailings/censorship/Surfwatch.Censorship.Against.Lesbigay.WWW.Pages.html; e-mail from Chris Kryzan, cited in note 17.

25. See Meeks & McCullagh, cited in note 17.

26. See Meeks & McCullagh, cited in note 17.

27. See Meeks & McCullagh, cited in note 17. Indeed, Cyber Patrol apparently blocks all of the alt.support groups (including, e.g., alt.support shyness and alt.support.depression), along with such groups as alt.war.vietnam and alt.fan.frank-zappa (e-mail from Declan McCullagh to the fight-censorship mailing list, October 4, 1996, on file with author); *Cyber Patrol: The truth*: http://www.canucksoup.net/CYBERWHY.HTM

28. Meeks & McCullagh, cited in note 17.

29. Meeks & McCullagh, cited in note 17.

system, Wallace fears, would likely force him to choose between the unsatisfactory alternatives of labeling the work as suitable for all ages, on the one hand, or "lump[ing it] together with the Hot Nude Women page" on the other.[33]

It seems to me that at least some of the rating services' problems in assigning ratings to individual documents are inherent. It is the nature of the process that no ratings system can classify documents in a perfectly satisfactory manner, and this theoretical inadequacy has important real-world consequences.

Consider first how a ratings system designer might construct a ratings algorithm. The designer might provide an algorithm made up entirely of simple, focused questions, in which each question has a relatively easily ascertainable "yes" or "no" answer. (Example: "Does the file contain a photographic image depicting exposed male or female genitalia?") Alternatively, the designer might seek to afford evaluators more freedom to apply broad, informal, situationally sensitive guidelines so as to capture the overall feel of each site. (Example: "Is the site suitable for a child below the age of 13?")

In jurisprudential terms, the first approach relies on "rules"; the second, on "standards."[34] The RSACi system makes the greatest attempt to be rule-based, seeking to enunciate simple, hard-edged categories, with results turning mechanically on a limited number of facts.[35] Other rating systems rely more heavily on standards. The SafeSurf questionnaire, for example, requires the self-rater to determine whether nudity is "artistic" (levels 4 through 6), "erotic" (level 7), "pornographic" (level 8), or "explicit and crude" pornographic (level 9).[36] Specs for Kids has its raters distinguish between sites that refer to homosexuality (a) "im-

---

30. See *Wired News,* December 10, 1996. Available [Online]: http://www.wired.com/news/story/901.html. CYBERSitter has threatened to block all 2500 domain names hosted by Peacefire's Internet service provider if the provider refuses to delete Peacefire's site.

31. J. Wallace (1996), *Why I will not rate my site* [Online]. Available: http://www.spectacle.org/ cda/rate.html#report

32. http://www.spectacle.org/695/ausch.html

33. Wallace, cited in note 31.

34. See D. Kennedy (1976), Form and substance in private law adjudication, *Harvard Law Review, 89,* 1685, 1685, 1698; P. Schlag (1985), Rules and standards, *UCLA Law Review, 33,* 379, 380, 383-398; K. Sullivan (1992), The Supreme Court, 1991 term—Forward: The justices of rules and standards, *Harvard Law Review, 106,* 22, 58-59; J. Weinberg (1993), Broadcasting and speech, *California Law Review. 81,* 1101, 1167-1169. See also C. R. Sunstein (1995), Problems with Rules, *California Law Review, 83,* 953, 956.

35. For (slightly different) listings of the RSACi categories, see *Rating the Web,* http://www.rsac.org/why.html; J. Miller et al., *Rating services and rating systems (and their machine readable descriptions),* rev. 5, last modified May 5, 1996, http://www.w3.org/pub/WWW/PICS/services.html, at Appendix B.

36. See *SafeSurf rating system:* http://www.safesurf.com/classify/index.html

partial[ly]," or (b) discuss it with "acceptance or approval," or (c) "active[ly] promot[e]" it or "attempt...to recruit the viewer."[37]

Fuzzier classifications require more judgment on the part of the evaluator. People with different outlooks and values may disagree as to where the lines fall. (With respect to the Specs treatment of references to homosexuality, people will disagree as to whether the categories are even coherent.) The categories work only within a community of shared values, so that evaluators can draw on the same norms and assumptions in applying the value judgments embedded in the standards.

This distinction follows the more general rules-standards dichotomy in law, which focuses on the instructions that lawmakers give law appliers in a variety of contexts.[38] Legal thought teaches that rules and standards each have disadvantages. A problem with standards is that they are less constraining; relatively speaking, a standards-based system will lack consistency and predictability.[39] Rules become increasingly necessary as the universe of law appliers becomes larger, less able to rely on shared culture and values as a guide to applying standards in a relatively consistent and coherent way.

It is for this reason that the designers of RSACi attempted to be rule-like. They contemplate that the universe of ratings evaluators will include every content provider on the Web; that group can claim no shared values and culture. To accomodate that heterogeneous group, RSAC offers a rules-based questionnaire that (it hopes) all can understand in a similar manner. This, RSAC explains, will "provide...fair and consistent ratings by eliminating most of the subjectivity inherent in alternative rating systems."[40]

Rules, though, have their own problems. They direct law appliers to treat complex and multifaceted reality acording to an oversimplified schematic.[41] The point of rules, after all, is that by simplifying an otherwise complex inquiry, they "screen...off from a decisionmaker factors that a sensitive decisionmaker would otherwise take into account."[42] They may thus generate results ill-serving the policies behind the rules.[43] At best, a rule-based filtering system will miss nuances; at worst, it will generate absurd results (as when America Online, enforcing a rule forbidding certain words in personal member profiles, barred subscribers from identifying themselves as "breast" cancer survivors).[44]

---

37. *Specs glossary*: http://www.newview.com/cust/ss_sg_lvl3a_cf_fcs.html

38. See sources cited in note 34.

39. See Kennedy, cited in note 34, pp. 1688-1689; Sullivan, cited in note 34, pp. 62–63; see also F. Schauer (1988), Formalism, *Yale Law Journal, 97,* 509, 539-540; Sunstein, cited in note 34, pp. 972–977.

40. *Before you begin*: http://www.rsac.org/images/spmenu.map?300,0

41. See Kennedy, cited in note 34, p. 1689.

42. Schauer, cited note 39, p. 510.

43. See Schauer, cited in note 39, pp. 534–537; Sunstein, cited in note 34, pp. 992–993.

Given this theoretical critique, one might think that the challenge facing ratings system designers is to devise really good rules-based systems, ones that track reality as well as possible, minimizing the difficulties I have noted. That is what RSAC claims to have done in RSACi.[45] I think the product of any such effort, though, necessarily will be flawed. Over the next few pages, I try to explain why.

Let us return to the choices facing a ratings system designer who constructs blocking software. So far in this chapter, I have glossed over the most basic question confronting the designer: What sort of material should trigger ratings consequences? Should children have access to material about weapons making? hate speech? artistic depictions of nudity? Again, the designer can take two different approaches. First, the designer can decide all such questions, so that the home user need only turn the system on and all choices as to what is blocked are already made for the user. CYBERSitter adopts this approach.[46] This has the benefit of simplicity, but seems appropriate only if members of the target audience are in basic agreement with the rating service (and each other) respecting what sort of speech should and should not be blocked.

Alternatively, the designer can leave those questions for the user to answer. The ratings system designer need not decide whether to block Web sites featuring bomb-making recipes, or hate speech. The designer can instead design the system so that the user has the power to block those sites if the user chooses. Microsoft's implementation of the RSACi labels, thus, allows parents to select the levels of adult language, nudity, sex and violence that the browser will let through.[47] Cyber Patrol allows parents to select which of the 12 CyberNOT categories to block.

Either approach, though, imposes restrictions on the categories chosen by the ratings system designer. If the system designer wishes to leave substantive choices to parents, he or she must create categories that correspond to the different sides of the relevant substantive questions. That is, if the designer wishes to leave

---

44. See R. A. Knox (1995, December 1), "Women go on line to decry ban on "breast," *Boston Globe*, p. A12. This incident was likely the result of string-identification software. String-identification programs are excellent examples of rules-based filtering systems.

45. In fact, RSACi is seriously flawed for reasons having nothing to do with rules and standards. The original RSAC rating system was designed for video games. RSACi carries over the categories and language of the earlier video-game rating system even where they are completely inappropriate. Thus, for example, RSACi's definition of "aggressive violence" on a web page excludes acts of nature "such as flood, earthquake, tornado, hurricane, etc., unless the act is CAUSED by Sentient Beings or Non-sentient Objects in the game or where the game includes a character playing the role of 'God' or 'nature' and the character caused the act." See *RSACi ratings dissected*, http://www.antipope.demon.co.uk/charlie/nonfiction/rant/rsaci.html. A striking consequence of RSAC's approach is that the system nowhere acknowledges a distinction between images and text.

46. See text accompanying note 16.

47. See *Using content advisor*: http://microsoft.com/ie/most/howto/ratings.htm

users the choice whether to block sites featuring hate speech, he or she must break out sites featuring hate speech into a separate category or categories. If the designer wishes to leave the user the choice whether to block sites that depict explicit sexual behavior but nonetheless have artistic value, he or she must categorize those sites differently from those that do not have artistic value. On the other hand, if the system designer makes those substantive decisions, making personal value choices as to what material should and should not be blocked, he or she must (of course) create categories that correspond to those value choices.

The problem is that many of these questions cleave on lines defined by standards. Many users, for example, might like to block "pornography," but to allow other, more worthy, speech, even if that speech is sexually explicit. SafeSurf responds to that desire when it requires self-raters to determine whether nudity is "artistic," "erotic," "pornographic," or "explicit and crude" pornographic. It gets high marks for attempting to conform its system to user intuitions, but its lack of rulishness means problems in application. Similarly, Specs' distinction between "impartial reference," "acceptance or approval," and "active promotion" of homosexuality may well correspond to the intuitions of much of its target audience, but will hardly be straightforward in actual application. The problem increases with the heterogeneity of the service's audience: The more heterogeneous the audience, the more categories a rating system must include, to accommodate different user preferences.

With this perspective, one can better appreciate the limitations of RSAC's attempt to be rule-bound. For one thing, RSACi ignores much content that some other ratings systems class as potentially unsuitable, including speech relating to drug use, alcohol, tobacco, gambling, scatology, computer hacking and software piracy, devil worship, religious cults, militant or extremist groups, weapon making, and tattooing and body piercing, and speech "grossly deficient in civility or behavior."[48] For many observers (myself included), RSACi's limited scope is good news. Software that blocks access to controversial political speech is not a good thing, even if parents install it voluntarily. My point, though, is that RSACi had to confine its reach if it was to maintain its rule-bounded nature.

The problem appears as well in connection with the categories RSACi does address. Consider RSACi's treatment of sex. It divides up sexual depictions into "passionate kissing," "clothed sexual touching," "non-explicit sexual activity," and "explicit sexual activity; sex crimes."[49] But note what RSACi, in contrast to some

---

48. All of these areas of speech trigger Cyber Patrol blocking (except for tattooing and body piercing, which constitute a CyberNOT only to the extent they result in "gross depictions"). See *CyberNOT list criteria*: http://www.microsys.com/cyber/cp_list.htm. Tattooing and body piercing are specifically blocked by Specs (in a category, called "subjects of maturity," that lumps them in with "illegal drugs, weapon making...[and] some diseases." *Specs glossary*: http://www.newview.com/cust/ss_sg_lvl3a_cf_fcs.html.

49. *Rating the Web*: http://www.rsac.org/why.html

other ratings systems, does not attempt. It does not seek to distinguish "education-al" materials from other depictions. It does not seek to distinguish "artistic" depic-tions from others. It does not seek to distinguish "crude" depictions from others. There is no way, consistent with rulishness, that it can adequately distinguish the serious or artistic from the titillating. It achieves rule-boundedness, and ease of ad-ministration, at the expense of nuance; it achieves consistent labeling, but in cate-gories that do not correspond to the ones many people want.

In sum, rating system designers face a dilemma. If a ratings service seeks to map the Net in a relatively comprehensive manner, it must rely on a relatively large group of evaluators. Such a group of evaluators can achieve fairness and consis-tency only if the ratings system uses simple, hard-edged categories relying on a few easily ascertainable characteristics of each site. Such categories, though, will not categorize the Net along the lines that home users will find most useful, and will not empower those users to heed their own values in deciding what speech should and should not be blocked. To the extent that ratings system designers seek to allow evaluators to consider more factors, in a more situationally specific man-ner, to capture the essence of each site, they will ensure inconsistency and hidden value choices as the system is applied.

UNRATED SITES

Blocking software can work perfectly only if all sites are rated. Otherwise, the software must either exclude all unrated sites, barring innocuous speech, or allow unrated sites, letting in speech that the user would prefer to exclude. What are the prospects that a rating service will be able to label even a large percentage of the millions of pages on the World Wide Web? What are the consequences if it cannot?

Consider first rating services associated with individual manufacturers of block-ing software, such as CYBERSitter and Cyber Patrol. These services hire raters to label the entire Web, site by site. The limits on their ability to do so are obvious. For one thing, as the services get bigger, and hire more and more employees to rate sites, their consistency will degrade. For another, no service could be big enough to rate the entire Web. Too many new pages come online every day. The content associated with any given page is constantly changing. Further, some of the sites most likely to be ephemeral are also among the most likely to carry sexually ex-plicit material. A ratings service simply can't keep tabs on every college freshman who gets to school and puts up a Web page, notwithstanding that college freshmen are of an age to be more interested in dirty pictures than most. A ratings service certainly can't keep tabs on every Web page put up by a college freshman in Osa-ka, say, or in Amsterdam. So any such rating service must take for granted that there will be a huge number of unrated sites.

As a practical matter, enabling access to unrated sites is not an option for these rating services; it would let through too much for them to be able to market themselves as reliable screeners. The only effective answer, in the eyes of industry members, is to block all unrated sites.[50]

What about self-rating approaches, like those of SafeSurf and RSACi? These services have the potential for near-universal reach, because they can draw on the services of an effectively unlimited number of evaluators. Although the evaluators will be a diverse group (to say the least), rating service designers can try to cope with that diversity by constructing rule-bound questionnaires. Although some evaluators may misrepresent their sites, rating services can try to devise enforcement mechanisms to cope with that as well. On the other hand, self-rating services will not achieve that potential unless content providers have a sufficient incentive to participate in the ratings process in the first place. That incentive is highly uneven.

Mass-market commercial providers seeking to maximixe their audience reach will participate in any significant self-rating system, so as not to be shut out of homes in which parents have configured their browsers to reject all unrated sites.[51] Many noncommercial site owners, though, may not participate. They may be indifferent to their under-18 visitors, and may not wish to incur the costs of self-rating. For the owner of large archives containing many documents, supplying a rating for each page may be a time-consuming pain in the neck. If self-rating services choose to charge content providers a fee to participate, that will provide another disincentive. RSAC's business plan for RSACi contemplates that it will charge a fee to Internet content providers, just as it charges a fee to video-game manufacturers who participate in its video-game self-rating system.[52] RSAC, though, has not yet announced any details of its charging plans; a ratings service with hopes of global reach might allow noncommercial sites to rate themselves gratis.

It may be that the only way to ensure participation in an a self-rating system even in a single country (let alone internationally) would be for government to

---

50. Microsoft, thus, cautions Internet content providers that "For a rating system to be useful, the browser application must deny access to sites that are unrated" (*The PICS standard*, http://www.microsoft.com/intdev/sdk/ docs/ratings/ratng001.htm). Other observers say the same. See W. Andrews, (1996, July 8), Site-rating system slow to catch on, *Web Week*, (quoting Compuserve representative Jeff Shafer). Available: http://www.webweek.com/96July8/comm/rating.html; *Specs FAQs: Quick Quest: General info*, http://www.newview.com/cust/ss_qq_lvl3a_reg.html; e-mail from A. Oram, O'Reilly and Associates, to telecomreg mailing list, May 21, 1996, (on file with author).

51. See Lewis, cited note 4; RSAC, *Rating the Web*: http://www.rsac.org/why.html

52. See Recreational Software Advisory Council, *RSACi: RSAC on the Internet: Business plan* (on file with author); *How to register software titles*, http://www.rsac.org/register.html; see also *Microsoft Press Release*, February 28, 1996: http://www.microsoft.com/corpinfo/press/1996/feb96/rsacpr.htm ("To encourage widespread rating of Internet content, RSAC will make its rating application available for no charge for the first year it is available on the Internet").

compel content providers to self-rate. Such a requirement, though, seems dubious
as a policy matter. The drafters of such a law would face the choice of forcing content providers to score their sites with reference to a particular rating system specified in the law, or allowing them leeway to choose one of a variety of PICS-
compliant ratings systems. Neither approach seems satisfactory. The first, mandat-
ing use of a particular rating system, would freeze technological development by
eliminating competitive pressures leading to the introduction and improvement of
new searching, filtering and organizing techniques. It would leave consumers unable to choose the rating system that best served their needs. The second would be
little better. Some government organ would have to assume the task of certifying
particular self-rating systems as adequately singling out material unsuitable for
children. It is hard to imagine how that agency could ensure that every approved
system yielded ratings that were in fact useful to most parents, while nonetheless
maintaining a healthy market and allowing innovation. In any event, a mandatory
self-rating requirement would likely be held unconstitutional.[53]

The result, though, is that child-configured lenses will show only a limited, flat-
tened view of the Net. If many Internet content providers decline to self-rate, the
only "safe" response may be to configure blocking software to exclude unrated
sites. The plausible result? A typical home user, running Microsoft Internet Explorer set to filter using RSACi tags (say), would have a browser configured to ac-

---

53. See McIntyre v. Ohio Elections Commission 115 S. Ct. 1511 (1995); Riley v. National Federation of the Blind, 487 U.S. 781 (1988); see also Hurley v. Irish-American Gay,
Lesbian and Bisexual Group, 115 S. Ct. 2338, 2347-2348. Even if the characterization of
speech according to the taxonomy of a particular rating system were deemed factual and
value-neutral, requiring a speaker to rate her speech would be subject to exacting scrutiny
under the rule of these cases, because it would require her to incorporate into her speech a
"statement...she would otherwise omit." (*McIntyre*, 115 S. Ct. at 1519-1520).

Mandatory self-rating triggers heightened scrutiny on other grounds as well. The Court has
repeatedly recognized the impermissibility of requiring a speaker to associate herself with
particular ideas she disagrees with. See Hurley; Pacific Gas & Electric Co. v. Public Utilities
Comm'n, 475 U.S. 1, 15 (1986); Wooley v. Maynard, 430 U.S. 705, 714 (1977); West Virginia
State Bd. of Educ. v. Barnette, 319 U.S. 624, 642 (1943). Requiring self-rating does that, because ratings are not value-neutral; mandatory self-rating compels the speaker to associate
herself with the values and worldview embodied in the rating taxonomy.

I am doubtful that a self-rating requirement could survive exacting scrutiny. Even without a self-rating requirement, parents can restrict their children's access to sexually explicit
sites by using blocking programs, and instructing the software to block all unrated sites. A
self-rating requirement would be helpful to parents only in that it would enable them to limit their children's access in such a way that the kids could also view an uncertain number of
additional sites, not containing sexually explicit material, whose providers would not otherwise choose to self-rate. In light of the first amendment damage done by a compelled self-
rating requirement, accomplishing that goal does not appear to be a compelling or overrid-
ing state interest.

cept duly rated mass-market speech from large entertainment corporations, but to block out a substantial amount of quirky, vibrant individual speech from unrated (but child-suitable) sites. This prospect is disturbing.

The Internet is justly celebrated as "the most participatory form of mass speech yet developed."[54] A person or organization with an Internet hookup can easily disseminate speech across the entire medium at low cost; the resulting "worldwide conversation" features an immense number of speakers and "astonishingly diverse content."[55] As Judge Dalzell noted in his *ACLU v. Reno* opinion, the Internet vindicates the First Amendment's protection of "the 'individual dignity and choice' that arises from 'putting the decision as to what views shall be voiced largely into the hands of each of us,'" because "every minute [Internet communication] allows individual citizens actually to make those decisions."[56] But this prospect is threatened if widespread adoption of blocking software ends up removing much of the speech of ordinary citizens, leaving the viewer little to surf but mass-market commercial programming. One hardly needs the Internet for that; we get it already from the conventional media.

In sum, blocking software could end up blocking access to a significant amount of the individual, idiosyncratic speech that makes the Internet a unique medium of mass communication. Filtering software, touted as a speech-protective technology, may contribute to the flattening of speech on the Net.

CHILDREN, ADULTS AND BLOCKING SOFTWARE

You may protest that I am making much of little here. After all, blocking software is intended to restrict children's access to questionable sites. It won't affect what adults can see on the Net—or will it? It seems to me that, in important respect, it will. The desire to restrict children's access has spurred the recent development of filtering technology. I am doubtful, though, that widespread adoption of the software will leave adults unaffected.

In a variety of contexts, we can expect to see adults reaching the Net through approaches monitored by blocking software. In the home, parents may set up filters at levels appropriate for their children, and not disable them for their own use.[57] Indeed, they may subscribe to an Internet access provider that filters out material at the server level (so that nobody in the household can see "objectionable" sites except by establishing an Internet access account with a new provider).[58] If,

---

54. ACLU v. Reno, 929 F.Supp. 824, 833 (E.D. Pa. 1996) (opinion of Dalzell, J.).

55. ACLU v. Reno, at 877, 883.

56. ACLU v. Reno, at 881–882 (quoting Leathers v. Medlock, 499 U.S. 439, 448–449 [1991]).

57. This concern is salient in connection with blocking programs, such as Microsoft's Content Advisor, that block any access to restricted sites through the computer on which the program is installed, unless the program is disabled.

as seems likely, future versions of the PICS specifications support the transmission
of filtering criteria to search engines, then users running Internet searches will not
even know which sites otherwise meeting their criteria were censored by the
blocking software.[59]

Other people get their Internet connections through libraries; indeed, some policymakers tout libraries and other community institutions as the most promising
vehicle for ensuring universal access to the Internet.[60] The American Library Association takes the position that libraries should provide unrestricted access to information resources; it characterizes the use of blocking programs as
censorship.[61] This policy, however, is not binding on member libraries. It is likely
that a significant number of public libraries will install blocking software on their
public-access terminals, including terminals intended for use by adults; indeed,
some have already done so.[62]

Still other people get Internet access through their employers. Corporations too,
though, wary of risk and wasted work time, may put stringent filters in place.
Some large companies worry about the possibility of being cited for sexual harassment by virtue of material that came into the office via the Net. Even more are concerned about sports and leisure information that they feel may detract from
business productivity.[63]

In sum, we may see home computers blocked for reasons of convenience, library computers blocked for reasons of politics, and workplace computers

58. See, e.g., *BESS.NET: The service,* http://demo.bess.net/about_bess/the_service.html.
Safesurf is now providing technology—the Safesurf Internet Filtering Solution—that allows any ISP to offer parents this easy option. See R. Aguilar, *Marketplace site filters criticized* (October 18, 1996): http://www.news.com/News/Item/0,4,4609,00.html

59. When a user running blocking software seeks to conduct a search using an Internet
search engine such as Alta Vista, the software will transmit the user's filtering rules to the
search engine. The search engine will tailor its search so as not to return any sites excluded
by the filter. Participants at the recent PICS Developers' Workshop agreed that this was the
preferable approach, in part because it would be undesirable for users to get search results
like "here's the first 10 responses, but 9 of them were censored by your browser" (PICS Developers' Workshop Summary, cited in note19; see also PICS, http://www.w3.org/pub/
WWW/PICS, [Frequently Asked Questions]).

60. See, e.g., G. Chapman. (1996, June 3). Universal service must first serve community,
Los Angeles Times, p. D1. See generally R. H. Anderson et al., (1995). *Universal access to
e-mail: Feasibility and societal implications,* Rand Corp., Santa Monica, CA, chap. 3 (discussing pros and cons of locating devices for e-mail access in the home, at work, in schools,
and in libraries, post offices, community centers, and kiosks).

61. See *QUESTIONS AND ANSWERS: Access to electronic information, services, and
networks: An interpretation of the Library Bill of Rights,* http://ala1.ala.org:70/0/alagophx/
alagophxfreedom/electacc.q%26a; *Access to electronic information, services, and networks: An interpretation of the Library Bill of Rights,* http://ala1.ala.org:70/0/alagophx/
alagophxfreedom/electacc.fin

blocked for reasons of profit. (Even one university temporarily installed blocking software in its computer labs, in aid of a policy "prohibit[ing] the display in public labs of pornographic material unrelated to educational programs."[64]) The result may be that large amounts of content may end up off-limits to a substantial fraction of the adult population.

There are limits to this—sex sells. Many home Internet users will be loathe to cut themselves off from the full range of available speech. Most online services and Internet access providers, while attempting to make parents feel secure about their children's exposure to sexually explicit material on the Net, will still host such material for adults who wish to view it. It seems safe to conclude, though, that blocking software will have the practical effect of restricting the access of a substantial number of adults.

CONCLUSION

Across the world, governments and industry are turning to filtering software as the answer to the problem of sexually explicit material on the Internet. In the United Kingdom, service providers and police have endorsed a proposal recommending that Internet service providers require users to rate their own Web pages, and that the providers remove Web pages that their creators have "persistently and deliberately misrated."[65] The European Commission has urged the adoption of similar codes of conduct to ensure "systematic self-rating of content" by all European content providers.[66] Some U.S. companies have been leaning the same way: Compuserve has announced that it will "encourage" its users and other content providers to self-rate using RSACi.[67]

---

62. I do not want to overplay this point. Many libraries have decided not to install blocking software. See, e.g., *Ann Arbor [Michigan] District Library Internet use policy* (on file with author). Other libraries have installed the software only on terminals intended for use by children. I am grateful to Linda Mielke, President of the Public Library Association and Director of the Carroll County (Maryland) Public Library, Kathleen Reif, Director of the Wicomoco County (Maryland) Free Library, and Naomi Weinberg, President, Board of Trustees, Peninsula Public Library (Lawrence, New York) for educating me on these issues.

63. See R. Retkwa, (1996, September). Corporate censors, *Internet World*, p. 60; see also, e.g., *Microsystems announces Cyber Patrol proxy* (July 31, 1996): http://www.microsys.com/prfiles/proxy796.htm

64. The university was the University of Arkansas at Monticello. See e-mail from Carl Kadie to fight-censorship mailing list, October 22, 1996; e-mail from Tyrone Adams to amend1-L mailing list, October 22, 1996; e-mail from Stephen Smith to Jonathan Weinberg (October 22, 1996) (all on file with author).

65. *Illegal material on the Internet*, http://dtiinfo1.dti.gov.uk/safety-net/r3.htm. Under the initial proposal, users were to rate with RSACi; the proponents apparently have backed away from that. The proposal also recommends that ISPs take steps to support rating and filtering of Usenet newsgroups. For an explanation of the mechanics of the Usenet proposal, see *Turnpike news ratings*: http://www.turnpike.com/ratings

Ratings, though, come at a cost. It seems likely that a substantial number of adults, in the near future, will view the Net through filters administered by blocking software. Intermediaries—employers, libraries, and others—will gain greater control over the things these adults read and see. Sites may be stripped out of the filtered universe because of deliberate political choices on the part of ratings service administrators, and because of inaccuracies inherent in the ratings process. If a ratings service is to categorize a large number of sites, it cannot simultaneously achieve consistency and nuance; the techniques it must rely on to achieve consistency make it more difficult to capture nuance, and make it less likely that users will find the ratings useful. The necessity of excluding unrated sites may flatten speech on the Net, disproportionately excluding speech that was not created by commercial providers for a mass audience.

This is not to say that ratings are bad. The cost they impose, in return for the comforting feeling that we can avert a threat to our children, is surely much less than that imposed (say) by the Communications Decency Act. Ratings provide an impressive second-best solution. We should not fool ourselves, though, into thinking that they impose no cost at all.

ACKNOWLEDGEMENTS

A later, expanded version of this chapter was published as J. Weinberg (1997), Rating the Net, *Hastings Communications and Entertainment Law Journal 19*, 453. I am grateful to Jessica Litman, whose comments greatly improved this chapter. I am indebted to the participants in the 1996 Telecommunications Policy Research Conference, in particular Paul Resnick and Rivkah Sass, for their perspectives on this material.

---

66. See Communication to the European Parliament, the Council, the Economic and Social Committee and the Committee of the Regions (October 16, 1996). Available [Online]: http://www2.echo.lu/legal/en/internet/content/communic.html. A recent EU working group document recommends research into new rating systems, not RSACi, so as to "take account of Europe's cultural and linguistic diversity" and "guarantee respect of [users'] convictions" (*Report of working party on illegal and harmful material on the Internet*, http://www2.echo.lu/legal/en/internet/content/ wpen.html).

67. See *RSAC press release, Compuserve to rate Internet content by July 1* (May 9, 1996). Available [Online]: http://www.rsac.org/press/960509-1.html

IV

TELECOMMUNICATIONS
AND POLITICS

13

Bell Had A Hammer: Using the First Amendment to Beat Down Entry Barriers

Susan Dente Ross
Washington State University

> A major First Amendment ruling in [the telecommunications] area...has the potential fundamentally to transform this area of the law in a single stroke.[1]

In 1992, the application of the First Amendment to telecommunications seemed revolutionary. For most of the preceding century, the telephone had been regulated like the telegraph, or the trucking industry, or any carrier of others' goods. The dominant constitutional paradigm was oversight of interstate commerce, not protection of free speech.[2] The telephone was regulated as a vehicle of interpersonal commercial trade, not as a medium of mass communication, and certainly not as a member of "the press" protected by the First Amendment.

But by 1996, that vehicle had reached a crossroads. Telephone companies had assaulted regulation that confined them to serve as pure vehicles with a barrage of lawsuits claiming a First Amendment right to provide content as well.[3] Telephone companies chaffed at the restricted role of transporter and moved to embrace a dual function as both content supplier and carrier.[4] To effect this shift in status, telephone companies claimed they were being unconstitutionally deprived of their right to speak by regulations the government claimed merely constrained the economic structure of the communications industry.[5]

Telephone company First Amendment arguments arose en masse virtually overnight[6] and followed closely on a variety of market changes that bade ill for the

---

1. M. K. Kellogg, J. Thorne, & P. W. Huber. (1992). *Federal Telecommunications Law, 719.*
2. See Interstate Commerce Act of 1887, ch. 104, 24 Stat. 379, 49 U.S.C. §§ 10101-11917 (1988), and the 1910 Mann-Elkins Act, 36 Stat. 539 (1910) (establishing federal regulatory principles on common carriers and applying them to railroads).

continued growth of traditional telephone company services and revenues, and on technological developments that enabled telephone readily to transport video programming.[7] Assertions of First Amendment rights seemed calculated to expand the economic market of telephone companies. The message telephone companies wanted to speak to their network of customers was lucrative cable video.[8]

An early effort to analyze telephone company First Amendment assertions, this research examines the historical context of the recent federal court rulings that prompted both the Federal Communications Commission (FCC) and Congress to further expand the ability of telephone companies to supply video and to control the messages they send over their wires. The court decisions demonstrate that the courts based rapid rulings on economic considerations but framed their holdings on First Amendment grounds. The courts' incompatible fact-finding and reasoning exacerbate problems of First Amendment jurisprudence. At the same time, the court decisions pushed policymakers and Congress to proceed rapidly with deregulation without careful scrutiny.

---

3. *Chesapeake and Potomac Tel. Co. v. United States*, 830 F. Supp. 909 (E.D. Va. 1993), aff'd, *United States v. Chesapeake and Potomac Tel. Co.*, 42 F.3d 181 (4th Cir. 1994), cert. granted, 115 S. Ct. 2608 (1995)[hereinafter *C&P*]; *BellSouth Corp. v. United States*, 868 F. Supp. 1335 (N.D. Ala. 1994); *Ameritech Corp. v. United States*, 867 F. Supp. 721 (N.D. Ill. 1994); *US West v. United States*, 855 F. Supp. 1184 (W.D. Wash. 1994), aff'd, 48 F.3d 1092 (9th Cir. 1994); *NYNEX Corp. v. United States*, Civil No. 93-323-P-V (D. Me Dec. 8, 1994); and *Pacific Telesis Group v. United States*, 48 F.3d 1106 (9th Cir. 1994). See also USTA wins summary judgment; U.S. Dist. Court says all telcos can offer video programming, *Comm. Daily*, January 30, 1995, p. 1; USTA, OPATSCO, NCTA win lawsuit to lift cable-phone ownership ban, *Bureau of National Affairs, Daily Report for Executives*, January 30, 1995, p. A19; Bell Atlantic files brief, *Comm. Daily*, October 20, 1995, p. 2.

4. See generally, D. Brenner. (1988). Cable television and the freedom of expression, *Duke Law Journal*, 329 (concluding that the First Amendment does not prohibit exclusive franchising or access requirements imposed on cable).

5. See generally, J. W. Emord. (1989). The First Amendment invalidity of FCC ownership regulations. *Catholic University Law Review, 38*; E. M. Reilly. (1994). The telecommunications industry in 1993: The year of the merger, *CommLaw Consp., 2*, 95; and E. T. Werner. (1991). Something's gotta give: Antitrust consequences of telephone companies' entry into cable television. *Fed. Com. Law Journal, 43*, 215.

6. But see B. W. Rein, J. C. Quale, J. R. Bayes, & J. S. Logan. (1982). The constitutionality of the FCC's television-cable cross-ownership restrictions. *Fed. Com. Law Journal, 34*, 1 (summarizing an earlier petition for rulemaking to the FCC on behalf of a television station and asking for elimination of the ban on television/cable cross-ownership as an unconstitutional restriction of television companies' First Amendment rights).

7. See generally, A. C. Barrett. (1993). Shifting foundations: The regulation of telecommunications in an era of change, *Fed. Com. Law Journal, 46*, 39 (providing an overview of the logic and utility of current telephone-cable alliances).

These cases suggest the susceptibility of the court system to the power and influence of businesses seeking to advance their rational economic interests. Expediency led telephone companies to assert First Amendment claims, and the courts and policymakers bowed to those claims without demanding sufficient and appropriate evidence to justify revolutionary change in the application of the First Amendment to common carriers.[9] These court decisions empowered telephone companies to batter down economic regulations with the First Amendment, muddied First Amendment jurisprudence, increased uncertainty about the regulation of changing telecommunications markets, and undermined clear policy deliberation and fact-finding.[10]

Although the Supreme Court chose not to address the question of the extent of a telephone common carrier's First Amendment rights,[11] lower court rulings eliminate a regulatory distinction between speakers and carriers and may open the door to more intrusive carrier regulation of all speakers. Based on speculative conclusions about predicted economic failures in the video marketplace, such rulings also establish precedents for free speech arguments to circumvent the empirical requirements imposed on economic challenges to structural regulations.[12] At the very least, these decisions reinforce and expand the application of intermediate, rather than strict, scrutiny to regulation of speech transmitted through new technologies.[13]

---

8. The Supreme Court has ruled that video programming is protected speech. See e.g., *Los Angeles v. Preferred Communications, Inc.*, 476 U.S. 488, 494 (1986) (comparing cable communications with traditional First Amendment speakers: newspapers, books, pamphlets, and public speakers); *Leathers v. Medlock*, 111 S. Ct. 1438, 1442 (1991) ("Cable...is engaged in 'speech' under the First Amendment"). See generally L. Levy. (1985). *Emergence of a Free Press* (containing a thoroughly documented discussion of the original meaning of the First Amendment and the conclusion that the amendment always allowed regulation of some forms of speech). But see *Telephone company-cable television cross-ownership rules*, CC Docket No. 87-266, First Report and Order, and Second Further Notice of Inquiry, 7 F.C.C.R. 300 (1991); Memorandum Opinion and Order on Reconsideration, 7 F.C.C.R. 5069 (1992). (Despite FCC disclaimers, the video programming distributed by telephone companies is indistinguishable from, and in some cases identical to, that distributed by cable operators.)

9. In re Telephone Company-Cable Television Cross-Ownership Rules, Fourth Further Notice of Proposed Rulemaking, CC Docket No. 87-266 (FCC No. 95-20) (January 12, 1995). See also R. M. Pepper. (1986). *Through the looking glass: Integrated broadband networks, regulatory policies, and institutional change*, 4 F.C.C.R. 1306 (1988) (presenting an insightful overview of the regulatory issues posed by new and merging telecommunications technologies).

10. M. D. Director and M. Botein. (1994). Consolidation, coordination, competition, and coherence: In search of a forward looking communications policy. *Fed. Com. Law Journal*, *39*, 229.

11. *United States v. Chesapeake & Potomac Tel. Co.*, 1996 U.S. LEXIS 1551; 64 U.S.L.W. 4115 (February 27, 1996).

This chapter begins with an examination of the history of the cross-ownership ban on video programming and repeated telephone company challenges to the ban on the basis of alleged changes in the video marketplace. A summary of the seminal First Amendment challenge to the Cable Act ban is followed by an overview of the wave of First Amendment rulings subsequently handed down by federal courts. The subsequent section places the Cable Act cases into the historical context of sparse First Amendment case law involving telephone common carriers, of economic regulation of telephone common carriers, and of the well-established trifurcated First Amendment jurisprudence of the media. For further context, the author then identifies a number of non-Constitutional challenges to structural regulation of telephone companies that arose as the Cable Act cases made their way through the courts. A discussion of the implications of the First Amendment rulings precedes some brief concluding remarks.

THE CROSS-OWNERSHIP BAN ON VIDEO PROGRAMMING

One focus of telephone company efforts to expand their markets during the 1990s consisted of First Amendment assaults[14] against the 1984 Cable Act's video programming ban, which prohibited dominant local telephone companies from offering cable television in their telephone service area.[15] The Cable Act ban essentially codified a rule first adopted by the FCC in 1970 out of fear that huge, powerful telephone companies would dominate the then-fledgling cable industry.[16] The FCC adopted the rule to eliminate both the opportunity and the incentive

---

12. But see *Minneapolis Star & Tribune Co. v. Minnesota Comm'r of Revenue*, 460 U.S. 575, 585 (1983)(holding that regulatory distinctions between media are presumptively invalid unless justified by "some special characteristic"); *Turner Broadcasting Sys. v. United States*, 114 S. Ct. 2445, 2458 (1994) (holding that speaker or medium partial regulations are always subject to heightened scrutiny).

13. See, e.g., *Turner Broadcasting Sys.*, 114 S. Ct. at 2467. Intermediate scrutiny as applied in *United States v. O'Brien*, 391 U.S. 367 (1968) has been established as the appropriate standard of review for regulation of cable. When compared to strict scrutiny, the intermediate O'Brien test requires that regulation (a) address an "important or substantial," rather than a compelling, government interest, and (b) be narrowly tailored, rather than the least intrusive, means of achieving that interest. See *Ward v. Rock Against Racism*, 491 U.S.781, 799 (1989) (rearticulating the second *O'Brien* prong to require only that regulation does not "burden substantially more speech than is necessary to further the government's legitimate interests"). Some justices argued, however, that speech-related regulation of cable must be assessed under strict scrutiny standards. 114 S. Ct. at 2476 (O'Connor, J., dissenting). See also *Central Hudson Gas & Elec. Corp. v. Public Service Com.*, 447 U.S. 557, 575-577 (1980) (Blackmun, J., concurring) (warning against the expanded application of the intermediate standard); and p. 584 (Rehnquist, J., dissenting) (arguing that state created monopolies enjoy no First Amendment protection).

14. *C&P et al.*, cited in note 3.

for telephone companies to discriminate against independent cable video operators in favor of their own video affiliates.[17]

Telephone companies began challenging the video programming ban virtually
from its inception,[18] but the ban survived more than a decade of FCC and congressional scrutiny,[19] including a formal FCC request that Congress eliminate the
ban.[20] In 1992, the FCC asked Congress to repeal the ban as a means to "promote
[the Commission's] overarching goals...by increasing competition in the video
marketplace, spurring the investment necessary to deploy an advanced infrastruc-

---

15. 47 U.S.C. § 533(b)(1)(2) (1988). Section 533(b) provides that:
   (1) It shall be unlawful for any common carrier...to provide video programming di-
       rectly to subscribers in its telephone service area, either directly or indirectly
       through an affiliate owned by, operated by, controlled by, or under common control
       with the common carrier.
   (2) It shall be unlawful for any common carrier...to provide channels of communica-
       tions or pole line conduit space, or other rental arrangements, to any entity which
       is directly or indirectly owned by, operated by, controlled by, or under common
       control with such common carrier, if such facilities or arrangements are to be used
       for, or in connection with, the provision of video programming directly to sub-
       scribers in the telephone service area of the common carrier.
In addition, the 1984 Cable Act defines a cable system as a "set of closed transmission
paths" that transmit, receive and control the signal "designed to provide cable service which
includes video programming" [47 U.S.C. § 522(6) (1988)]. The act's definition of cable excludes any "facility of a common carrier which is subject, in whole or in part, to the provisions of subchapter II of this chapter" unless the common carrier is providing "video
programming directly to subscribers" [47 U.S.C. § 522(6)]. A cable operator is "any person
or group of persons...who otherwise controls or is responsible for, through any arrangement, the management and operation of such a cable system" [42 U.S.C. § 522(4)(B)]. A
cable service, under the act, is "the one-way transmission to subscribers of (i) video programming, or (ii) other programming service" [42 U.S.C. § 522(5)].

16. In re Applications of Telephone Companies for Section 214 Certificates for Channel
Facilities Furnished to Affiliated Community Antenna Television Systems (Memorandum
Opinion and Order), 22 F.C.C.2d 746 (1970).

17. As cited in note 17. (The 1978 Pole Attachments Act codified a portion of the FCC
rule and prevented discriminatory pricing for use of telephone poles).

18. See *United States v. Western Elec. Co.*, 592 F. Supp. 846, 850-51 (D.D.C. 1984).
Telephone company waiver requests to the Modified Final Judgment's line-of-business restrictions began January 26, 1984, with a request from Bell Atlantic to lease equipment.
That was followed by a January 27, 1984, BellSouth motion to offer software programs and
related services; a February 8, 1984, Pacific and Nevada Bell motion to enter into foreign
businesses; a February 15, 1984, NYNEX request to provide office equipment; a February
24, 1984, BellSouth request to provide communications services and equipment to NASA;
a March 20, 1984, US West request to engage in real estate transactions. Also see citation
in note 3, p. 850.

ture, and increasing the diversity of services made available to the public."[21] Congress failed to act.

Evidence suggests that from the beginning neither the FCC nor Congress viewed the ban, or its elimination, from a First Amendment perspective. The dominant issue was one of promoting competition and increasing regulatory efficacy in a dynamic communications environment.[22] In the 1980s and 1990s, support for the ban diminished because of a new attitude at the FCC. Renewed faith in the ability of competition to efficiently achieve public policy objectives[23] undermined established doctrine that the "exceptional grant of power to private enterprises justifies extensive oversight on the part of the state to protect the ratepayers from exploitation of the monopoly power."[24]

The policy debate between unfettered competition and control of monopolies masked a similar dialectic tension between free speech and the economic objectives of telephone/cable regulation. As Columbia University economist Eli Noam noted, "Common carriage is not only a free speech matter. The reason for common car-

---

19. FCC Office of Plans and Policy: FCC Policy on Cable Ownership, A Staff Report, 1981, at 162. See also, Commissions' Rules Concerning Carriage of Television Broadcast Signals, 1 F.C.C.R. 864 (1986); Telephone Company-Television Cross-Ownership Rules, 2 F.C.C.R. 5092 (1987); Telephone Company-Cable Television Cross-Ownership Rules, 3 F.C.C.R. 5849 (1988); In re Telephone Company-Cable Television Cross-Ownership Rules, 7 F.C.C.R. 5781 (1992) (Second Report and Order, Recommendation to Congress, and Second Further Notice of Proposed Rulemaking). Also see, e.g., Competitive Issues in the Cable Television Industry: Hearings Before the Subcomm. on Antitrust, Monopolies and Business Rights of the Senate Comm. on the Judiciary, 100th Cong., 2d Sess. (March 17, 1988); Cable Television Regulation: Hearings Before the Subcomm. on Telecommunications and Finance of the House Comm. on Energy and Commerce, (Parts 1 and 2), 101st Cong., 2d Sess. (March 1, April 19, and May 9, 1990); Cable Instructional TV and S. 1200 Communications Competitiveness and Infrastructure Modernization Act of 1991: Hearings Before the Subcomm. on Communications of the Senate Comm. on Commerce, Science and Transportation, 102d Cong., 2d Sess. (February 28, 1992); and Effects of Telecommunications Mergers: Hearing before the Subcomm. on Antitrust, Monopolies and Business Rights of the Senate Comm. on the Judiciary, 103d Cong., 1st Sess. (October 27, 1993). See also Cable Television Consumer Protection and Competition Act of 1992, Pub. L. No. 102-385, § 2, 106 Stat. 1460 (1992) (reforming telecommunications regulation but leaving intact the telephone/cable cross-ownership ban); and proposed telecommunications reform bills, see H.R. 2437, 101st Cong., 1st Sess. (1989); H.R. 1504, 103d Cong., 1st Sess. (1993); S. 1200, 102d Cong., 1st Sess. (1991); and S. 1068, 101st Cong., 1st Sess. (1989).

20. In re Telephone Company-Cable Television Cross-Ownership Rules, 7 F.C.C.R. 5781, 5784 (1992) (Second Report and Order, Recommendation to Congress, and Second Further Notice of Proposed Rulemaking).

21. As cited in note 20, p. 5847. For Department of Justice support of the repeal, see Reply Comments of the U.S. Department of Justice, In re Telephone Company-Cable Television Cross-Ownership Rules, CC Docket No. 87-266, p. 44.

riage, whether in transportation or communication, is generally to reduce transaction costs in the use of infrastructure and hence to benefit its development."[25] As early as 1973, the FCC noted the dual role of the telephone/cable cross-ownership rules, saying they were intended to foster both "increased competition in the economic marketplace" and "increased competition in the marketplace of ideas."[26]

Fifteen years later, and four years after Congress codified the ban, a working paper from the FCC Office of Plans and Policy suggested potential conflict between

---

22. See, e.g., S. Rep. No. 92, 102d Cong. (1991) (noting 11 hearings on cable television between 1989 and 1991); Oversight of Cable TV: Hearings Before the Subcomm. on Communications of the Senate Comm. on Commerce, Science and Transportation, 101st Cong., 1st Sess. (November 16 & 17, 1989) (examining competition in the video programming industry); Cable Television Consumer Protection Act: Hearings Before the Subcomm. on Telecommunications and Finance of the House Comm. on Energy and Commerce, 101st Cong., 2d Sess. (March 29 & April 4, 1990) (exploring the ineffectiveness of cable rate regulation in response to consumer complaints of gouging); Cable Television Consumer Protection and Competition Act of 1990: Conference Report to Accompany H.R. 5267 of the Committee on Energy and Commerce, 101 Cong., 2d Sess., H.R. Conf. Rep. No. 682, 101st Cong., 2d Sess. (1990) (favoring amendment of the Communications Act of 1934 to increase consumer protection and industry competition); Cable Television Consumer Protection and Competition Act of 1992: Hearings Before the Subcomm. on Telecommunications and Finance of the House Comm. on Energy and Commerce, H. Rep. No. 628, 102d Cong., 2d Sess. (1992); The Communications Act of 1994: Hearings on S. 1822 Before the Senate Comm. on Commerce, Science and Transportation, H. Rep. No. 367, 103d Cong., 2d Sess., (1994) (favoring enactment of a bill to foster development of the nation's telecommunications infrastructure). See also Cable Competition Act, S. 1068, 101st Cong., 1st Sess. (1989); Cable Consumer Protection Act, S. 905, 101st Cong., 1st Sess. (1989); Cable Television Consumer Protection Act of 1990, S. 1880 (1990); Cable Television Consumer Protection Act of 1991, S. 12 (1991); S. 1200, 102d Cong., 1st Sess. (1991) (permitting telephone provision of cable service and video programming); Communications Act of 1994, S. 1822, 103d Cong., 2d Sess. (1994) (deregulating telecommunications and increasing competition and investment); S. 2111, 103d Cong., 2d Sess. (1994) (deregulating telecommunications and encouraging development of a national infrastructure); S. 1883, 103d Cong., 2d Sess. (1994) (allocating funds to promote and develop telecommunications infrastructure); Communications Competitiveness and Infrastructure Modernization Act of 1991, H.R. 2546, 102d Cong., 1st Sess. (1991); H.R. 3626, 103d Cong., 2d Sess. (1994) (superseding the Modified Final Judgment and broadly amending the Communications Act of 1934).

23. R. Hundt. (1994)., Toward regulation that fosters competition. *Fed. Com. Law Journal, 39,* 265. See also E. M. Noam, Principles for the Communications Act of 2034: The superstructure of infrastructure. *Fed. Com. Law Journal, 39,* 317 (arguing that competition between common and private carriers distorts the market and is unstable).

24. A. Kahn. (1971). *The Economics of Regulation* (p. 113–171).

25. E. M. Noam. (1994). Principles for the Communications Act of 2034: The superstructure of infrastructure. *Fed. Com. Law Journal, 39,* 317, 320.

common carrier regulation and the First Amendment.[27] Robert Pepper asked, in part, whether First Amendment protection of cable would invalidate regulatory constraint of telephone company broadband networks if telephone companies provided cable-like content through their lines.[28] Pepper presciently questioned whether the potential that common carrier regulations would become unconstitutional when a carrier provided content should foreclose telephone company entry into content.[29]

Although Congress failed to pass reform legislation or to repeal the Cable Act telephone/cable cross-ownership ban until 1996, other policy arenas forged policies that effectively eliminated structural barriers between telephone and cable.[30] The FCC proceeded to develop a policy called video dial tone to permit a limited role for telephone companies to provide video.[31] At the same time, courts eliminated barriers to local telephone provision of cable imposed by the 1982 Modified Final Judgment,[32] and telephone companies overturned the Cable Act ban on First Amendment grounds.[33] As FCC Commissioner Andrew Barrett noted, "the courts [were taking] the lead in rearranging the telecommunications industry."[34]

---

26. Memorandum Opinion and Order in Docket No. 18397, 39 F.C.C.2d 377, 391 (1973).

27. R. M. Pepper. (1988). *FCC Office of Plans and Policy, Through the looking glass: Integrated broadband networks, regulatory policies, and institutional change.* 4 F.C.C.R. 1306.

28. As cited in note 27

29. As cited in note 27.

30. See H.R. 3636, 103d Cong., 2nd Sess. (1994) (approving immediate sweeping communications deregulation); 103 H. Rep. 560, 103d Cong., 2nd Sess., National Communications Competition and Information Infrastructure Act of 1994 (finding that diversity in telecommunications is best advanced through unfettered market operation). But see 103 S. Rep. 367, 103d Cong., 2nd Sess., Communications Act of 1994 (September 14, 1994) (urging incremental changes and retention of local and national cross-ownership rules to protect diversity in the telephone industry).

31. See Reporting Requirements on Video Dialtone Costs and Jurisdictional Separations for Local Exchange Carriers Offering Video Dialtone Services, Memorandum Opinion and Order, 1995 FCC LEXIS 6460, September 29, 1995; Reporting Requirements on Video Dialtone Costs and Jurisdictional Separations for Local Exchange Carriers Offering Video Dialtone Services, Order Inviting Comments, *Fed. Reg., 60,* 35548 (June 23, 1995). See also Telephone Company/Cable Television Cross-Ownership Rules, Sec. 63.54–63.58, Further Notice of Proposed Rulemaking, First Report and Order and Second Further Notice of Inquiry, 7 F.C.C.R. 300 (1991), recon. 7 F.C.C.R. 5069 (1992), aff'd, National Cable Television Asso. v. FCC, 33 F.3d 66 (D.C. Cir. 1994); Telephone Company/Cable Television Cross-Ownership Rules, Sec. 63.54-63.58, Second Report and Order, Recommendation to Congress, and Second Further Notice of Proposed Rulemaking, 7 F.C.C.R. 5781 (1992), aff'd, Memorandum Opinion and Order on Reconsideration and Third Further Notice of Proposed Rulemaking, 10 F.C.C.R. 244 (1994), appeal pending sub nom., *Mankato Citizens Tel. Co. v. FCC,* No. 92-1404 (D.C. Cir., filed 1992).

THE SEMINAL CASE

The initial constitutional challenge to the Cable Act's ban on video programming came when Chesapeake and Potomac Telephone Co. of Virginia (hereafter C&P Telephone) brought suit in December 1992 against Alexandria, VA. The city had cited the Cable Act's cross-ownership ban as grounds for its denial of the telephone company's request for a cable franchise to provide a competitive cable video system.[35] C&P Telephone challenged the ban as an unconstitutional denial of its right to free speech.[36]

Both the federal district and the circuit courts ruled that the so-called cross-ownership ban unconstitutionally restricted C&P Telephone's First Amendment right to free speech.[37] Both courts subjected the ban to the intermediate scrutiny test articulated in *United States v. O'Brien*[38] and found the ban failed to overcome *O'Brien*'s requirements that content-neutral regulations of speech[39] (a) further an important or substantial governmental interest, (b) that the interest be unrelated to the limitation of expression of views, and (c) that the incidental limitation of free expression be no greater than necessary to achieve the governmental interest.[40] Both courts accepted the government's interest as important and unrelated to the content of speech, and found the ban failed the third prong because it was unconstitutionally overbroad.

A RISING TIDE OF COURT DECISIONS

By 1995, an array of federal courts had ruled on telephone companies' facial and as-applied constitutional challenges to the statutory ban on telephone/cable cross-

---

32. *United States v. Western Elec. Co.*, 767 F. Supp. 308 (D.D.C. 1991), stay vacated, 1991-1992 Trade Cas. (CCH) sec. 69,610 (D.C. Cir.), review denied sub nom. *American Newspaper Publishers Asso. v. United States*, 112 S. Ct. 366 (1991) (lifting the ban and permitting the regional Bell companies to offer cable and other information services).

33. *C&P*, cited in note 3. Some speculate that C&P was the regional Bell to initiate this series of parallel lawsuits because it could bring suit in the 4th Circuit "rocket docket," known for the speed with which it renders judgments.

34. A. C. Barrett. (1993). Shifting foundations: The regulation of telecommunications in an era of change. *Fed. Com. Law Journal, 46,* 39, 53.

35. *C&P*, cited in note 3.

36. *C&P*, cited in note 3. (citing Cable Communications Policy Act of 1984 Pub. L. No. 98-549, § 533(b)).

37. *C&P*, cited in note 3.

38. 391 U.S. at 377.

39. See *Turner Broadcasting Sys.*, 114 S.Ct. 2445 (rejecting the premise that speech regulations that favor a particular group or speaker are necessarily content-based and defining as content-neutral all speech regulations not directed to advance or suppress a specific point of view).

40. 391 U.S. at 377.

ownership. All held the ban unconstitutional.[41] The federal courts also concurred in applying intermediate scrutiny and in finding the ban fatally overbroad.

The government had argued in favor of the ban as essential to promote competitive and diverse local media ownership, and to prevent telephone company anticompetitive practices. In a representative opinion, district court Judge Sharon Lovelace Blackburn found that although these government interests might be sufficiently substantial to support regulation, the government had failed to prove that the ban advanced that goal with no greater burden on speech than necessary.[42]

Judge Blackburn refused to defer to congressional judgment about the need for the ban because of vast changes in the cable television market since the ban's 1984 enactment and because the congressional record failed to show any independent fact-finding when Congress codified the FCC's established cross-ownership ban.[43] Instead of deference, Judge Blackburn said the Supreme Court's decision in *Turner Broadcasting System, Inc. v. FCC*[44] required courts to conduct independent evaluations of the facts.[45]

Every judge ruling in the Cable Act cases, whether through summary judgment or independent evaluation of the facts, struck down the ban, which one judge called "draconian."[46] The courts focused their rulings on the overbreadth of the ban and sidestepped the thorny issue of reconciliation of common carrier regulation[47] and judicial precedent that categorically separates common carriers from content control.[48] Only the U.S. Court of Appeals for the 4th Circuit, in affirming the lower court's *C&P* ruling, mentioned the tension between established common carrier principles and the assertion of First Amendment rights by telephone companies. The appeals court first noted that strict scrutiny of speech regulation is unwarranted when differential treatment of speakers is justified by some special characteristic of the medium being regulated.[49] The court reasoned that intermediate scrutiny was applicable to *C&P* because the regulatory discrimination

---

41. *C&P* et al., cited in note 3.

42. *BellSouth Corp.*, cited in note 3.

43. *BellSouth Corp.*, cited in note 3. See also Ameritech Corp., cited in note 3, pp. 743–747 (asserting the court's duty to independently assess the necessity of the ban to achieve its stated goals).

44. *Turner Broadcasting Sys.*, 114 S. Ct. 2445.

45. See, e.g., *Home Box Office, Inc. v. FCC*, 567 F.2d 9 (D.C. Cir. 1977) (holding that rules that implicate First Amendment freedoms must be supported with factual evidence, not unsupported theory, hypothesis, or speculation).

46. *C&P*, 1993 U.S. Dist. LEXIS 11822 at *64.

47. 47 U.S.C. § 202 prevents common carriers "from exercising editorial control over the communications they transmit" (*C&P*, 42 F.3d 181).

48. *C&P*, 1994 U.S. App. LEXIS 32985 (4th Cir. 1994) at *47.

49. *C&P*, as in note 48, *46-47 (citing *Minneapolis Star and Tribune*, 460 U.S. at 581; and *Arkansas Writers' Project v. Ragland*, 481 U.S. 221, 231 [1987]).

against telephone companies as video speakers was content-neutral and was justified by the unique economic and physical characteristics of video provision.[50] The
court said regulation might be justified because cable service involves a physical
bottleneck that provides an incentive for telephone companies to monopolize the
information flowing through the bottleneck.[51] Moreover, the court said, content-
neutral laws are constitutionally suspect "only in certain circumstances."[52] Then
the court concluded:

> Although common carriers are not members of "the press" insofar as 47
> U.S.C. § 202 precludes them from exercising editorial control over the
> communications they transmit, the foregoing would nevertheless seem
> applicable to Section 533 (b), which restricts a class of speakers from
> joining the press by operating, with editorial control and within certain
> areas, cable systems.[53]

This lone attempt to reconcile Title II regulation and editorial control is extremely muddled. It appears to say that although Title II does not define telephone companies as protected speakers and, indeed, proscribes their exercise
of editorial control over the messages they carry, the Cable Act ban nevertheless restricts this "class of speakers" from exercising certain types of editorial
control. The court ruled this restriction unconstitutional.[54] The remainder of
the First Amendment rulings in favor of telephone companies preferred to
avoid this imbroglio altogether, and none of the rulings confronted precedent
upholding the constitutionality of similar cross-ownership bans applied to
clearly established First Amendment speakers.[55]

THE ABSENCE OF PRECEDENT

The Sparse History of Telco First Amendment Claims

Although the FCC had spent years attempting to balance First Amendment and
common carrier doctrines, references to FCC debate or even more generally to
common carrier principles were notably absent from federal court decisions affirming telephone companies' First Amendment right to provide video telephony.[56] This omission eliminated a critical thread of policy because prior to the mid-
1990s almost no common law precedent existed to support the assertion of editorial control by a common carrier. Indeed, prior to the 1993 ruling of the U.S. Dis-

---

50. *C&P*, as in note 48, at *48.
51. *C&P*, as in note 48, at *43. See also at *38, *39.
52. *C&P*, as in note 48, at *46 (citing *Leathers v. Medlock*, 449 U.S. at 444).
53. *C&P*, as in note 48, at *47.
54. *C&P*, as in note 48, at *47.
55. *C&P et al.*, cited in note 3. See also notes 61–65 and accompanying text.

trict Court in *C&P*, few courts had ever been asked to consider the extent of First Amendment protection enjoyed by a traditional common carrier when it also functioned in part as a private speaker.[57]

A handful of cases and FCC rulings suggests that a First Amendment speaker may function in part as a common carrier.[58] However, neither the courts nor the FCC had explored the implications of the reverse: allowing a regulated common carrier to assert autonomous First Amendment control over a portion of its capacity.[59]

In general, courts have attempted to avoid ruling on the constitutionality of structural regulations imposed on communications industries and to rest holdings on statutory grounds whenever possible.[60] However, in 1977, the D.C. Circuit Court ruled in *National Citizens Committee for Broadcasting v. FCC*[61] that the newspaper/broadcast cross-ownership ban, similar to that imposed on telephone

---

56. Policy and Rules Concerning Rates for Competitive Common Carrier Services and Facilities, Notice of Inquiry, 77 F.C.C.2d 308 (1979); Policy and Rules Concerning Rates for Competitive Common Carrier Services and Facilities, First Report and Order, 85 F.C.C.2d 1 (1980); Policy and Rules Concerning Rates for Competitive Common Carrier Services and Facilities, Further Notice of Proposed Rulemaking, 84 F.C.C.2d 445 (1981); Policy and Rules Concerning Rates for Competitive Common Carrier Services and Facilities, Second Report and Order, 91 F.C.C.2d 59 (1982); Policy and Rules Concerning Rates for Competitive Common Carrier Services and Facilities, Third Report and Order, 48 Fed. Reg. 46,791 (1983); Policy and Rules Concerning Rates for Competitive Common Carrier Services and Facilities, Fourth Report and Order, 95 F.C.C.2d 554 (1983); Policy and Rules Concerning Rates for Competitive Common Carrier Services and Facilities, Fifth Report and Order, 98 F.C.C.2d 1191 (1984); Policy and Rules Concerning Rates for Competitive Common Carrier Services and Facilities, Sixth Report and Order, 99 F.C.C.2d 1020, overturned sub nom., *MCI Telecommunications v. FCC*, 765 F.2d 1186 (D.C. Cir. 1985).

57. *C&P*, cited in note 3. See also Northwestern Indiana Tel. Co. v. FCC, 872 F.2d 465 (D.C. Cir. 1989), cert. denied, 493 U.S. 1035 (1990) (declining to address an issue raised on appeal by telephone companies to have the cross-ownership ban declared unconstitutional); BellSouth et al., cited in note 3.

58. See, e.g., 46 F.C.C.2d at 763-764; 51 F.C.C.2d at 967-959; FCC First Report and Order in Docket No. 18397, 20 F.C.C.2d 201, 207 (1969) (holding that designation as a speaker and as a common carrier are not mutually exclusive). See also *United States v. Southwestern Cable Co.*, 392 U.S. 157, 172-173(1968) (holding that cable partakes of "characteristics both of broadcasting and of common carriers but with all of the characteristics of neither"); and *Frontier Broadcasting Co. v. FCC*, 24 F.C.C. 251, 254 (1958) (holding that one-way cable services are not engaged in common carriage because the content is not under the control of the subscriber); but see *National Asso. of Regulatory Utility Comm'rs v. FCC*, 533 F.2d 601, 610-611 (1976), hereafter *NARUC II* (holding that two-way cable systems are common carriers if customers have explicit or implicit discretion over content).

59. See, e.g., In re Telephone Company-Cable Television Cross-Ownership Rules, Fourth Further Notice of Proposed Rulemaking, CC Docket No. 87-266 (FCC No. 95-20) (January 12, 1995).

companies by the Cable Act, did not violate the newspaper's First Amendment rights because it "neither mandates nor prohibits what may be published...[and] is an attempt to enhance the diversity of information heard by the public."[62] The court called a constitutional challenge to the rule "ironic,"[63] and in dicta that may prove prescient in the case of telephone company assertions of First Amendment rights, warned that regulated separation of media protected the full editorial autonomy of newspapers.[64]

> It may be that newspapers cannot truly be free of government interference so long as they operate government licensed broadcast stations. An unsavory fact of life is that government has the power to regulate expression by a "raised eyebrow" reminding the broadcaster of the triennial government renewal process. A newspaper opens itself up to similar intimidation by affiliation with a broadcast station.[65]

This language suggests that at least one judge believed full First Amendment rights were best protected by identifying and protecting only pure speakers, as distinguished from broadcasters or other electronic media. In this theory, the extension of First Amendment protection to nonpure speakers, such as telephone companies, erodes the unequivocal nature of the protection of free speech. This interpretation suggests one of several reasons why courts generally had shunned First Amendment challenges raised by common carriers.[66]

In another case, the D.C. Circuit Court, venue of many telephone company regulatory challenges, suggested that it legally was "constrained to turn a deaf ear to these [First Amendment] complaints" of telephone companies.[67] In a rather typical response, district court Judge Harold Greene said a 1987 First Amendment challenge to restrictions of the Modified Final Judgment[68] was without

---

60. See, e.g., *Northwestern Indiana Tel. Co. v. FCC*, 824 F.2d 1205 (D.C. Cir. 1987); and *Northwestern Indiana Tel. Co. v. FCC*, 872 F.2d 465 (D.C. Cir. 1989). See also *General Tel. Co. of the Southwest v. United States*, 449 F.2d 846 (5th Cir. 1971) (affirming against a due process challenge FCC requirements that banned telephone companies from providing cable antenna television services unless they first offered independent CATV operators access to carriers' telephone poles).

61. 555 F.2d 938 (D.C. Cir. 1977).

62. As cited in note 61, at 954–955. Note that while apt in many respects, this case is distinguishable at least on the grounds that broadcast precedents apply uniquely to that medium characterized by spectrum scarcity. See *Turner Broadcasting Sys.*, 114 S. Ct. at 2457.

63. As cited in note 61, at 955.

64. As cited in note 61, at 955.

65. As cited in note 61, at 955.

66. See, e.g., *United States v. Western Elec. Co.*, 673 F. Supp. 525, 585–586 (D.D.C. 1987); and *United States v. Western Elec. Co.*, 774 F. Supp. 11, 12 n. 2 (D.D.C. 1991) (noting that the First Amendment "argument adds nothing to the Regional Companies' claim of injury").

67. *United States v. Western Elec. Co.*, 846 F.2d 1422 (D.C. Cir. 1988).

68. 552 F. Supp. 131 (1982).

merit:[69] "These [telephone] companies, which have never been publishers, thus cannot bootstrap their own failure to make the showing necessary for the relief of their obligations under an antitrust decree into an infringement of their First Amendment rights."[70]

The district court cited *FCC v. Midwest Video Corp.*[71] and *Columbia Broadcasting System, Inc. v. Democratic National Committee*[72] as establishing the principle that "common carriers are quite properly treated differently for First Amendment purposes than traditional media."[73] Both cases, however, are readily distinguishable from the telephone company First Amendment challenges to structural regulations and court decrees. In *Midwest Video*, the Supreme Court held that the FCC could not impose common carrier-type access requirements on cable operators because "Congress has restricted the Commission's ability to advance objectives associated with public access at the expense of...journalistic freedom."[74] Similarly, in *CBS*, the Court affirmed that broadcasters enjoyed First Amendment protection and could not be required to accept paid editorial announcements.[75]

The cases cited by Judge Greene to establish that speech protection may not be extended to common carriers instead represent the principle that common carrier regulation may not be imposed on speakers. Indeed, a portion of *Midwest Video* more relevant to telephone company First Amendment claims may be found in the concurrence[76] in which Justice William Brennan argues for "a comprehensive re-examination of the statutory scheme as it relates to...new development, so that the basic [communications] policies are considered by Congress and not left entirely to the Commission and the courts."[77]

A case more analogous to telephone company First Amendment challenges to the Cable Act ban was presented in *Central Hudson Gas & Electric Corp. v. Public Service Commission of New York*.[78] The question before the *Central Hudson* court was whether "a state-created monopoly, which is the subject of a comprehensive regulatory scheme, is entitled to protection under the First Amendment." The Supreme Court answered yes, and ruled that the utility had a constitutional right to promote its services through advertising.[79]

69. *Western Elec. Co.*, 673 F. Supp. 525.
70. *Western Elec. Co.*, 673 F. Supp. 586.
71. 440 U.S. 689 (1979).
72. 412 U.S. 94 (1973).
73. *Western Elec. Co.*, 673 F. Supp. 586.
74. 440 U.S. at 707.
75. As cited in note 70.
76. *United States v. Midwest Video Corp.*, 406 U.S. 649, 676 (1972).
77. As cited in note 76.
78. See note 14 (ruling 8 to 1 that a ban on promotional advertising by the state's electrical utility company did not pass intermediate scrutiny and was unconstitutionally overbroad). But see the same at 583 (Rehnquist, J., dissenting).

In a lone dissent, Justice William Rehnquist disagreed strongly. Justice Rehnquist wrote: "When the source of the speech is a state-created monopoly such as this, traditional First Amendment concerns, if they come into play at all, certainly do not justify the broad interventionist role adopted by the Court today."[80] In arguing against extension of the First Amendment to the utility company, Justice Rehnquist said the Court's ruling "could invite dilution, simply by a leveling process, of the force of the Amendment's guarantee."[81]

Nonetheless, beginning in August 1993, a string of federal court rulings unanimously held that the First Amendment protected state-regulated telephone utilities from the federal Cable Act ban.[82]

The Critical but Unclear Category of Common Carriage

Although classification as a First Amendment speaker historically has been antithetical to imposition of common carrier obligations,[83] the uncertain rights and responsibilities of common carriers might have offered courts hearing the telephone cases a nexus to delineate the intersection of the two concepts. Neither statute nor common law stipulates a clear definition of common carriage or describes the specific rights and obligations of common carriers.[84] The Communications Act of 1934, which established the obligations of communications common carriers, circularly defines a common carrier as "any person engaged as a common carrier for hire."[85] The FCC offered a similarly unenlightening definition when it said a common carrier was "any person engaged in rendering communication services for hire to the public."[86]

The common law is no more helpful. In *National Association of Regulatory Utility Commissioners v. FCC*, a case cited by numerous courts struggling with common carrier doctrine, the D.C. Circuit "define[d] a common carrier firm as a firm that engages in common carriage."[87] The *NARUC* court also offered the functional definition that common carriage arises from "holding oneself out to serve the public indiscriminately."[88] Thus, subsequent court and FCC decisions focused on whether a firm affirmatively held itself out to offer nondiscriminatory service to like customers.[89]

In general, courts and regulators avoided defining "common carrier" and focused instead on determining the conditions that justified regulation. Regulators and courts held that regulatory control of common carriers was necessary to min-

79. As in note 14.
80. As cited in note 14 at 585.
81. As in note 14 at 589.
82. *C&P et al.*, as cited in note 3; see also Bell Atlantic Files Brief, *Communications Daily*, October 20, 1995, at 2 (quoting the Bell Atlantic brief which states that "every one of the 16 federal judges who has considered the [video programming ban] has concluded that it is invalid under the First Amendment").

imize disruption of public property; to assure the greatest service to the greatest number of citizens; and to control monopoly power and prevent abusive business practices.[90] Thus, it is hardly surprising that the 1934 Communications Act contains a recurrent theme that communications carriers should be regulated to serve the public interest, convenience, and necessity.[91]

Recognizing this quasi-public character, the Supreme Court historically afforded telephone common carriers First Amendment protection inferior to either print

---

83. Common carriers transport communications but do not control the content originated by other speakers. 47 U.S.C. § 201 (1988). The U.S. Code states that a "common carrier" shall be defined by common law. 18 U.S.C. § 660 (1988). Statutes, however, mandate that communications common carriers, such as telephone and telegraph, shall provide access to anyone who can pay and shall not alter the content of the senders' messages. 47 U.S.C. § 201 (1988). Also see *National Asso. of Regulatory Utility Comm'rs. v. FCC*, 525 F.2d 630 (D.C. Cir. 1976), cert. denied, 425 U.S. 992 (1976)[hereinafter *NARUC I*] (common carriers may not discriminate between two like customers) and *NARUC II*, 533 F.2d 601(common carriers do not control the content they transmit). In contrast, speakers such as newspapers do control the content they disseminate and are not required to provide access to all comers. See *Miami Herald Publ. Co. v. Tornillo*, 418 U.S. 241 (1974) (ruling it unconstitutional to require newspapers to provide a right of reply). See Burch, Common carrier communications by wire and radio: A retrospective. *Fed. Com. Law Journal*, *37*, 85, 102 (1985) (FCC raises novel issue of legal definition of common carrier); but see R. S. Homet Jr. (1985). "Getting the message": Statutory approaches to electronic information delivery and the duty of carriage, *Fed. Com. Law Journal*, *37*, 217; P. Nichols. (1987). NOTE: Redefining "common carrier": The FCC's attempt at deregulation by redefinition, *Duke Law Journal*, *501*. For recent unsuccessful First Amendment challenges by telephone companies, see, e.g., *United States v. Western Elec. Co.*, No. 87-5288 (D.C. Cir. 1990) (court's refusal to rule on First Amendment claim); and *Northwestern Indiana Tel. Co.*, as cited in note 57 (court's refusal to consider whether a ban against cable television programming by telephone companies violates the First Amendment) contra *C&P et al.*, as cited in note 3.

84. The U.S. Code states that a "common carrier" shall be defined by common law [18 U.S.C. § 660 (1988)]. Statutes, however, mandate that communication common carriers, such as telephone and telegraph, shall provide access to anyone who can pay and shall not alter the content of the senders' messages [47 U.S.C. § 201 (1988)]. See also *NARUC I*, 525 F.2d at 641(holding that common carriers may not discriminate between two like customers); hereafter *NARUC I*; and *NARUC II*, 533 F.2d 601(deciding that common carriers do not control the content they transmit).

85. 47 USC § 153 (h) (1988).

86. 47 CFR § 21.1 (1974). Historically, communications common carriers have been required to: (a) offer their services to the general public, (b) permit subscribers to control the messages they send, and (c) engage in interstate commerce. See *Midwest Video Corp.*, 440 U.S. 689; and *Frontier Broadcasting Co.*, 24 F.C.C.R. 251.

87. R. W. Poole. (1985). Unnatural Monopolies (p. 43 note) citing 525 F.2d 633 (D.C. Cir. 1976).

media or other electronic carriers.[92] With few exceptions, free speech rights to communicate over the telephone wires were the exclusive province of the individual users of the telephone,[93] and extensive telephone regulation was upheld as a reasonable means to advance the First Amendment right of telephone users to have nondiscriminatory, near-universal, interconnected service.[94] Common law holds principally that telephone common carriers must: (a) offer their services to the general public, (b) permit subscribers to control the messages they send, and (c) engage in interstate commerce.[95] To protect the citizen's right of free speech, regulation generally barred both the telephone system operator and the government from control of telecommunications content.[96]

88. 525 F.2d 630, 641 (D.C. Cir. 1976).

89. See, e.g., *NARUC I*, cited in note 83, at 641-642; *NARUC II*, cited in note 58, at 609; *Wold Communications Inc. v. FCC*, 735 F.2d 1465 (D.C. Cir. 1984).

90. See FCC Final Report and Order, 21 F.C.C.2d 307 (1970). See also *NARUC I*, cited in note 83, at 640-641. See also 49 U.S.C. §§ 301-327 (1970) and *American Trucking Asso. v. United States*, 101 F. Supp. 710 (D. Ala. 1951) (upholding trucking regulation against constitutional challenge). For communications common carriers the concept of operation in the public interest, convenience, and necessity dominated regulation. See In re Sec. 214 Applications with the FCC, e.g., 49 F.Reg. 21333 (May 21, 1984).

91. 47 U.S.C. §§ 151, 214, 310(d) (1988). See also *Munn v. Illinois*, 94 U.S. (4 Otto) 113 (1877) (sustaining legislative policy on rates for grain elevators and introducing the leitmotif of the "public interest").

92. See W. J. Baumol & J. G. Sidak. (1994). Toward competition in local telephony (pp. 18-20) (examining the economic and competitive disadvantages of current common carrier regulations).

93. Limited but notable exceptions to the autonomy of telephone callers include obscenity and harassment.

94. See, e.g., J. A. Barron. (1967). Access the press—A new First Amendment right. *Harvard Law Review, 80*, (1967) (asserting that the First Amendment legitimately advances the free speech rights of the individual citizen against the power of media owners); L. Bollinger. (1976). Freedom of the press and public access: Toward a theory of partial regulation of the mass media. *Michigan Law Review, 75*, 1.

95. See *Midwest Video Corp.*, cited in note 76; and *Frontier Broadcasting Co.*, cited in note 58. The U.S. Code states that a "common carrier" shall be defined by common law [18 U.S.C. § 660 (1988)]. Statutes, however, mandate that communication common carriers, such as telephone and telegraph, shall provide access to anyone who can pay and shall not alter the content of the senders' messages [47 U.S.C. § 201 (1988)]. See also *NARUC I*, 525 F.2d at 641; *NARUC II*, cited in note 58. See also FCC, Report and Order, Industrial Radiolocation Service, 5 FCC 2d 197 (October 5, 1966) at 202 ("the fundamental concept of a communications common carrier is that such a carrier makes a public offering to provide, for hire, facilities by wire or radio whereby all members of the public who choose to employ such facilities may communicate or transmit intelligence of their own choosing").

96. See W. J. Baumol & J. G. Sidak, cited in note 92 (examining the economic and competitive disadvantages of current common carrier regulations).

Then, in response to changing market conditions in the 1980s, the FCC attempted to redefine common carriage by focusing on the market power of a communications carrier[97] rather than on the elusive quasi-public nature of telephone service. This controversial discretionary "deregulation by redefinition" eased the regulatory burden on small or competing telephone suppliers but did nothing to clarify the First Amendment status of telephone common carriers or to establish a consistent definition of common carriage.[98]

Despite the interpretive flexibility afforded by the vague definition of common carriage, none of the courts hearing the Cable Act cases explored the definition as a means to clarify the ambit of the First Amendment or to support the simultaneous exercise of free speech and common carriage responsibilities by telephone video providers.

The Trifurcated First Amendment

Although the uniform extension of broad First Amendment protection to telephone companies in the telephone Cable Act cases represented a revolution in First Amendment jurisprudence,[99] the courts failed to explain their abandonment of well-established precedent that afforded different media different levels of constitutional protection.

---

97. Policy and Rules Concerning Rates for Competitive Common Carrier Servs. and Facilities Authorizations Therefor, Notice of Inquiry, 77 F.C.C.2d 308, 359 (1979); Policy and Rules Concerning Rates for Competitive Common Carrier Servs. and Facilities Authorizations Therefor, Further Notice of Proposed Rulemaking, 84 F.C.C.2d 445, 465 (1981); FCC Major Matters Report 40 (1982). See also *Cox Cable Communications Inc.*, 102 F.C.C.2d 110, 121–122 (1985) (applying the market power analysis to determine carrier status); *International Competitive Carrier Policies*, 102 F.C.C.2d 812, 829-839 (1985) (using a similar market power definitional strategy).

98. P. Nichols, cited in note 83. See also *General Tel. Co. v. FCC*, 413 F. 2d 390 (D.C. Cir. 1969); and *General Tel. Co.*, cited in note 60 (reasoning that although CATV systems were neither broadcasters nor common carriers, the common carrier status of telephone companies involved in CATV service was determinative).

99. For an example of non-First Amendment appeals, see, e.g., *General Tel. Co. v. United States*, 449 F.2d 846 (5th Cir. 1971). For recent unsuccessful First Amendment appeals, see, e.g., *United States v. Western Elec. Co.*, No. 87-5288 (D.C. Cir. 1990) (court's refusal to rule on First Amendment claim); and *Northwestern Indiana Tel. Co. v. FCC*, 872 F.2d 465 (D.C. Cir 1989) (court's refusal to consider whether a ban against cable television programming by telephone companies violates the First Amendment). But First Amendment appeals have been successful against other cross-ownership bans. See, e.g., *FCC v. National Citizens Committee for Broadcasting*, 436 U.S. 775 (1978) (newspaper/ broadcast cross-ownership rules); and *Marsh Media, Ltd. v. FCC*, 798 F.2d 772 (5th Cir. 1986) (broadcast/ cable cross-ownership rules).

Historically, the courts, Congress, and the FCC applied the First Amendment's prohibition against restraint of a free press[100] differently to each communications medium and developed a trifurcated system of First Amendment jurisprudence.[101] Either through analogy to or distinction from established media,[102] media were placed on a constitutional scale wherein telephony was virtually devoid of First Amendment protection, print media were sacrosanct,[103] and broadcast and a broadening array of electronic technologies were somewhat free and somewhat regulated.[104] Although telephone and telegraph services fell within common carriage,[105] by specific exemption, broadcast did not,[106] and the carrier status of relatively newer technologies, such as cable, was muddled.[107]

Under this tradition, telephone companies enjoyed no editorial control of their wires because the telephone was treated as an essential utility, not a speaker.[108] For print, the theory was that government intervention was unnecessary to effect an open-market exchange of ideas.[109] When speech was delivered by radio, television, or cable, regulation could infringe upon the owner's editorial freedom to protect the First Amendment rights of the audience.[110]

---

100. U.S. Const., Amend. 1. ("Congress shall make no law...abridging the freedom of speech, or of the press."). Although interesting, the ultimate resolution of the debate over Constitutional intent is less important to this chapter than is the day-to-day interpretation and application of that document to telecommunications. For relevant discussions of the philosophical underpinnings of free press in America see, e.g., E. D. Cohen. (1992). Philosophical issues in journalism; L. Levy. (1985). Emergence of a free press; J. H. Altschull. (1990)., From Milton to McLuhan: The ideas behind American journalism.

101. See, e.g., *Kovacs v. Cooper*, 336 U.S. 77, 97 (1949); *Southeastern Promotions Ltd. v. Conrad*, 420 U.S. 546, 557 (1975) ("Each medium of expression...must be assessed for First Amendment purposes by standards suited to it."); and *United States v. Western Elec. Co.*, 673 F. Supp. 525, 586 (D.D.C. 1987) ("Common carriers are quite properly treated differently for First Amendment purposes than traditional news media," citing *Midwest Video Corp.*, cited in note 76). See also D. Brenner. (1991). Telephone company entry into video services: A First Amendment analysis. *Notre Dame Law Review, 67,* 1, 111; A. J. Campbell. (1991). Publish or carriage: Approaches to analyzing the First Amendment rights of telephone companies. *North Carolina Law Review, 70,* 1071.

102. For a discussion of how this premise has been called into question see, e.g., I. de Sola Pool. (1983). Technologies of freedom. Also for a general discussion of the objectives of FCC licensing, see D. H. Ginsburg, M. H. Botein, & M. D. Director. (1991). Regulation of the electronic mass media: Law and policy for radio, television, cable and the new video technologies (Vol. 2, p. 158).

103. See *Tornillo*, cited in note 83 (ruling it unconstitutional to require newspapers to provide a right of reply). See contra, Policy and Rules Concerning Rates for Competitive Common Carrier Servs. and Facilities Authorizations Therefor, Notice of Inquiry, 77 F.C.C.2d 308 (1979); FCC Final Report and Order, 21 F.C.C. 2d 307 (1970); P. Nichols, cited in note 83.

From the outset, broadcasting was licensed as a limited public resource, and licensing justified further regulatory intervention necessary to assure fair and responsible use of a scarce, public commodity.[111] Courts extended the broadcast regulatory model to varying degrees to other electronic media[112] such as of cable, which the courts ruled to be both like and unlike broadcast.[113] However, courts likened video programming to newspaper content, a form of speech strictly protected by the First Amendment.[114] Thus, video telephony found a convergence of the three branches of First Amendment jurisprudence: the absolute protection of video programming, the intermediate protection of cable service, and the low protection of telephony.

With the advent of converging technologies, differential regulation of telephone companies and cable operators increasingly was seen as a source of inequity, com-

---

104. See, e.g., A. Campbell, cited in note 101 (discussing the trifurcated regulatory system that distinguishes among newspapers, broadcast, and cable); and J. A. Barron, cited in note 94 (asserting that regulation is necessary to assure public access to monopolistic media). Also note that certain types of speech, such as obscenity and libel, present separate regulatory rationales, and their regulation may not pose constitutional questions. See, e.g., *Miller v. California*, 413 U.S. 15 (1973) (holding that a legitimate government interest exists sufficient to prohibit dissemination of obscene material); and compare *Red Lion Broadcasting Co. v. FCC*, 395 U.S. 367 (1969) (affirming a government right to regulate access to scarce, licensed air waves), and Tornillo, cited in note 83 (holding that government-mandated access to newspaper columns violated the First Amendment). See also W. W. Van Alstyne. (1984). Interpretations of the First Amendment (presenting a graphic interpretation of the meaning of the First Amendment).

105. Subchapter II, 47 U.S.C. §§ 201 et seq. (1994) (outlining the services and charges of wire or radio communication common carriers such as telegraph and telephone). See also Mann-Elkins Act, Pub. L. No. 61-218, @7, 36 Stat. 539, 544 (1910). See also *Parks v. Alta California Telegraph Co.*, 13 Cal. 422, 424-425 (1859). Conventional wisdom holds that telephone companies held themselves out as common carriers in a quid pro quo for protected monopoly status. But see R. S. Homet Jr., cited in note 83, 245 (arguing against this oft-repeated position and maintaining that telephone functioned as a common carrier prior to its monopoly status because of its logical linkage to other vital carriers of business materials, i.e., trains and telegraph). Much common carrier regulation is designed to offset market imperfections. Common carriers, such as railroads and telegraphs, historically were viewed as natural monopolies because the high cost of installation and the limited customer base made it economically impossible for competitors to enter the market. Although competition now exists in long-distance telephone service, Regional Bell Operating Companies continue to exert market power over local telephone service within their operating areas. See also I. de Sola Pool, cited in note 102.

106. Communications Act of 1934 @3(h), 47 U.S.C. @153 (h); see also *Red Lion Broadcasting Co. v. FCC*, 395 U.S. 367 (1969).

107. See *NARUC I*, cited in note 83; *Turner Broadcasting Sys.*, cited in note 13.

108. I. de Sola Pool, cited in note 102, at pp. 95–98.

109. *Tornillo*, cited in note 83.

110. *Red Lion*, cited in note 106.

petitive disadvantage, and conflict.[115] Telephone companies responded by wielding the First Amendment hammer to bludgeon away at disparate regulation,[116] and the courts obligingly acknowledged telephone company First Amendment rights.

Some observers believed this extension of First Amendment rights to telephone companies would have a far-reaching impact on all First Amendment speakers.[117] Mark Director and Michael Botein called the redefinition of telephone company rights "a dramatic step with possibly cataclysmic effects on the entire market."[118] Legal and media scholars argued that the rulings "frayed [the] fibers of social policy, economic reality, and constitutional constraint."[119] The tatters of trifurcated First Amendment jurisprudence seemed ill suited to cover the ongoing transformation of the media. Instead, some argued that coherent telecommunications policy should rest on the consistent application of antitrust law.[120]

SHIFTING TELECOMMUNICATIONS POLICY

However, antitrust rulings in telecommunications also were under assault during the 1990s. The regional Bell companies continued to attack and wear down the

---

111. *Red Lion*, cited in note 106. See also Turner Broadcasting Sys., cited in note 13. Red Lion, 395 U.S. at 389 (holding that the First Amendment permits the government to require a broadcaster to "share his frequency" to create an open marketplace of ideas). See also *FCC v. League of Women Voters*, 468 U.S. 364 (1984) (holding that spectrum scarcity permits broadcast programming requirements to serve the public interest). But see *CBS v. Democratic National Committee*, 412 U.S. 94, 124(1974) (holding that broadcasters have the right to determine how best to create open discussion and to deny access to individual speakers).

112. *Turner Broadcasting Sys.*, cited in note 13.

113. An appeal of *Turner Broadcasting System v. FCC*, 819 F. Supp. 32 (D.D.C. 1993), argued before the Supreme Court in January 1994, focused in part on the appropriateness of "pigeonholing any communications industry in[to] a First Amendment pecking order." T. Mauro. (1994, January 10). Cable industry case expected to be a landmark, *Recorder*, 5; L. Greenhouse. (1994, January 13). New law regulating cable TV gets skeptical response from high court, *New York Times*, p. A12. *Turner* was remanded to the circuit court for fact finding and application of the intermediate level of scrutiny on June 27, 1994, (114 S.Ct. 2445).

114. *Red Lion*, 395 U.S. at 389; see also *Turner Broadcasting Sys.*, 114 S. Ct. at 2456.

115. *NARUC I*, cited in note 83.

116. *C&P*, cited in note 3.

117. M. K. Kellogg, J. Thorne, & P. W. Huber, cited in note 1.

118. Director & Botein, cited in note 10.

119. Director & Botein, cited in note 10.

120. See, e.g., W. E. Lee. (1995). The First Amendment, economic power, and judicial review, *23d Annual Telecommunications Policy Research Conference*, Solomons, MD, September 30–October 2.

constraints imposed by the Modified Final Judgment that broke up AT&T in 1982.[121] Early in 1995, Judge Harold Greene[122] of the federal district court in Washington, DC, held that the Modified Final Judgment[123] ban on regional Bell provision of long-distance services did not apply to one Bell company's delivery of video programming.[124] Telephone companies said the economies of scale in national broadband networks were critical to the economic viability of telephone video efforts.[125] Some observers expected similar court rulings to allow all local Bell companies to establish nationwide video networks, but it was Congress' passage of the Telecommunications Act of 1996, not court action, that effectively extended Judge Greene's ruling to all telephone companies nationwide.[126]

The courts clearly were not alone in reassessing established regulation of telephone services. The FCC and Congress also encouraged deregulated telephone entry into video.[127] In the mid-1990s, amid nascent competition between telephone and cable operators, the FCC argued that "differences in regulatory treatment create[d] artificial barriers" and impeded introduction of new technologies that served the public interest.[128] Entry barriers and uncertainty about regulatory mechanisms created "disincentives for the offering of new services and...impediments to investment in the multichannel marketplace."[129] So the FCC initiated inquiries to determine how best to deregulate video telephony to minimize regulatory disparities between telephone and cable and to enhance opportunities for new services.[130]

In 1995, after more than a decade of scrutiny of its policy of separation of video and telephony, the FCC began to develop and expand what it called video dial tone rules, which permitted telephone entry into video programming as well as deliv-

---

121. *Western Elec. Co.*, cited in note 66.

122. *Western Elec. Co.*, cited in note 66. Judge Greene is charged with oversight of the rules governing the postdivestiture Bell companies.

123. As in note 122.

124. M.Landler. (1995, March 18). Phone companies clear TV hurdle. *New York Times,* p. 1-1.

125. Landler, cited in note 124.

126. Pub. L. No. 104-104.

127. See, e.g., S. 652 §202, 104th Cong., 2d Sess. (1995) and H.R. 1555 §201, 104th Cong., 2d Sess. (1995) (both would immediately remove the Cable Act's cross-ownership ban).

128. R. B. Chong. (1994). Trends in communications and other musings on our future. *Fed. Com. Law Journal, 39,* 213, 218. See also Director & Botein, cited in note 10.

129. See generally A. C. Barrett. (1993). Shifting foundations: The regulation of telecommunications in an era of change, *Fed. Com. Law Journal, 46,* 39, 56.

130. In re Telephone Company-Cable Television Cross-Ownership Rules, CC Docket No. 87-266, Fourth Further Notice of Proposed Rulemaking (FCC No. 95-20) (January 12, 1995). As used herein, video telephony is a broad term that subsumes video dial tone common carrier systems and video systems operated by telephone companies that offer both programming and delivery.

ery.[131] The FCC issued a blanket waiver of the Cable Act ban for most telephone systems.[132] Cable companies promptly raised challenges to the FCC's video dial tone rules and waivers, but courts generally deferred to the Commission's power and expertise.[133]

During that same period, congressional debate over the Telecommunications Act evinced a desire to broadly deregulate electronic communications firms to encourage the economic benefits of competition. Aside from the requisite number of references to diversity of voices, congressional debate did not reflect a desire to deregulate as a means to enable telephone companies to advance First Amendment interests in public discourse.[134] Policymakers instead argued that telephone competition would counteract the market power of cable monopolies[135] and speed deployment of a national broadband telecommunications network.[136]

ANALYSIS

In the 1990s, the FCC, the courts and Congress initiated a sea change in telephone's regulatory status. Suddenly, telephone providers were transformed from passive, nondiscriminatory conduits to active speaker/conduits in the video marketplace, but policymakers failed to determine or provide guidance on how to rec-

---

131. In re Telephone Company-Cable Television Cross-Ownership Rules, CC Docket No. 87-266, Fourth Further Notice of Proposed Rulemaking (FCC No.95-20) (January 12, 1995). See also Telephone Company-Cable Television Cross-Ownership Rules, CC Docket No. 87-266, First Report and Order, and Second Further Notice of Inquiry, 7 F.C.C.R. 300 (1991); Memorandum Opinion and Order on Reconsideration, 7 F.C.C.R. 5069 (1992); Telephone Company-Cable Television Cross-Ownership Rules, CC Docket No. 87-266, Second Report and Order, Recommendation to Congress, and Second Further Notice of Proposed Rulemaking, 7 F.C.C.R. 5781 (1992); Memorandum Opinion and Order on Reconsideration and Third Further Notice of Proposed Rulemaking (FCC No. 94-269)(November 7, 1994). See, e.g., In re Telephone Company-Cable Television Cross-Ownership Rules 3 F.C.C.R. 5849 (1988) (Further Notice of Inquiry and Notice of Proposed Rulemaking) (holding that greater participation of telephone companies in providing cable services pursuant to appropriate safeguards resulted in greater competition in cable television service, and therefore, in greater public interest benefits to consumers (for early scrutiny of the cross-ownership ban)).

132. Commission Announces Enforcement Policy Regarding Telephone Company Ownership of Cable Television System, Public Notice, 1995 FCC LEXIS 1813 (March 17, 1995).

133. *National Cable Television Asso. v. FCC*, 33 F.3d 66 (D.C. Cir. 1994).

134. See, e.g., In re Telephone Company-Cable Television Cross-Ownership Rules, CC Docket No. 87-266, Fourth Further Notice of Proposed Rulemaking (Jan. 12, 1995).

135. *Turner Broadcasting Sys.*, cited in note 13.

136. Clinton Administration Information Infrastructure Task Force, *The National Information Infrastructure*: Agenda for action (September 15, 1993).

oncile the common carrier obligations of telephone companies with the newly
asserted First Amendment rights. The difficult and unresolved policy decisions lay
"in distinguishing, as reasonably as possible, among the expressive and
non-expressive activities of operators" to constrain anticompetitive practices
while freeing speech.[137] Indeed, "not all [telephone] activities are First Amendment fungible. Some [telephone] activity has characteristics that should invoke
First Amendment protection, but much does not."[138]

Nonetheless, technological innovation implemented during the 1990s smudged
bright-line regulatory distinctions between media and telephony, and blurred the
separation of speech and economic activity.[139] Although First Amendment jurisprudence long had distinguished among speakers and applied different regulation
to each according to its unique characteristics,[140] such distinctions became increasingly impractical as technological convergence and economic mergers erased
the "special characteristics"[141] that distinguished one medium from another. Economic or technological distinctions between video telephony and cable systems
seemed arbitrary, speculative, or capricious.

Yet outmoded regulatory barriers to telephone entry into video delivery and programming dissolved not because the regulations no longer advanced a legitimate
government economic objective, but because the courts accepted telephone companies' constitutional claims.[142] Lower courts consistently upheld telephone company First Amendment rights while ignoring common carrier precedent and
failing to require the showing demanded by the Supreme Court to support inter-
media regulatory distinctions and shifts in economic regulation.[143] On one hand,
these rulings may support the telephone company claim that there existed no log-

---

137. D. Brenner. (1988). Cable television and the freedom of expression. *Duke Law
Journal, 1988*, 331.

138. J. A. Barron. (1989). On understanding the First Amendment status of cable: Some
obstacles in the way, *George Washington Law Review, 57*, 1495, 1504.

139. In re Applications of Telephone Companies for Section 214 Certificates for Channel
Facilities Furnished to Affiliated Community Antenna Television Systems (Memorandum
Opinion and Order), 22 F.C.C.2d 746 (1970) (holding that no telephone common carrier
subject to the Communications Act could supply CATV service to the viewing public in its
area unless a waiver of the rules had been granted under specified conditions).

140. See, e.g., A. Campbell, cited in note 101 (detailing the theory and application of trifurcated First Amendment jurisprudence).

141. *Minneapolis Star and Tribune*, 460 U.S. at 585.

142. 47 U.S.C. § 151 (1988) (empowering the FCC to regulate interstate communications by wire and broadcast). Title II of the Communications Act of 1934 gives the FCC
broad authority to regulate interstate common carriers to prevent discriminatory service,
pricing, or quality.

143. See, e.g., *Turner Broadcasting Sys.*, cited in note 13; *Quincy Cable Television Inc.
v. FCC*, 768 F.2d 1434 (D.C. Cir. 1985); *Minneapolis Star and Tribune*, cited in note 13;
Red Lion, cited in note 106.

ical basis for the regulatory distinctions. On the other, they may demonstrate little more than "modish deference to even the faintest mention of the First Amendment."[144] Then, once the courts removed the barrier, the FCC simply followed the courts' lead, and Congress endorsed the new status quo.

The unanswered question at the core of the constitutional metamorphosis of telephone companies remains which, if any, telephone company activities truly are imbued with First Amendment rights. This determination is critical to the future force and direction of First Amendment jurisprudence. For although the First Amendment should arise as an obstacle every time regulation affects expressive activities, it should present no barrier to laws of general application unrelated to speech. Use of the First Amendment to overturn economic regulations should concern both economists and First Amendment scholars.

Of greatest concern in the telephone Cable Act cases is the courts' uniform avoidance of the nettlesome issues. The courts failed to provide guidance on how to distinguish between economic and expressive activities, or to suggest mechanisms to replace the historical carrier/speaker dichotomy that would permit more logical determinations of the rights and responsibilities of various members of the electronic press. The courts failed to offer a useful definition for common carriers that would help clarify which communications entities would qualify and under what conditions, or to determine whether common carrier status is a self-imposed condition or a regulatory mandate. The courts failed to determine what evidence is necessary to justify regulatory distinctions, or to establish the degree of deference that should be given to historical or contemporary administrative judgment when the two conflict. The courts consistently failed to demonstrate how extension of First Amendment protection to video telephony conformed with precedent.

These unresolved issues become increasingly thorny, and critical, as powerful businesses merge or splinter, and as technologies intermingle once separate services. Yet the uncertainty inherent within an industry undergoing rapid and extensive transformation increases the difficulty of fact-finding that should underlie rational, legal decision making. What data are available about a developing technology or a proposed service are largely speculative, predictive, and incomplete. The decisions that spring from these data may transform or handicap the development of even unforeseen capacities.

Thus, courts, regulators, and Congress face a conundrum. Decisions of potentially enormous impact must be made within an historical framework ill-suited to contemporary conditions, and based on partial information supplied by parties with vested interests.[145] Moreover, technological, economic, and regulatory uncertainty increase the likelihood of error. This is not a new problem, but the problem becomes increasingly significant as the size of the telecommunications

---

144. Barron, cited in note 138.
145. L. Garcia. (1996, September). The failure of telecom reform, *Telecommunications*.

industry and its centrality to public discourse and economic exchange grow. Thus, it has been said that today "the social and economic implications of controlling and communicating knowledge are of unprecedented concern."[146]

Therefore, caution is advisable. Yet the path of caution is by no means certain. The cautious libertarian would choose to err on the side of underregulation, preferring to apply regulation responsively, only as a corrective to demonstrable market failures or imperfections. The cautious regulatory advocate, however, would prefer to impose regulation more proactively to address the reasonable potential that the market will not operate properly or will not address important externalities.

Neither strategy is unequivocally preferable, particularly because current conditions in the telecommunications industry disfavor monolithic solutions. The accelerating change and increasing complexity of the communications system make the potential for useful central guidance increasingly unlikely.[147] Micromanagement, as embodied by established statutory, regulatory, and constitutional distinctions predicated on the form of communications, becomes both impractical and counterproductive in this volatile communications arena. On the other hand, "hybrid [regulatory] solutions that try to assure the coexistence of common and private carriage will not be stable in a dynamic environment."[148]

Pervasive experimentation, technological and economic uncertainty, and rapid change in telecommunications belie the ability of policymakers to achieve a single optimal, unified regulatory policy. Although congressional inaction on numerous telecommunications deregulatory proposals has been criticized as the failure of special-interest politics, it also reflects the enormous complexity of the task and lack of experience upon which to make sound decisions. The inability of the Telecommunications Act to resolve fundamental problems should surprise no one.

In reality, ad hoc telecommunications decision making—such as court and FCC video telephony rulings in the early 1990s—may afford the best opportunity to incorporate experience and learning into the policy process. Ad hoc policy may be erratic and inconsistent, and it may increase regulatory uncertainty and slow deployment of new technologies. Yet case-by-case decision making enables policy to respond to real change in the industry and in societal goals. The piecemeal dismantling of the Cable Act's cross-ownership ban and other regulatory impediments to video telephony allowed for the development of empirical evidence on the effects of telephone entry into various cable markets.

Yet the value of this ad hoc policy as an essential precursor to the development of broad policy largely was undermined by the FCC's unwillingness to free tele-

---

146. E. G. Krasnow. (1989). Mass media regulation and the constitution: Foreword: The First Amendment identity crisis, *Catholic University Law Review, 38,* ix.

147. E. M. Noam. (1994). Principles for the Communications Act of 2034: The superstructure of infrastructure, *Fed. Com. Law Journal, 39,* 317, 320–321.

148. As cited in note 147, at 321.

phone video proposals from its cumbersome licensing process. This licensing hindered the rapid entry of telephone companies into the video market with various and evolving technologies, and hampered the development of empirical evidence of the effects of telephone company entry. Such evidence would have provided a foundation for a Supreme Court ruling on the cross-ownership ban and for reasoned legislative decisions by Congress.[149]

A separate and equally pressing concern arises from telephone companies' successful use of the First Amendment to gain economic goals in the Cable Act cases. At the core, these cases involve equity and competition, not speech. Fairness dictates elimination of the cross-ownership ban in "a world in which broadcasting, cable, and telephone operations [no longer] are distinct businesses separate from each other."[150] Moreover, the ban lost its justification when the government found that the ban no longer was necessary to prevent anticompetitive conduct.[151] Yet because the courts and Congress refused to eliminate the Cable Act ban when presented with these economic arguments, the telephone companies battered the ban down with the First Amendment hammer.

Such use of the First Amendment shifts the burden of proof of the [in]validity of regulation from the telephone companies to the government. If telephone companies are speakers, this is appropriate. The obligation of showing should fall to the government when it seeks to regulate speech. Moreover, when regulations single out a particular message or speaker, as with the cross-ownership ban, such reg-

---

149. See, e.g., Telephone Company/Cable Television Cross-Ownership Rules, Sec. 63.54-63.58, Fourth Report and Order, 1995 FCC LEXIS 5418 (Aug. 11, 1995) (streamlining rather than waiving the Sec. 214 filing process for telephone video dialtone systems). See also Telephone Company/Cable Television Cross-Ownership Rules, Sec. 63.54-63.58, Public Notice, Supplemental Comments Sought on Possible Grant of Blanket Sec. 214 Authorization, 1995 FCC LEXIS 2216 (April 3, 1995); Accounting and Reporting Requirements for Video Dialtone Service, Letter, 77 Rad. Reg. 2d (P&F) 766 (April 3, 1995); Pleading Cycle Established for Comments on NCTA's Motions to Dismiss Sec. 214 Applications, Public Notice, 1994 FCC LEXIS 5888 (Nov. 23, 1994); Blanket Sec. 214 Authorization for Provision by a Telephone Common Carrier of Lines for Its Cable Television and Other Non-Common Carrier Services Outside Its Telephone Service Area, 49 Fed. Reg. 21,333 (May 21, 1984); Applications of Telephone Companies for Sec. 214 Certificates for Channel Facilities Furnished to Affiliated Community Antenna Television Systems, 21 F.C.C.2d 307 (Jan. 28, 1970).

150. Chong, cited in note 129.

151. See, e.g., Telephone Company/Cable Television Cross-Ownership Rules, Sec. 63.54-63.58, Further Notice of Inquiry and Notice of Proposed Rulemaking, 3 F.C.C.R. 5849, 5849 paragraph 1, 5853 paragraph 20 (1988) (concluding that elimination of the cross-ownership ban would serve the public interest); Telephone Company/Cable Television Cross-Ownership Rules, Sec. 63.54-63.58, Second Report and Order, Recommendation to Congress, and Second Further Notice of Proposed Rulemaking, 7 F.C.C.R. 5781 (August 14, 1992) (urging Congress to eliminate the cross-ownership ban).

ulations should be subject to strict scrutiny. Yet it is unclear whether telephone companies should be viewed as speakers, given the speculative nature of telephone companies' involvement in video programming.[152]

To the extent that telephone companies are not speakers but are instead—or simultaneously—common carriers, heightened scrutiny is ill applied. The First Amendment is inapposite to the economic aspects of telephone company operations, where the government enjoys a presumption in favor of reasonable regulation. In fact, ample legal precedent exists to deny the First Amendment arguments of telephone companies in the absence of explicit demonstration of the deprivation of real, not conjectural, speech.[153]

The Cable Act cases reasonably could have rested on either thread of jurisprudence: strict scrutiny of speaker-based regulations, or ordinary scrutiny of generally applicable laws unrelated to constitutional rights. The courts chose neither. Instead, they refused to commit to either course and fudged their way through with intermediate scrutiny, which affords broad latitude in interpretation and application. The Cable Act courts took advantage of the imprecision of intermediate scrutiny to exercise judicial activism.

If the cross-ownership ban should have been eliminated, it logically should have been rejected either as an impermissible constraint of a particular class of speakers or as an unreasonable economic regulation. Instead these cases commingled constitutional and economic arguments. They defy logic, undermine established standards of proof, and muddle both First Amendment and economic jurisprudence.

Despite (or because of) their substantial flaws, these cases deserve careful scrutiny. They embody a redirection of telecommunications policy and have the potential to revolutionize the communications landscape. They suggest the willingness of policymakers to embrace court- and Commission-made law. They suggest the power of the elite to use the courts to gain advantage in a rent-seeking environment and the willingness of the courts to ignore precedent and to overcome, or circumvent, legislative inertia.[154]

Through the Cable Act cases and the Telecommunications Act of 1996, the telephone companies achieved a substantial degree of deregulation. They gained the ability to enter new economic markets and to compete with long-distance carriers, cable operators, and an array of other communications entities. But the time was ripe for such change. It is possible then that renewed economic-based assault on

---

152. See generally Barrett, cited in note 129. See also In re Application of the New Jersey Bell Telephone Company for Authority pursuant to Section 214 of the Communications Act of 1934, as amended, to construct, operate, own, and maintain facilities to provide video dialtone service within a geographically defined area in Dover Township, Ocean County, N.J., 9 F.C.C.R. 3677 (1994) (approving the first commercial video dial tone system, which was virtually indistinguishable in content or function from a standard cable system).

153. *Turner Broadcasting Sys.*, cited in note 13.

154. P. W. MacAvoy. (1994). Telecommunications in transition. *Yale Journal on Regulation, 11*, 115.

the cross-ownership ban also could have attained this goal without wreaking havoc with First Amendment jurisprudence.

Instead the telephone companies partially redefined themselves as First Amendment speakers without delineating the parameters of their speech function. Consequently, these cases muddy the definition of speech and may establish a precedent for invasive regulation of constitutionally protected speakers. What telephone companies championed as an expansion of First Amendment rights may, therefore, "boomerang"[155] against all speakers.

CONCLUSION

Use of the First Amendment to open markets to competitive entry is efficient but foolhardy. It commingles legitimate economic and speech objectives, and confounds coherent application of distinct principles to achieve separate ends. Rhetorical laxity that enshrines marketplace metaphors into First Amendment doctrine is in part to blame, but the failure of federal courts directly to address the issue before them—the extent of First Amendment protection enjoyed by common carriers—is equally problematic. Better legal policy either would have applied strict scrutiny to these First Amendment questions or rejected the constitutional arguments entirely because the ban regulated the economic, not speech, activities of a common carrier.

Instead, the FCC, the courts, and Congress demonstrated a desire to forge ahead with a procompetitive policy that enables telephone companies to provide video content. This may be good economics, but it makes little sense under established First Amendment doctrine. Thus, these cases demonstrate the ability of the courts and the Commission to abandon precedent and to fashion new law without the public accountability imposed on Congress. The efficiency of this approach comes at the price of unresponsiveness to express public desires.

ACKNOWLEDGEMENTS

The author thanks Tom Aust, Sandy Berg, Tim Brennan, Bill Chamberlin, Wendy Gordon, Tom Hazlett, and Tom Krattenmaker for helpful insights on earlier versions of this chapter. All errors or omissions are the author's own.

---

155. Barron, cited in note 138.

Voting on Prices:
The Political Economy of Regulation

Gerald R. Faulhaber
Wharton School, University of Pennsylvania

Economists have long recognized that regulation is an imperfect solution to the problem of market failure. The term regulatory failure is often used to denote the inefficiencies inherent in government intervention in the marketplace. Policymakers are thus confronted with a practical problem of comparative institutions: Do the inefficiencies of regulation outweigh the inefficiencies of the asserted market failure (say, of monopoly)? In this chapter, we develop a simple stylized model of a (potential) monopolist offering two services, one more widely demanded than the other. We compare aggregate surplus from unregulated monopoly with aggregate surplus from a median voter model of price setting in a (perfectly) regulated monopoly. We find that:

1. Median voter pricing can not only yield lower surplus than monopoly pricing, it can actually yield negative surplus for one of the services.
2. Whether or not median voter pricing is surplus-inferior to monopoly pricing depends upon the elasticities of demand in the two service markets.
3. The more widely demanded both services become, the less is the inefficiency from median voter pricing.

In addition, the model offers hypotheses regarding the recent evolution of local and toll service pricing in the telephone industry. We provide some empirical evidence from this industry that confirms these hypotheses.

At the risk of overgeneralizing, the development of regulatory economics over the past quarter century has followed two main themes: a regulatory models theme and a Chicago School political economy theme. The first theme has focused on current industrial organization (IO) models applied to regulation, often expanding the theoretical frontiers of IO. The second theme has focused on self-interested interactions in the political marketplace of regulators and various constituencies. Each of these themes has developed substantial insight into the problems of regulation, but each has borrowed surprisingly little from the other. One objective of this chapter is to apply models consistent with the first theme to the concepts of the second.

The salient history of each theme is reviewed briefly, assessing the strengths and weaknesses of each. The regulatory models theme has had two phases: natural monopoly theory and agency theory. The first phase was ascendant during the 1970s and early 1980s, starting with Baumol and Bradford's (1970) popularization of Ramsey pricing, through cross-subsidy analysis to sustainability and the theory of contestable markets (Baumol, Panzar, & Willig 1982). The central problem addressed by this work is the welfare and equilibrium analysis of market structure in which firms are characterized by scale and scope economies and marginal cost pricing does not cover total cost. Regulators are implicitly assumed to maximize welfare subject to a budget constraint, and Ramsey pricing is the answer. Contestability theory was a pre-game-theoretic analysis of market structures with such firms, with normative implications about the appropriate scope of regulation.

The second phase of the regulatory models theme, agency theory, has been ascendant throughout the 1980s and 1990s, starting with the early chapter of Vogelsang and Finsinger (1979) through the work of Laffont and Tirole (1993) and many others, well -summarized in Laffont (1994). It is without question the reigning paradigm of this theme of regulatory economics. The central problem addressed by this work is the asymmetry of information between regulated firms and the regulators, the strategic use of this asymmetry by firms, and appropriate responses by regulators. Agency theory goes beyond natural monopoly theory, in its recognition that regulation is far from perfect, which is a major focus of this chapter. However, the agency theory approach to regulation has limitations:

- Regulatory agency theory generally is focused on one very specific imperfection: asymmetric information. No other regulatory failures are encompassed.
- There is nothing inherent in agency theory that is uniquely applicable to public, or government, control of firms. This approach to regulation thus limits the analysis to those problems of public regulation that are similar to, say, firm managers trying to control workers, or firm shareowners trying to control firm managers.
- This line of research generally maintains the polite but fantastical fiction that regulators are seeking economic efficiency in their control of the industry, but are hampered in this pursuit by asymmetric information.

It is precisely these weaknesses of the regulatory models theme that are the strengths of the Chicago School theme. This theme was pioneered by Stigler in the early 1960s and carried on by Peltzman, Becker and many others, well -represented in the collection of articles in Stigler (1988). The central problem addressed by this work is "demanders" of regulations (industry constituencies) using political influence to enlist "suppliers" of regulations (regulators and legislators) in order to capture market rents. In this view, regulation is not about achieving efficient outcomes (in the sense given earlier), but rather redistributing rents using the coercive power of government.[1] Unfortunately, this powerful concept has no central organizing analytic model to fully realize it. Most of the applications of this concept (see, e.g., Stigler, Peltzman, & Linneman in Stigler, 1988) have used reduced form or ad hoc models applicable to their view of the industry being studied. Becker's (1983) attempt at a more general model of political influence has had little influence on the literature since its appearance. Closest in spirit to this chapter is Peltzman (1980), which uses a stylized majority-voting approach to model government tax and transfer policies. This model is used to guide the empirical analysis of cross-country tax and transfer policies to confirm the author's hypotheses. Similarly, in this chapter, we also use a stylized majority-voting approach to model the (rather narrower issue of) political choice of prices charged by a regulated utility. As in Peltzman, we subject this model to empirical test, using data from the telephone industry, 1960–1993.

Curiously, almost no efforts appear in the literature to span these two themes. A notable exception is the attempt by Laffont and Tirole (1993) to bridge this gap (Chap. 11, Regulatory Capture). They developed an interesting model of interest group politics within the overall framework of agency theory; essentially, it is an agency model with (enforceable) side payments. Although it is an important step forward, it still suffers from all three of the previous objections to agency theory.

In contrast, this chapter seeks to apply a simple *political economy* model to the problem of regulatory pricing, in order to compare the relative efficiency of an unconstrained monopoly market with that of voter-responsive regulators wishing to maximize their likelihood of (re)election. The political process takes center stage in this model, rather than being modeled as just another agency (or natural monopoly) problem.

This is not the first economics paper to examine voting behavior and economic outcomes. Several papers in public finance have addressed how a progressive income tax in a two-period model can emerge from a median voter model (Roberts, 1977 and Creedy & Francois, 1993, among many others). Spann (1977) considered the political choice between collective consumption and individual consump-

---

1. The Virginia School, associated with Buchanan and Tullock, takes an even less rosy view of political influence, claiming that potential rents are dissipated by rent-seeking behavior on the part of constituencies. Little of this literature is focused directly on regulation.

tion. Closely related to the present work is Ye and Yezer (1992) and following chapters that examine regulatory pricing of freight movements in spatial monop- olies; their results appear to be highly specialized to their application, however, and few general results emerge. Beard and Thompson (1996) use very similar techniques to those of this chapter to analyze political choices over the form of two-part tariffs. This excellent article is quite close in spirit to this work, although focused on a somewhat different set of pricing issues.

Perhaps the most advanced work on relating political economy with economic outcomes is Grossman and Helpman (1995) from the literature on trade. This chapter presents a game-theoretic model in which various interest groups "buy" influence (via campaign contributions) with policymakers in their home country, which then enter into either negotiations or a trade war with a second country (whose interest groups also buy influence in their home country). This quite am- bitious and insightful model yields important empirical predictions (which the au- thors do not test) regarding the outcomes of trade wars verses trade talks and the relevance of in-country political power of interest groups. Theirs is not a decisive- voter model, with the concomitant "winner-take-all" discontinuous payoffs. Rath- er it focuses on campaign contributions, which in their model result in direct utility to politicians.

The second section of this chapter develops the political economy model of reg- ulation, specifying how the public regulation game is played and the resulting in- centives of regulators. The third section, the results of this model are presented, including a comparison of political pricing with monopoly pricing. The fourth sec- tion, model hypotheses of regulatory objectives are developed and tested using telephone usage data from 1960 to 1993. A fifth section concludes the chapter.

THE MODEL

We consider a monopolist[2] that produces two services, a "mass" service M that ev- eryone is likely to consume and a "specialized" service S that is consumed (per- haps intensely) by part of the population. An example from telephony would be local service as service M and toll service as service S; an example from informa- tion networks would be e-mail and local library text access as service M and full graphics worldwide capability as service S.

Demand Consumers are indexed by θ, $0 \leq \theta \leq 1$, with individual demand func- tions $q_x(p;\theta)$.

---

2. To keep the analysis uncluttered, we do not specify the source of the monopolist's market power. It is closest to the spirit of the analysis to think of the monopolist as owning an exclusive franchise.

We assume:
1. $q_x(p_x;\theta) = \lambda_x(\theta) \cdot Q_x(p_x)$;individual demand functions are separable in p and θ, demands for the two products are independent, and income effects are ignored. Consumer θ consumes the fraction $\lambda_x(\theta)$ of total demand for service $x=M, S$, so

$$\int_0^1 \lambda_x(\theta)d\theta = 1$$

2. $\lambda_M(\theta) > 0$, for all θ; everyone consumes the mass service.

The special assumption that each individual's demand function is a multiple of the aggregate demand function is by no means innocuous. It implies that all consumers with a positive λ consume at any price at which anyone consumes. In this formulation, the model is not appropriate for addressing regulatory questions of "universal service" or network externalities.

We use the convention that $\lambda(\theta) = \lambda_S(\theta)/\lambda_M(\theta)$ is an increasing function of θ;[3] consumers are indexed according to their ratio of specialized to mass demand.

Consumer θ's (indirect) utility is:

$$U(p_M, p_S;\theta) = U_M(p_M;\theta) + U_S(p_S;\theta) = \int_{p_M}^{\infty} \lambda_M(\theta) \cdot Q_M(z)dz + \int_{p_S}^{\infty} \lambda_S(\theta) \cdot Q_S(z)dz \quad (1)$$

with $U'_M = -\lambda_X(\theta)Q_X(p_X)$

and aggregate consumer surplus is:

$$W(p_M, p_S) = W_M + W_S = \int_{p_M}^{\infty} Q_M(z)dz + \int_{p_S}^{\infty} Q_S(z)dz, \quad (2)$$

with $W'_X = -Q_X(p_X)$

Total producer plus consumer surplus is:

$$T = W_M + \Pi_M + W_S + \Pi_S \quad (3)$$

with $W'_X + \Pi'_X = (p_x - c_x)Q'_x$

---

3. This condition need not uniquely determine the ordering of the index. If we further assume that $\lambda'_S \geq 0, \lambda'_M < 0,$ then the ordering is unique.

Production

We assume that the production technology exhibits constant returns to scale and no economies of scope, thereby abstracting from the problems of "natural monopoly" theory. The cost function is therefore $C(Q_M, Q_S) = c_M \cdot Q_M + c_S \cdot Q_M$, where the marginal cost coefficients are positive constants. The monopolist firm may choose whether or not to produce.

Regulation

There is a government that (it is assumed) has chosen to regulate this monopolist. Candidates from each of two parties may run for an elective office of regulator, who then controls the prices charged by the monopolist. These candidates only care about being elected and have no preferences over prices. The candidate receiving a majority of votes wins. Candidates make promises during their election campaign about the level of prices they will permit the monopolist to charge if they are elected. All citizen/consumers vote in this election.

This majority-voting model of regulation should not be taken literally. It is meant to focus attention on price determination through political forces (rather than market forces), abstracting away from the rich institutional structure that characterizes regulation. What is important is that voter support for a particular policy increases the likelihood it will be adopted.

Information and Commitment

We assume that there is perfect information: consumers, regulators, and firms know the full model, thereby abstracting away from agency theory. Additionally, we also assume platform promises made by candidate regulators prior to election are binding commitments that the regulators have the power to enforce. Neither of these assumptions reflects the way the world works; our purpose here is to focus attention on the functioning of regulation in a politically-driven context without such imperfections, the impacts of which have been studied extensively elsewhere.

There are, of course, many ways in which private parties can influence lawmakers and regulators: campaign contributions, either in cash or in kind, providing information, outright bribery, "get-out-the-vote" campaigns, political education efforts, or smear tactics. In this model, we abstract away from these interesting questions, and focus exclusively on how the political system "ought" to work: informed voters expressing their preferences in elections.

In sum, this stylized model focuses on examining how regulation works if it were to meet its idealized "design specifications": no lack of voice on the part of customer/citizens, no information asymmetries, no commitment problems, and no rent-seeking behavior. A case can be made that taking into account these real-world complications are likely to reduce the social performance of regulation.[4]

Characterizing Regulated Prices

The familiarity of the majority-voting approach to political economy problems suggests that we can forgo the development of a formal model of the game in the interest of brevity. The price outcome $(p_M^V, p_S^V) = \mathbf{p}^V$ of the game is characterized as follows:

1. $\Pi(\mathbf{p}^V) \geq 0$; the assumptions of (a) perfect information and (b) the monopolist does not have to produce imply that citizen/consumers would see candidate platforms that promised unrealistically low prices as "pie in the sky," and correctly perceive that no output would be produced unless the firm has nonnegative prices.

2. $\mathbf{p}^V$ is undominated. That is, there exists no price vector $\hat{\mathbf{p}}$ such that (a) $\Pi(\hat{\mathbf{p}}) \geq 0$ and (b) $\hat{\mathbf{p}} < \mathbf{p}^V$. Clearly, if such a price vector existed, all citizen/consumers would prefer it, so it wins unanimously over $\mathbf{p}^V$, proving the assertion by contradiction.

3. $\Pi(\mathbf{p}^V) = 0$. By continuity of Π, this is a necessary condition for $\mathbf{p}^V$ to be undominated. Therefore a candidate has a choice of only one price variable, with the zero profit constraint determining the other, that is, $p_S = p_S(p_M)$. Not all prices p_M are feasible; only prices in the interval $[p_M^{min}, p_M^{\Pi}]$ are undominated, where $p_M^{min} = p_S^{-1}(p_S^{\Pi})$. Within this interval, it is easy to show that a lower mass price requires a higher specialized price: $p_S' < 0$. Hence, individual welfare can be written as a function of the single variable p_M for example, $U(p_M, p_S(p_M); \theta) = U(p_M); \theta)$.

4. $U(p_M); \theta)$ is single-peaked in p_M for all θ under the assumption that the profit function is concave on the interval $[p_M^{min}, p_M^{\Pi}]$.

5. There exists a unique majority voting equilibrium at the price that maximizes the median voter's θ^V utility: $p_M^V = \arg\max_{p_M}[U(p_M; \theta^V)]$. This is simply the median voter theorem of Downs (1957); results 3 and 4 are the conditions for which the theorem is true. We refer to this as *the median voter price*.

---

4. Becker (1983) and others of the Chicago School would argue that rent-seeking on the part of interested parties in the political marketplace leads (under special assumptions) to efficient redistribution, and therefore this activity is welfare-enhancing. This is not a universally-held view, but because it is not the point of this chapter, we need make no judgment on this issue.

The median voter price-pair is

$$p = \arg\max_{p_M} [U(\mathbf{p}; \theta^V) s.t. \Pi(\mathbf{p}) = 0]$$

$$= \left(\frac{c_M}{1 + \frac{\mu - \lambda_M(V)}{\mu \varepsilon_M}}, \frac{c_S}{1 + \frac{\mu - \lambda_S(V)}{\mu \varepsilon_S}} \right)$$

where μ is the Lagrangian multiplier on the constraint and ε_X is price elasticity of demand, and

$$\varepsilon_X = \frac{p_X}{Q_X} \cdot \frac{dQ_X}{dp_X} < 0$$

This immediately leads to the following results:

Proposition 1

If $\lambda_M(\theta^V) = \lambda_S(\theta^V)$, then the median voter price-pair is

$$p_M = c_M, \ p_S = c_S, \text{ and } p_S < c_S, \text{ and } \lambda_M(\theta^V) = \lambda_S(\theta^V) = \mu.$$

Proof

With $\lambda_M(\theta^V) = \lambda_S(\theta^V)$, if both are greater than μ then both prices are less than their respective marginal costs, yielding negative profits, from Equation 4. Similarly, if both are less than μ, both prices are greater than their respective marginal costs, yielding positive profits, again from Equation 4. Thus, only when both are equal to μ, $\lambda_M(\theta^V) = \lambda_S(\theta^V) = \mu$, are profits zero. ∎

Thus if the median voter has the same proportion of mass and specialized service as does the population as a whole, then the solution will be efficient prices.

However, this is most likely not an accurate representation of such markets. Rather, the median voter is likely to consume a greater fraction of one service over another; our choice of terminology for these services leads us to identify the mass service as of greater use to the median voter than specialized service. The next proposition shows that the median voter's preferred price pair will reflect this consumption allocation favoring mass service.

Proposition 2

If $\lambda_M(\theta^V) > \lambda_S(\theta^V)$, *then median voter pricing results in*

    (i) $\lambda_M(\theta^V) > \mu > \lambda_S(\theta^V)$

    (ii) $p_M < c_M$ *and* $p_S < c_S$.

Proof

Obvious from Equation 4; in order to assure zero profits, it cannot be the case that both λ values are greater than μ, or that both λ values are less than μ. ■

Clearly, regulation driven by median voter pricing caters to the consumption preferences of the median voter, extracting monopoly profits from S to subsidize M. The extent to which the regulator can lower p_M depends on the profits extracted from S.

Efficiency of Median Voter Pricing

These two propositions merely restate the well-known result that only if the median voter has the same preferences as the mean voter will majority rule result in efficient outcomes. This "well-known result" depends, of course, on defining efficiency in terms of aggregate surplus. The problems involved with this definition of efficiency are also well-known, in the absence of compensation. If we adopt the more careful but much weaker Pareto criterion for efficiency, then in this model *any* price-pair for which $\Pi(\mathbf{p}) = 0$ is efficient, because any other price-pair makes one of the consumer groups worse off. In this latter view, regulation results in transfers among groups of citizen/consumers, and normative conclusions regarding their desirability cannot be reached on the basis of a scalar measure such as aggregate surplus. In this chapter, we take the somewhat agnostic view that both aggregate surplus and the distribution of surplus are important and worthy of analysis, as evidenced by the results that follow. However, at the risk of abusing or confusing terminology, we reserve the term efficient prices to mean prices that maximize aggregate surplus. In this context, it is only this definition of efficiency that has discriminatory power.

It is no surprise that allocating resources by voting is not efficient, and it is hardly a surprise that voters will opt for subsidies that are in their favor. Even in this case of perfect information and regulators that are perfectly responsive to voters, inefficient outcomes result—hence, "regulatory failure."

The more interesting question is to compare the efficiency cost of regulatory failure to that of market failure. In the context of this model, we ask how median voter prices compare to monopoly prices, which are

$$p_X^{\Pi} = \frac{c_X}{1 + \dfrac{1}{\varepsilon_X}}$$

The problem of above-marginal cost pricing is well understood in the literature and constitute the efficiency case against monopoly. What is not as clear is the effect of *below* marginal cost pricing as occurs in the market for service M under regulation. Can below-marginal cost pricing in the mass market lead to lower aggregate surplus than monopoly pricing? Can prices be driven so low that negative surplus is generated? If we contrast this possibility with the case of monopoly pricing, in which surplus is never negative, we may find that the regulatory cure is worse than the monopoly disease, a problem to which we turn our attention in the next section.

MODEL RESULTS

The principal theoretical results of this chapter are:
1.   Median voter pricing can result in total surplus which is less than that achieved under monopoly,
2.   Median voter pricing can result in a negative total surplus for the mass service,
3.   An exogenous increase in the amount of specialized service consumed by the median voter leads to an increase both in p_M^V and in total surplus, whether this increase comes about through an increase in total demand or from a change in the distribution of demand.

Efficiency

In order to address the first two issues, we first note that below-cost pricing can only be supported by extracting profit from service S; the greater the potential profit from S, the lower the price for the mass service M can be. We have, from the zero proof constraint,

$$\left(p_M^V - c_M\right)Q_M\!\left(p_M^V\right) + \left(p_S^V - c_S\right)Q_S\!\left(p_S^V\right) = 0$$

$$\Rightarrow p_M^V = c_M + \frac{\Pi_S}{Q_M\!\left(p_M^V\right)} \qquad (5)$$

$$\Rightarrow \frac{\partial p_M^V}{\partial \Pi_S} = \frac{-1}{Q_M(p_M^V)} < 0$$

In order to explore the relative inefficiencies of median voter regulation verses monopoly, we examine an interesting special case: (a) The median voter consumes none of the specialized service: $\lambda_S(\theta^V) = 0$; (b) The demand functions are well-approximated by the semi-log form, $Q_X(p) = a_X \cdot p^{\varepsilon x}$; and (c) We focus on the mass market, assessing efficiency loss as parametric in p_M^V, which is determined largely by profits in the S market, as shown in equation (5).

The first assumption yields the straightforward result that the median voter price for S is the profit-maximizing price, since that yields the lowest possible price for M:

$$\lambda_S(\theta^V) = 0 \Rightarrow p_S^V = \frac{c_S}{1+\frac{1}{\varepsilon_S}} = p_S^{\Pi}$$

We can now compare the total surplus achieved by the two institutions:

$$T^{\Pi}(p_M^{\Pi}, p_S^{\Pi}) \lessgtr T^V(p_M^V, p_S^V)$$

$$\Rightarrow Q_M(p_M^{\Pi}) \cdot \left[p_M^{\Pi} - c_M - \frac{p_M^{\Pi}}{(\varepsilon_M + 1)} \right] + Q_S(p_S^{\Pi}) \cdot \left[p_S^{\Pi} - c_S - \frac{p_S^{\Pi}}{(\varepsilon_S + 1)} \right] \gtrless$$

$$Q_M(p_M^V) \cdot \left[p_M^V - c_M - \frac{p_M^V}{(\varepsilon_M + 1)} \right] + Q_S(p_S^V) \cdot \left[p_S^V - c_S - \frac{p_S^V}{(\varepsilon_S + 1)} \right]$$

Rearranging terms yields

$$Q_M(p_M^{\Pi}) \cdot \left[p_M^{\Pi} - c_M - \frac{(\varepsilon_M + 1)}{\varepsilon_M} \right] \gtrless Q_M(p_M^V) \cdot \left[p_M^V - c_M - \frac{(\varepsilon_M + 1)}{\varepsilon_M} \right]$$

The last step follows from the fact that with $\lambda_S(\theta^V) = 0$, the total surplus from the S market is the same under both institutions, so the surplus comparison depends only on p_M^V.

By taking the semilog approximation of the demand curves, elasticities are held constant, and therefore can be treated as exogenous parameters. With this approximation, the surplus comparison is:

$$T_M^V \gtrless T_M^{\Pi} \Rightarrow \left(\frac{p_M^V}{c_M} \right)^{\varepsilon_M} \left[\frac{p_M^V}{c_M} - \frac{\varepsilon_M + 1}{\varepsilon_M} \right] \gtrless \left(\frac{\varepsilon_M}{\varepsilon_M + 1} \right)^{\varepsilon_M} \left[\frac{\varepsilon_M}{\varepsilon_{M+1}} - \frac{\varepsilon_M + 1}{\varepsilon_M} \right]$$

Defining the relative price as $\hat{p}_M^V = \frac{p_M^V}{c_M}$, we have

$$T_M^V \gtrless T_M^\Pi \Rightarrow (\hat{p}_M^V)^{\varepsilon_M}\left[\hat{p}_M^V - \frac{\varepsilon_M+1}{\varepsilon_M}\right] \gtrless \left(\frac{\varepsilon_M}{\varepsilon_M+1}\right)^{\varepsilon_M}\left[\frac{\varepsilon_M}{\varepsilon_M+1} - \frac{\varepsilon_M+1}{\varepsilon_M}\right] \quad (6)$$
$$\text{and}$$
$$T_M^V \gtrless 0 \Leftrightarrow \hat{p}_M^V \gtrless \frac{\varepsilon_M+1}{\varepsilon_M} \quad (7)$$

The right-hand side of Equation 6 is always positive, so the set of relative prices with positive mass surplus is larger than the set of prices in which total surplus is greater with median voter regulation over monopoly.

Denote by $v(\varepsilon_M)$ the relative price at which Equation 6 holds at equality and by $z(\varepsilon_M)$ the relative price at which Equation 7 holds at equality. Then these two equations can be represented graphically as in Fig. 14.1. This is captured in the following proposition:

Fig. 14.1
Relative Efficiency of Median-Voter Regulation
versus Unregulated Monopoly

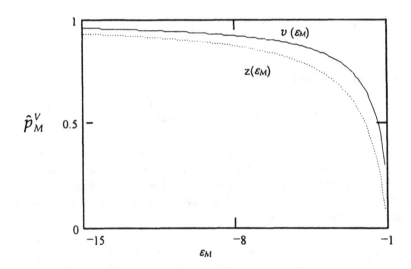

Proposition 3

Assuming (a) semilog demand and (b) the median voter consumes no specialized service, then median voter pricing may lead to

- Lower total surplus than monopoly pricing
- Negative total mass service surplus

depending on profits available from S *and the elasticity of demand of M.*

Proof

See Fig. 14.1. ■

The intuition behind Fig. 14.1 is straightforward. A higher valued specialized service leads to greater extraction of profit (and greater surplus loss) in order to subsidize lower prices for mass service, moving down the vertical axis of the figure. A more elastic mass service leads to greater quantity response to lower prices, and so greater consumption at prices below cost, moving to the left on the horizontal axis. Both a higher valued specialized service and a more elastic mass service lead to greater inefficiencies, and a greater likelihood that regulation performs worse than monopoly or even results in a negative total mass service surplus.

Fig. 14.1 also has important policy implications. If the mass service has fairly inelastic demand, then the relative price $\hat{p}_M^V$ can be quite low and still have median voter prices superior to monopoly prices. However, even a small increase in $-\varepsilon_M$ greatly narrows the scope of median voter regulation to outperform monopoly. In the case of telephone service, the mass service is access and local calling, which has a very low estimated elasticity. This would suggest that the long-standing subsidy from toll service to local, although it had efficiency costs, was still superior to monopoly. However, for new information services this need not be the case. For example, if e-mail and text access to local information providers is considered the mass service, it is as likely as not to be reasonably price-sensitive. In this case, if regulation ended up subsidizing a highly price sensitive service, regulation could be less efficient than pure monopoly, or even generate negative surplus from the mass service.

Rent Distribution

It is obvious that in the special case in which the median voter consumes none of the specialized service, regulation simply redistributes the rents extracted from the consumers of S from the monopoly to the consumers of M. The previous example shows that this may result in a reduction in total surplus, but that example does not address the equally important question of the distribution of the surplus.

There are two approaches to understanding rent distribution: The first, more classical welfare economics approach is to note that using aggregate surplus as an efficiency measure is making an implicit choice regarding a social welfare function, in which all consumers are equally weighted. Suppose that the welfare weights of mass service consumers and specialized service consumers were different; would that change the implications of Fig. 14.1? A moment's reflection should convince the reader that using *weighted* aggregate surplus, with differing weights for mass and specialized consumers will shift the location of the curves in Fig. 14.1, but will not alter the basic structure of the model. Placing more weight on mass service shifts the curves down and to the left, whereas placing more weight on specialized services shifts the curves up and to the right. More precisely, for any mass versus specialized weights, there exists an unbounded region in Fig. 14.1 in which weighted aggregate surplus is reduced under regulation as compared to monopoly. Fore example, even if the surplus from specialized service has a relatively low social weight. In Fig. 14.1, a huge (unweighted) specialized surplus translates to a very low (possibly negative) relative price for the mass service.

The second approach to understanding rent distribution is simply to identify which players get what rents. It is to this that we now turn. Our next example demonstrates how surplus is divided up among M consumers, S consumers, and the monopolist F, still under the assumption of $\lambda^V = 0$. The distribution of rents depends on (a) the elasticities ε_M and ε_S, and (b) the relative magnitudes of the surplus available from M and S. For concreteness, we use $\varepsilon_M = -1.2$ and $\varepsilon_S = -5$ in this example, and we normalize the total surplus available with competition to be unity (from Equation 2, $W_M + W_S = 1$). The interpretation of greater W_S is that the specialized service is of greater value to its consumers relative to the consumption value of service M.

In Fig. 14.2, the surplus frontiers for groups M and S, for competition (as the benchmark), and for both median voter pricing and monopoly pricing are shown. The heavy black line with slope -1 is simply the surplus achievable under competition, defined to satisfy $W_M + W_S = 1$. The thin black line is the frontier with median voter pricing. It lies almost everywhere inside the competitive frontier; each S consumer achieves only about 40% of the surplus a competitive regime would yield. The thin shaded line represents the frontier of direct surplus each group gets under monopoly pricing. Clearly, this lies inside the median voter pricing frontier, because both groups now face monopoly prices. Under this regime, each S consumer obtains 70% of the surplus a competitive regime would yield (and each S consumer again achieves 40%, as earlier). However, this does not take account of the profits of the monopolist. In order to include these profits, we assume that each group owns shares of the monopolist and thus receives a share of the profits, either as dividends or prospective growth in earnings. For illustrative purposes, we assume that profits are split among the two groups in the same proportion as the competitive surplus.[5] This

---

5. Varying the assumption of how profits are shared among the two groups does not alter the result that the two frontiers cross. However, note that we assume implicitly that profits are shared equally *within* each group, in order to simplify the exposition. Further, because this analysis is for illustrative purposes only, voter/consumers do not take the profit distribution into account when voting on prices.

yields the "Profit Max with Distribution" frontier, which crosses the median voter frontier, and lies outside it for larger values of W_S (as implied by Fig. 14.1).

To focus more sharply on distribution, the plotted points (C, V, Π) show the surplus distribution for each regime when the competitive surplus is split equally: $W_S = W_M = 0.5$, located at point C. V is the distribution of surplus for this example with median voter pricing and Π is the surplus distribution with monopoly pricing (assuming profits shared in proportion to competitive surplus share). Clearly, service M customers prefer median voting to monopoly, and even to competition. Service S consumers, on the other hand, prefer monopoly to median voting, but not to competition.

Figure 14.2

Rent Distribution—Monopoly versus Median Voter

Total Competitive Surplus = 1

Profit Share = Competitive Surplus Share

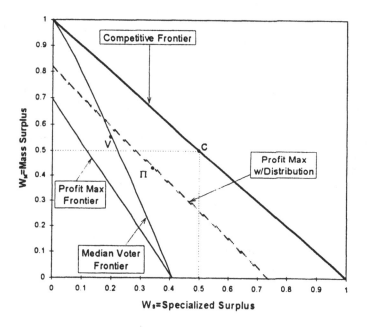

Changing Median Voter Consumption

The Fig. 14.1 and Fig. 14.2 examples illustrate starkly the potential negative efficiency effects if the median voter consumes none of the specialized service. However, Proposition 1 portrays a more optimistic picture, in which efficiency is achieved when the median voter consumes equal fractions of both services. We next analyze the effect of increasing the share of service S by the median voter on total surplus. We focus on the *relative* distribution of toll demand for the median voter, defined as $\lambda^V = \lambda_S (\theta^V)/\lambda_M (\theta^V)$.

Welfare for the median voter can be rescaled to be

$$U(p_M; \theta^V) = W_M(p_M) + \lambda^V W_S(p_S(p_M))$$

Proposition 4

Both p_M and total surplus increase as λ^V increases, for $0 < \lambda^V < 1$.

Proof: See Appendix. ∎

The interpretation is straightforward: The closer the demand pattern of the median voter is to the mean voter (and thus population as a whole), the greater is the total surplus that results from maximizing the median voter's utility. This proposition strengthens and extends our previous results to show that both aggregate surplus and price of the mass service are monotonic in λ^V.

The last proposition shows that exogenous increases in the amount of the specialized service consumed by the median voter lead to more efficient pricing. This relatively straightforward model result yields a very strong prediction regarding pricing behavior in telecommunications over the past 35 years. It is widely believed that toll telephone calls, considered a rare luxury in the 1950s, are now widely demanded by most consumers. Proposition 4 would then imply that initially, toll service would subsidize local service, but that as the demand pattern changed over this period, the subsidy would be reduced: Local prices would increase and toll prices would decrease. It is also widely believed that the policy changes of the 1970s and 1980s, which led to the rate restructuring of the 1980s and 1990s, have indeed moved toll and local prices closer to their respective costs. These results, therefore, demonstrate the link between the two, suggesting that changing demand patterns of the median voter (and not efficiency considerations) have led to increased competition and changing rates. It is to this empirical issue that we now turn.

EMPIRICAL RESULTS

The substantial changes that have occurred in the telecommunications industry over the past two decades constitute a natural experiment for testing this model. Conventional wisdom in the field is that for many years, toll service provided a large subsidy to local service. The AT&T divestiture constituted a significant shift away from regulated monopoly that supported this[6] subsidy, toward a more competitive model which will not support subsidies. Further, the data (shown later) support this conventional wisdom. Surprisingly, economists have not attempted to explain why this important and fundamental shift occurred at this particular time. The model developed in this chapter suggests a place to look for this explanation: If the preferences of the median voter changed significantly toward toll over this period, then the model predicts that the political system will respond by adopting a competitive regime that would favor reductions in the previous subsidies.

The hypothesis, then, is that shifts in the preferences of the median voter toward toll service are closely related (with appropriate lags) to shifts in toll and local prices produced by the recent more competitive market structure.[7] In order to test this hypothesis, we estimate the model[8] using toll and local telephone price and cost data, as well as data on the distribution of toll usage across the population, from 1960 to 1993.[9]

The overall strategy to test this hypothesis is

1. Derive equations for price-cost margins for each service, based on the model developed in the preceding text.
2. Construct series that represent prices, costs, revenues, and median voter preferences from actual telephone data.
3. Use the equations, the data, and constructed series to predict the price-cost margins.
4. Regress actual margins against predicted margins, using lags of order 0 through 10.

---

6. As well as a number of other subsidies, such as from urban-suburban users to rural users, business to residential users, and Bell companies to independent companies. Although these subsidies were (and are) politically important, the most economically important subsidy was that from toll to local, and it is this subsidy that is the focus of this paper.

7. Some analysts dispute that the shifts in toll and local prices were produced by increased competition. Taylor and Taylor (1993) asserted that the extensive rate rebalancing that occurred in the 1980s was due principally to aggressive reductions by the FCC in carrier access charges, followed by aggressive reductions in AT&T rates, again at the behest of the FCC. Whether regulators/legislators brought about this rebalancing directly or indirectly via competition does not matter to our hypothesis. The rebalancing occurred as a result of some aggressive public policy action.

8. It is as obvious to the author as it is to the reader that the assumptions of the model, especially that of constant returns to scale, are gross oversimplifications of telecommunications economics. This analysis is intended only as a first-order analysis of the model, designed merely to assess the relevance of this form of political economy modeling to real-world situations.

9. The relative paucity of the data required that various estimation methods be employed to "backcast" certain information that was not available over the sample period, that involves the use of assumptions and approximations which appear reasonable to the author but perhaps not to the reader.

The hypothesis is accepted or rejected based on the t-statistics and of the regression of the final step.

Predictive Model

The model to be estimated consists of the first-order condition for maximization of the median voter's utility plus the budget constraint (substituting T [toll] for the S [specialized service] and L [local] for M):

$$U'(p_L, p_T; 0.5) = 0 \Leftrightarrow \lambda(0.5) = -\frac{Q_L}{Q_T p_T'} = \frac{1 + \dfrac{p_T - c_T}{p_T} \varepsilon_T}{1 + \dfrac{p_L - c_L}{p_L} \varepsilon_L}$$

$$(p_T - c_T)Q_T + (p_L - c_L)Q_L = 0$$

where the last equality in the first equation can be obtained by noting that

$$p_T' = -\frac{\frac{\partial \Pi}{\partial p_M}}{\frac{\partial \Pi}{\partial p_S}} = -\frac{Q_M(1 + \frac{p_M - c_M}{p_M} \varepsilon_M)}{Q_S(1 + \frac{p_S - c_S}{p_S} \varepsilon_S)}.$$

It is somewhat more convenient to use margins m_X rather than prices, so the relevant equations are

$$\lambda(0.5) = \frac{1 + m_T \varepsilon_T}{1 + m_L \varepsilon_L} \qquad 0 = m_T R_T + m_L R_L \tag{8}$$

The data required to estimate this model consist of the distributional information, prices, unit costs, quantities, and elasticities.

Distributions of Toll and Local Usage

Because local usage typically carried a zero price for most consumers during most of this period, and telephone penetration was relatively constant at over 90% during this period, we assume that local service was uniformly distributed over the relevant population. For toll service, we note that our assumption that the distribution of demand does not depend on price plays a key role in this empirical analysis, in that the demand distribution can be recovered from revenue distribution. We examine two data sources:

1.   Toll revenues by income quintile,[10] annual, 1984–1993 (Lande, 1994),

2. Toll revenues by income percentiles, 1972–1973 (Bureau of Labor Statistics, 1974) and 1960–1961 (Bureau of Labor Statistics, 1963).

For each of these 12 years, the one-parameter cdf $\theta^{\alpha+1}$ was estimated. For this family of distributions, the uniform corresponds to $\alpha = 0$; the larger is α, the more skewed the distribution is toward higher θ, and therefore the more specialized is the service. Thus, α declining over time implies that toll service is becoming more of a mass service. The Table 14.1 shows the coefficients of each year's regression.

TABLE 14.1
Estimated Toll Distribution Parameter α

Year	α	Year	α
1960	0.882	1988	0.408
1972	0.694	1989	0.371
1984	0.356	1990	0.342
1985	0.387	1991	0.338
1986	0.382	1992	0.338
1987	0.387	1993	0.370

For each year, the regression coefficient had a t-statistics in excess of 5 and an $R^2 \geq 0.99$. A time trend was fit to these coefficients, yielding

$$\hat{\alpha}_t = 1.897 - 0.172 \cdot t, t = 60, 72, \ldots, 93 \qquad (9)$$

with both coefficients having a t-statistic in excess of 11, and an adjusted and $F = 140.81$. It should be noted that although the linear fit is quite good, it appears that the downward trend seemed to mitigate during the period of the 1980s and early 1990s. This suggests that the principal changes in toll distribution occurred prior to this period.

Fig. 14.3 shows the cumulative distribution functions, both estimated and empirical, for toll usage in 1960 and in 1993, showing the trend toward mass service. These results are used to provide estimates of the preferences of the median voter, for 1960–1993:

$$\hat{\lambda}_t = (\hat{\alpha}_t + 1) \cdot (0.5)^{\hat{\alpha}_t}$$

---

10. The use of income as the basis of the underlying distribution of usage is an approximation forced on us by what data are available. Although there may be reason to suspect that toll usage and income may be weakly correlated, it is highly unlikely to be a perfect correlation, nor is it likely that the correlation, if it exists, is stationary.

Figure 14.3
Cumulative Distributions (Fitted)

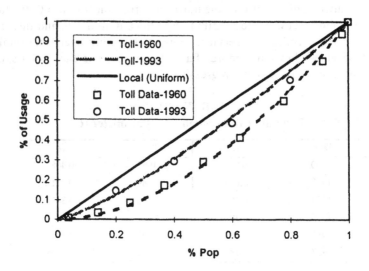

Quantities and Elasticities

The quantities of toll and local (number of calls) are taken from *Statistics of the Common Carriers* (Federal Communication Commission; 1960–1994). Toll and local elasticities are assumed constant over this period, following Taylor (1980, 1994). We assume $\varepsilon_L = -0.2$, $\varepsilon_T = -1.0$ based on these estimates.

Prices and Unit Costs

The consumer price index for telephone consists of an overall price index, available from 1960 to 1993, and separate indices for local and toll. Unfortunately, the separate indices are available only from 1978 to 1993. In order to estimate the separate indices from 1960 to 1977, it was noted that the relationship between the separate indices and the overall index was very stable from 1978 to 1983 (the period just prior to the AT&T divestiture); therefore, these relationships were estimated using a linear model (*t*-statistics over 80, $R^2 \geq 0.93$ for both local and toll) over this time period, and then backcast for the period 1960–1977 based on the overall telephone price index. The price series used, then, consists of actuals from 1978–1993 and backcast estimates from 1960–1977.

Unit cost data in the telecommunications industry are virtually nonexistent. It was therefore necessary to derive cost numbers based on two assumptions: (a) during the period prior to the AT&T divestiture, toll prices were estimated[11] to be about two to three times the marginal cost of toll; (b) assuming zero profits, the quantity-weighted unit costs must equal the overall price index. These assumptions imply relationships between the overall price index and the unit costs for local and toll, which we use to construct a unit cost series for the time period. The results of the price and unit cost derivations are shown in Fig. 14.4:

Figure 14.4
Constructed Price and Cost Indices

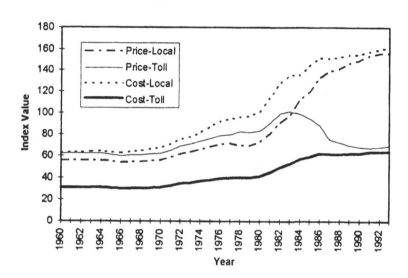

The price index series from 1978 onward (actual data) shows the extraordinary change in pricing policy that followed the AT&T divestiture, with toll prices declining and local prices rising. If the unit cost numbers are credible, this chart suggests that toll and local prices in the 1990s are rather close to their unit costs.

Predicted Margins and Estimation

Equations 8 can be rearranged to yield predicted toll and local margins for each year, based on the data just given:

$$\tilde{m}_L' = -\frac{1-\hat{\lambda}_t}{0.2 \cdot \hat{\lambda}_t + \frac{R_L}{R_T}} \qquad \tilde{m}_T = -\tilde{m}_L \cdot \frac{R_L}{R_T} \tag{10}$$

11. See Rohlfs (1979) for one of the few extant careful price and unit cost analyses.

The predicted margins are based on both the market revenues and the preferences of the median voter. Recognizing that political influence on pricing may not be instantaneous, we estimate a lagged model of actual margins verses predicted margins:

$$m_L = \beta_L + \sum_{t=0}^{10} \beta_t \tilde{m}_L(-t) + e_L \qquad m_T = \beta_T + \sum_{t=0}^{10} \beta_t \tilde{m}_T(-t) + e_T \qquad (10)$$

Because whatever causes political lags in setting local prices should also cause lags in setting toll prices, the coefficients of the lag terms are constrained to be equal. Further, we would also expect that the error terms of the two equations would be correlated, because the regulatory and market decisions in the two markets are linked. Therefore, system (10) was estimated using seemingly unrelated regression.

Conducting the unit root test on each of the two series m_L and m_T suggests that neither series is stationary. Taking first differences improved the situation, but nonstationarity cannot be rejected. However, taking second differences does yield series for which nonstationarity can be strongly rejected. Consequently, seemingly unrelated regressions were run on the levels, the first differences, and the second differences, applied on both sides of system (10). The results of all three regressions are given below in Table 14.2.

TABLE 14.2
Lagged Regression Results—Model versus Actual

	Level Coefficient	Std Deviation	First Difference Coefficient	Std Deviation	Second Difference Coefficient	Std Deviation
Toll constant	−0.41	(0.18)	0.23	(0.10)	0.00	(0.01)
Local constant	0.68	(0.09)	−0.02	(0.07)	0.00	22
Lag = 0	−184.96*	(19.50)*	−177.47*	(30.93)*	−177.01*	(28.39)*
Lag = 4	125.89	(32.51)	−25.27	(93.48)	−272.18	(84.12)
Lag = 6	316.16*	(36.83)*	106.18*	(33.72)*	99.67*	(33.06)*
Lag = 7	−308.40	(35.63)	82.67	(118.82)	201.46	(96.92)
Lag = 8	109.74	(40.96)	−65.94	(99.88)	−9.43	(78.48)
Lag = 10	−72.61*	(33.68)*	109.55*	(46.70)*	169.94*	(82.94)*
Toll adj R^2	0.86		0.95697		0.97240	
Local adj R^2	0.88		0.94884		0.94126	

* Lag coefficients that are significant in each of the regressions.

Each method of estimation leads to a very high R^2 and adjusted R^2, suggesting that the political model explains almost all the observed variation in toll and local price/cost margins. Most surprising is that differencing the estimation equation leads to a higher R^2, and differencing it twice leads to even higher R^2, which strongly suggests the stability of the relationship.

Closer examination reveals that the model does not have the strong predictive power that the t-statistics and R^2 suggests, for several reasons.

1. The sum of the lag coefficients for the second difference estimation is 12.453, and over 20 if the insignificant coefficient is dropped. On net, the magnitude of the actual margins (m) is over 12 times the magnitude of the predicted margins ($\tilde{m}$), although the theory suggests they should be equal.

2. There are very significant swings in the lag coefficients, from a positive 200 to a negative 272. One might expect a smooth decay process, yet there is no obvious trend in the coefficient estimates. These swings are not explained by the theory, and no obvious explanation for them comes to mind.

Thus, the regression results, which imply an impressive predictive power for the model, should be tempered by the unexplained magnitude and variation among the coefficients. It must also be recalled that many of the numbers used to derive the predicted margins are actually constructs rather than actual numbers, and these constructs are necessarily built upon somewhat patchy data, suggesting further caution in interpreting the strong regression results.

Nevertheless, it appears clear that the hypothesis that the political model explains the substantial variations in toll and local prices (and therefore price/cost margins) that have occurred in telecommunications is confirmed by the empirical analysis.

CONCLUSIONS

In this chapter, a political economy model of regulation has been developed, based on a median voter model of pricing two services of a regulated monopolist. This model is in sharp contrast to the dominant agency paradigm of regulatory economics, in that a political process takes center stage in determining economic outcomes, rather than asymmetric information (or natural monopoly).

We assume an idealized model of regulation, in which there are no information asymmetries, no bribery, no inefficiencies, and regulators are completely responsive to the wishes of the voters. The model may be viewed as an extreme case of "civics class" politics. In this model, specialized services, consumed by only some of the voting population, subsidize mass services, reflecting the preferences of the median voter. We compare this idealized regulation to unregulated monopoly, and find that for a broad range of parameter values, median voter regulation is welfare-inferior to monopoly, and may lead to negative surplus for the mass service. As the specialized service becomes more widely consumed, the subsidy from specialized to mass decreases: Specialized service price decreases and mass service price increases.

The very substantial price changes that occurred in telecommunications attendant to the AT&T divestiture were seen by many as wiping out the subsidy structure that had characterized the industry since before World War II. If this change was a response to changing preferences of the median voter, then this event constitutes a natural experiment against which to test the model. Using the rather limited data available, the hypothesis that the model explained the price changes as a result of changes in the distribution of toll demand was tested; the hypothesis was strongly confirmed.

We make no claim for the realism of either the economics or the politics of the model. Surely the median voter model is quite primitive, ignoring institutions, multi-issue politics, and agenda -setting. Further, the economic model of constant marginal costs and no fixed costs ignores important determinants of telecommunications technology. Nevertheless, bringing the two ideas together into a political economy model appears to have substantial explanatory power of an empirical phenomenon in a major U.S. industry that hitherto has been left unexplained by economists.

APPENDIX

Proof of Proposition 4:

The first-order condition for a maximum is

$$W_M' + \lambda^V \cdot W_S' p_S' = 0, \tag{11}$$

and the second-order condition is

$$W_M'' + \lambda^V \cdot [W_S'' \cdot (p_S')^2 + W_S' p_S''] < 0. \tag{12}$$

Total differentiation of Equation 11 yields

$$\frac{dp_M^V}{d\lambda^V} = -\frac{W_S' p_S'}{W_M'' + \lambda^V [W_S''(p_S')^2 + W_S' p_S'']} > 0 \tag{13}$$

Each term in the numerator is negative, and the denominator is negative, from Equation 12, so the expression can be signed, so that the price for service M is monotonic increasing in λ^V.

The derivative of total surplus is

$$T' = (p_M - c_M)Q'_M + (p_S - c_S)Q'_S p'_S \gtreqless 0$$

$$\text{as} \tag{14}$$

$$p_M \gtreqless c_M$$

Provided $\lambda^V < 1$, $p_M < c_M$, and surplus is increasing in p_M; therefore, combining Equations 13 and 14, we have that welfare is increasing in λ^V.■

ACKNOWLEDGEMENTS

The author would like to thank Profs. Franklin Allen and Robert Inman for their comments, as well as workshop participants at the Universidad Carlos III, Madrid, for theirs. J. Lande and others at the FCC provided crucial data, as did P. Srinigesh at Bellcore. Sue Kim provided excellent research assistance for the project.

REFERENCES

Baumol, W. & Bradford D. (1970). Optimal departures from marginal cost pricing, *American Economic Review, 60*, 265–83.

Baumol, W., Panzar, J., & Willig, R. (1982).*Contestable markets and the theory of industry structure* San Diego: Harcourt Brace Jovanovich..

Beard, T., & Thompson, H. (1996). Efficient verses "popular" tariffs for regulated monopolies. *J. Business, 69*, 75–87.

Becker, G. S. (1983). A theory of competition among pressure groups for political influence. *Quarterly Journal of Economics, XCVIII*, 371–400.

Bureau of Labor Statistics. (1963). *Consumer expenditure survey 1960–61*, Table 29A. Washington, DC: U.S. Department of Labor.

Bureau of Labor Statistics. (1974). *Consumer expenditure survey, 1972–73*, Table 1. Washington, DC: U.S. Department of Labor.

Creedy, J. & Francois, P. (1993). Voting over income tax progression in a two-period model. *Journal of Public Economics 50*, 291–298

Downs, A. (1957). *Economic theory of democracy.* New York: HarperCollins.

Federal Communications Commission. *Statistics of the common carriers*, 1960–1994. Washington, DC: U.S. Federal Communications Commission.

Grossman, G. & Helpman, E. (1995). Trade wars and trade talks. *Journal of Political Economy 104*, 675–708.

Laffont, J.-J. (1994). The new economics of regulation ten years after. *Econometrica 62*, 507–537.

Laffont, J.-J & Tirole, J. (1993). *A theory of incentives in procurement and regulation.* Cambridge: MIT Press: MA.

Lande, J. (1994). *The reference book: Rates, price indexes, and household expen-*

ditures for telephone service. Washington, DC: Industry Analysis Division, Common Carrier Bureau, Federal Communications Commission.

Peltzman, S. (1980). The growth of government. *Journal of Law & Economics, 23,* 209–87.

Roberts, K.W.S. (1977). Voting over income tax schedules. *Journal of Public Economics, 34,* 329–340.

Rohlfs, J. (1979). *Economically efficient bell-system pricing.* Bell Laboratories Economic Discussion Paper No. 138.

Spann, R. (1974). Collective consumption of private goods. *Public Choice, 20* (winter), 63–81.

Statistics of the common carriers, 1960–1994. Federal Communications Commission. Washington, DC: Author.

Stigler, G. (Ed.). (1988). *Chicago studies in political economy.* Chicago: University of Chicago Press.

Taylor, L. (1980). *Telecommunications demand.* Cambridge, MA: Ballinger Publishing Co.

Taylor, L. (1994). *Telecommunications demand in theory and practice.* Norwell, MA: Kluwer Academic Publishers.

Taylor, W.. & Taylor, L. (1993) Postdivestiture long distance competition in the United States. *American Economic Review 83,* 185–190.

Vogelsang, I. & Finsinger, J. (1979). A regulatory adjustment process for optimal pricing of multiproduct monopolies. *Bell Journal of Economics, 10,* 157–171.

Ye, M. & Yezer, A., (1992). Voting, spatial monopoly, and spatial price regulation. *Economic Inquiry, 30,* 29–39.

Restructuring the Telecommunications Sector in Developing Regions: Lessons from Southeast Asia

Heather E. Hudson
University of San Francisco

The importance of telecommunications to economic development is becoming widely recognized throughout the developing world. Less developed countries are striving to provide access to basic communications to their citizens, and industrializing economies face demands for new services and globally competitive prices. To encourage investment in technology and a market-driven approach to providing services, developing countries are adopting a variety of innovative policies and strategies. Although there are many paths to achieving universally accessible information infrastructure, there are also numerous potential pitfalls.

This chapter examines the telecommunications policies adopted by several Southeast Asian nations and the issues they are confronting as they open their doors to new technologies and services, and ultimately to new sources of information. It attempts to draw lessons from these experiences that may be relevant to other jurisdictions. Although these findings are likely to be most relevant to other developing countries, many are also relevant for industrialized countries that are grappling with strategies to encourage investment in telecommunications as a means to stimulate economic development.[1]

The restructuring process involves establishment of a regulatory authority in countries that typically have no tradition of independent regulation. Thus the new

---

1. In the United States, for example, many states have adopted incentive-based regulation to encourage upgrading of facilities in rural areas and extension of advanced services such as the Internet and videoconferencing to schools and health centers.

regulators often face conflicting mandates, as well as challenges in setting the rules for competition and tariff reform. Recognizing the importance of telecommunications for economic development, policymakers are looking for strategies to extend the public network, including designation of a carrier of last resort and incentives to encourage investment. Yet there may be inherent conflicts in information infrastructure policies because, despite encouraging investment, many governments seem wary of the consequences of increased access to information.

THE CONTEXT

Basic Indicators

The countries of Southeast Asia range from the economically dynamic city states of Singapore and Hong Kong[2] to the diversifying economies of major ASEAN members, to low income countries struggling to build their economies and infrastructure. In general, they inherited the PTT model of government monopoly, but most are moving toward privatization of the sector, with varying approaches to introducing competition. The countries highlighted in this analysis share some economic and demographic similarities, but differ in many respects, including the structure of their telecommunications sector (see Table 15.1).

TABLE 15.1
Basic Indicators of Selected Southeast Asian Countries

	World Bank Classification	Population 1993 (millions)	GDP/Capita 1993 (US$)	Teledensity 1994 (lines/100)
Singapore	High income	2.8	19,214	54.0
Hong Kong	High income	5.8	18,687	47.3
Malaysia	Upper middle	19.5	3,392	14.7
Philippines	Lower middle	66.2	817	1.7
Vietnam	Low income	72.9	181	0.6

Source: International Telecommunications Union (ITU, 1995).

Singapore and Hong Kong are the most developed, falling within the World Bank's high income classification, which also includes Japan and the other Organization for Economic Cooperation and Development (OECD) nations of North America, Europe, and Oceania. Singapore's trading economy is highly information intensive; its goal is to become an "intelligent island" that can serve as a regional commercial hub and lure international companies from Hong Kong.

2. Hong Kong was a British territory and became a Special Autonomous Region of China on July 1, 1997.

Official policy in Hong Kong is much more "laissez-faire," but it could already be considered an "intelligent island" based on its high utilization of telecommunications and information-intensive economy.

Malaysia is one of the fastest growing economies in the region, with much of its income derived from petroleum. It also has attracted high-technology industries that have set up assembly plants in the Penang free trade zone, and is intent on further diversifying its economy. The Philippines status as a low income country reflects the stagnation of the Marcos era, high population growth rates, and the continuing domination of the economy by a few powerful families, despite reforms by the Ramos government. Vietnam has become attractive to investors, apparently because of its economic potential and entrepreneurial spirit, but it remains among the world's poorest nations after decades of war, exacerbated by a centrally controlled economy and elderly Soviet-trained leadership.

RESTRUCTURING THE SECTOR: POLICIES AND PITFALLS

Approaches to Restructuring

Each of these countries has taken a different approach to organizing its telecommunications sector (see Table 15.2). Singapore, Malaysia, and Vietnam are evolving from a PTT structure carried over from their colonial administration, in which telecommunications was a government-operated monopoly, under the same jurisdiction as the postal service. Hong Kong's telecommunications were provided by Cable and Wireless, the company that provided telecommunications services for Britain's colonies. The Philippines inherited a primarily American model, with a large dominant private company (PLDT, the Philippines Long Distance Company) and some services in rural areas provided by national and municipal governments.

Hong Kong has taken the boldest steps to open telecommunications to competition. Regulation is the responsibility of the Office of the Telecommunications Authority (OFTA), which is generally modeled on the United Kingdom's OFTEL and Australia's AUSTEL. (In fact, the Director General was recruited from AUSTEL.) Following the Hong Kong government's policy of "positive nonintervention," OFTA has liberalized the sector into one of the most competitive in the world (Mueller, 1994). The extent of its future autonomy in telecommunications will depend on China's interpretation of the "one country, two systems" policy that will come into force with Hong Kong's reversion to China in July 1997.

The Cable and Wireless subsidiary Hong Kong Telecom (HKT) remains the sole provider of all international service, having been granted a 25-year monopoly in 1981 (before OFTA) that will not expire until the year 2006. HKT's local monopoly expired in 1995, after which OFTA introduced local competition. Three operators have been licensed to compete with HKT in the provision of fixed-network service. By linking with call-back operators, they have effectively introduced international competition, despite HKT's international monopoly.

TABLE 15.2
Structure of the Telecommunications Sector
in Selected Southeast Asian Countries

	Local	Trunk	International	Mobile	Value-added Services
Singapore	M-PP	—[a]	M-PP	C[b]	C
Hong Kong	C	—[a]	M-PP	C	C
Malaysia	M-PP	C	C	C	C
Philippines	M-P	C	C	C	C
Vietnam	M-G	M-G	M-G	M-G	M-G

Note: M, monopoly, C, competitive, G, government owned; P, private; PP, partially privatized.
[a] Domestic traffic is considered local within Singapore and Hong Kong.
[b] A second mobile carrier began operation in March 1997.

Hong Kong has the one of the highest concentrations of wireless communications in the world, with a cellular teledensity of 7.4 per 100 in 1994 (ITU, 1994); there is also almost one pager for every 4 inhabitants. Services are provided by four cellular operators on seven networks covering five analog and digital technologies. There are also 4 telepoint cordless telephone (CT2) licensees and 36 paging service operators (Asia-Pacific Telecommunity, 1995). Up to six operators are being licensed to provide personal communications services (PCS).

Singapore Telecom was corporatized in 1992, taking over the former commercial functions of the Telecommunications Authority of Singapore (TAS), which is now the regulator. Privatization began in 1993; Singapore Telecom is now the largest company listed on the Stock Exchange of Singapore. However, unlike Hong Kong, the government has been reluctant to allow competition. The company was given a 15-year monopoly by TAS in domestic and international telephone services and leased circuits. It was also given exclusive rights to provide cellular mobile service for 5 years, until 1997.

Among ASEAN nations, the Philippines is perhaps the most open market, with more than two dozen companies offering telecommunications services, but most are controlled by a small circle of ruling families. They form alliances with foreign companies for financial backing and technical expertise, but the alliances are volatile, with considerable turnover. The Philippine Long Distance Telephone Company (PLDT) is the dominant operator with 86% share of the market; the remaining 14% is the responsibility of some 50 other franchises. PLDT has been a fully private operator since 1928; its franchise extends until 2028.

Several private carriers have entered the market to provide local, long-distance, and value-added services (VAS). In 1995, seven record carriers provided domestic telex, facsimile, and leased-line services; four provided international services. There are also several private carriers for paging and data communications. Five companies have been granted provisional authority to establish national cellular networks.

In 1990, Malaysia partially privatized its former government operator, Telekom Malaysia, which was corporatized in 1984. Jabatan Telekom Malaysia, under the Ministry of Energy, Telecommunications, and Posts, was converted to a regulator. Telekom retains a local monopoly and has responsibility for rural services. However, competition has been introduced in cellular services, pay phones, long distance, and international services. The new entrants are all local companies, many with strong ties to Prime Minister Datuk Seri Mahathir Mohamad or other members of the ruling party UMNO (United Malays National Organization).

Vietnam has retained the PTT model, with telecommunications policy and operations still under government control. Telecommunications policies are set by the Directorate General of Posts and Telecommunications (DGPT), which was separated from the Ministry of Communications and Transport in 1992. A new entity, Vietnam National Posts and Telecommunications (VNPT), was established to operate the national network under the regulation of the DGPT. The VNPT established several subsidiaries including Vietnam Mobile Services (VMS), Vietnam Data Corporation (VDC), Vietnam Telecoms National (VTN), Vietnam Telecoms International (VTI), and the Vietnam Postal Service (VPS). VNPT is also responsible for transmission of radio and television.

Conflicting Mandates

Countries that are in transition from a PTT structure to a commercialized telecommunications sector often face difficulties in establishing an autonomous regulator. Typically, the regulatory staff is drawn from part of the operator that was responsible for tariffs, standards, and intergovernmental affairs. Thus, the regulators are really not at "arm's length" from the dominant carrier where they began their careers. Also, because the government generally remains a shareholder in the dominant carrier, it has a vested interest in ensuring that the carrier remains profitable. Thus, regulators have allegiances that conflict with their duties to serve the public interest and to set and enforce fair rules of competition.

These conflicts are evident in both Singapore and Malaysia. Neither the Telecommunications Authority of Singapore (TAS) nor Jabatan Telekom in Malaysia is truly independent of Singapore Telecom or Telekom Malaysia, where many of their former colleagues now work. TAS still interweaves regulation, operation, and policy making. Its function is not only to regulate but also to develop and promote the telecommunications industry in support of the National Information Technology 2000 (IT 2000) Plan to make Singapore an "intelligent island." TAS also receives about 60% of Singapore Telecom's surplus, and may receive revenue from licenses and administrative fees and may raise capital through stocks and bonds (Hukill & Jussawalla, 1991).

It would appear that TAS faces numerous potential conflicts of interest. However, Hukill (1993) concluded: "The close link between government, regulator and operator in Singapore must be seen not in terms of conflict of interest as might be the case in other countries, but as a tripartite strategy for development opportunity." As long as what is good for ST is good for the country, this may be true. But as more competition is introduced, and users develop a stronger voice, TAS may find it difficult to play several policy roles.

Although Malaysia has separated its regulatory agency, JTM, from its now partially privatized operator, JTM is really not independent of either the government or STM. As in Singapore, its employees were formerly employees of the government-run operator. JTM only recommends decisions; it can be overruled by the Minister or by the Cabinet. Also, the government still sees Telekom as an important element of its industrial policy and, with its majority share, wants Telekom to be commercially successful.

The licensing process is opaque, with no obvious criteria and evidence of political cronyism in selection of successful licensees. The terms may also not be fixed; in 1996, the government decided that too many licenses had been awarded during the previous administration and indicated that it wanted some licensees to buy out others, despite the fact that all had made investment plans and were building networks based on their license terms.

Setting the Rules for Competition

As competition is introduced, regulators must resolve issues such as network quality, network standards and compatibility, revenue sharing, and interconnection agreements. In the rush to open the sector, licenses may be issued before the "rules of the game" are established. The result may be bottlenecks if new carriers face barriers and delays in interconnecting with the dominant carrier. Also, policies must anticipate growth in demand. For example, dominant carriers must be required to supply additional trunks to carry traffic from new carriers, and numbering plans must allow for subscriber growth for new carriers and new services, particularly wireless networks.

In Malaysia, Jabatan Telekom faces these issues as it authorizes more competition. The new carriers cite problems with interconnection, numbering plans, procedures for obtaining construction permits and rights of way, and a prohibition on resale. Without the option of obtaining capacity from Telekom Malaysia at wholesale prices, new carriers are choosing to build their own backbone networks, which are highly capital intensive. However, the government is urging Malaysian industry to reduce imports in order to cut its foreign exchange deficits. Authorization of resale could help to reduce this deficit by creating an incentive to use sur-plus fiber and satellite capacity.

In the Philippines, Executive Order 59 issued by President Ramos in 1993 re-quired compulsory interconnection of authorized public telecommunications car-riers to create a nationalized integrated network and encourage greater private sector investment. This decision paved the way for other authorized carriers in-cluding small "mom-and-pop" operators to interconnect with the national back-bone of large carriers. The Philippines was also divided into 10 local exchange carrier (LEC) service areas.

The NTC has left PLDT and the local exchange carriers to negotiate among themselves, intervening only if invited or if the parties cannot agree within 90 days. Because PLDT's network includes 90% of all installed lines, the other car-riers have felt pressure to agree to PLDT's terms. PLDT had not designed its net-work for interconnection with numerous other carriers and had not budgeted for the transition. Also, PLDT acts as a bottleneck in high-demand areas where the new carriers depend on the PLDT network. In mid-1996, the new carriers cited back orders for hundreds of trunks in Manila from PLDT (personal interviews, Manila, June 1996). Other issues to be resolved included billing protocols and cost allocations among the carriers.

Tariff Reform

Tariff reform in many developing countries is a particularly sensitive issue. Poli-cymakers have traditionally viewed profits, particularly from international traffic, as a source of funds to extend and upgrade their infrastructure. However, compe-tition typically leads to lower rates, although it may create incentives for invest-ment by new operators. Where governments have not yet authorized competition, international call-back services have effectively created lower priced alternatives to the monopoly carrier. Call-back is thus viewed with great consternation by ad-ministrations that have authorized high international rates to generate revenue to extend and upgrade their domestic networks, or to keep local rates low.

Vietnam has one of the world's highest accounting rates, and perceives call-back as illegal and apparently a significant threat to VNPT's revenues (inter-view with DGPT, Hanoi, April 1996). In contrast, OFTA in Hong Kong has ef-fectively introduced international competition, in spite of HKT's monopoly until 2006, by licensing competitive local companies that are offering call-back access. In fact, the Hong Kong government encourages its departments to use call-back to save money.

The Philippines and Malaysia have authorized competitive international gateways. Like Vietnam, Singapore has retained its monopoly, but ST's tariffs are set to be competitive with a basket of tariffs from other countries. The strategy is apparently designed to protect ST's monopoly while responding to demands of its information-intensive economy for competitive international pricing. However, Singaporeans are also increasingly resorting to call-back to save on international calls.

Operators that retain a local monopoly may turn to their captive customers to make up for revenue lost through competition in domestic long-distance (trunk) and international services. For example, in March 1996, Telekom Malaysia received approval from Jabatan Telekom to increase local rates significantly. There was no previous notification that such a change was under consideration, nor any mechanism for users to express their views (interview with Consumers Association of Penang, March 1996). One group likely to be significantly affected was Internet users. In fact, the government-owned Internet access provider protested the increase, and a compromise was reached to introduce a new pricing structure for Internet access.

The Malaysian case illustrates the many conflicting goals of tariff policies. On the one hand, Telekom Malaysia, which is still partly government owned, was apparently losing revenues to call-back, and saw raising its local monopoly rates as an attractive solution. However, Malaysia also wants to be a leader in introducing new information services, including the Internet. In addition, the government owns the largest Internet provider! In addition, as there is no notification or hearings process, rates could simply be changed without any input from users or service providers.

In the long term, competition is likely to force rate reductions for both domestic and international services. Other sources of funding for investment will likely come from growth in volume of use rather than excessively high usage charges, and from policies designed to encourage investment (discussed later). Yet the lack of rules for fair competition and for participation in the policy process may penalize users in many countries for years to come.

EXTENDING THE PUBLIC NETWORK

The Carrier of Last Resort

When competition is introduced, regulators face the problem of how to ensure that services are provided to less profitable regions. In developing countries, these are typically rural areas. One model adopted by major industrialized countries, including the United Kingdom, Canada, and Australia, is the carrier of last resort. The dominant carrier is required to provide service to these areas, and may receive subsidies from the government or from a fund to which all carriers have contributed.

Malaysia has adopted the carrier of last resort approach; Telekom Malaysia (STM) is required to provide service in less profitable rural areas, while competing with carriers that apparently can cream skim the most lucrative business and urban customers. Yet reserving a rural monopoly for Telekom Malaysia provides no incentive for the company to reduce costs for rural services.

A test may come with the availability of the Measat satellite, which would be very suitable for providing services in isolated areas of Sarawak and Sabah. It is unclear whether Binariang, which owns Measat, can obtain authorization to serve these communities directly, or whether STM would be required to provide the service. STM could lease capacity from Measat, but appears to favor building its own more expensive terrestrial networks. If STM does build its own more expensive terrestrial network, it will be following in the path of Telecom in Australia (now Telstra), which rejected the opportunity to use Aussat in the 1980s to serve remote settlements in the Outback, instead building its own microwave network, a more costly and less flexible system.

Incentives for Investment

Another approach is to use incentives to encourage investment in rural areas. Many U.S. states have adopted policies that offer incentives to carriers to accelerate upgrading of their rural networks (see, e.g., Parker & Hudson, 1995). The Philippines has adopted an innovative strategy to create incentives to install telecommunications networks in unserved areas. Licenses for international gateways and domestic services now require that operators also undertake to install several hundred thousand lines in an unserved region. Executive Order 109 issued by President Ramos in 1993 requires a total of 5 million landlines from gateway and cellular telephone operators. Each cellular mobile telephone service (CMTS) operator is required to install a minimum of 400,000 local exchange lines. Similarly, each international gateway facility (IGF) operator is required to install a minimum of 300 local exchange lines per international switch termination and a minimum of 300,000 local exchange lines, within 3 years from the date of authority to operate and maintain local exchange carrier service (Asia-Pacific Telecommunity, 1995).

The Philippines policy of requiring licensees of international gateways and cellular systems to build networks in unserved parts of the country is a very innovative strategy to create incentives to invest in unserved areas. Although some observers have questioned the wisdom of this approach in that it may perpetuate internal cross subsidies, it shows promise as a model that may be emulated in other countries with large unserved territories. Some operators apparently see the new policy as a burden, whereas others welcome the opportunity, and anticipate profitable operations in their new franchise areas. Perhaps in a few years it will be possible to make a market in these rural franchise areas, so that rights to build and operate may be traded, with operators interested in expanding their franchise areas buying the obligations from those who find them a burden.

Vietnam is also encouraging foreign investment to build out its networks. Its new Foreign Investment Law, part of the constitutional reform of 1992, which adopted the guiding principle of "doi moi" or economic innovation, permits joint ventures and 100% ownership of assets. However, telecommunications is still regarded as important to national security; to retain government control, the DGPT has authorized only one form of foreign participation, the business cooperation contract (BCC), which is an agreement between a foreign and Vietnamese partner for "the mutual allocation of responsibilities and sharing of product, production or losses, without creating a joint venture enterprise or any other legal entity" (Ure, 1995, p. 71).

Vietnam's first BCC in telecommunications was with Telstra of Australia (formerly OTC) in 1988 to install Intelsat earth stations in Hanoi, Ho Chi Minh City, and Danang for international communications. International revenues are shared between Telstra and the DGPT, which uses part of these revenues to expand the domestic network (Ure, 1995, pp. 71-72). Operating companies from Singapore, Malaysia, Thailand, and Hong Kong are also participating in ventures in services such as cellular, paging, and payphones. As noted earlier, VNPT's international rates are among the highest in the world. It will likely be forced both to reduce these rates and to create additional incentives for investment to increase its teledensity significantly from its current level of less than one line per 100 inhabitants.

MISPLACED PRIORITIES? ACCESS TO TELEVISION VERSUS TELEPHONES

In developing countries, television is often more accessible than telephone service. In some countries, such as Vietnam and the Philippines in the Marcos era, this discrepancy reflects a deliberate policy to extend mass media for national unity or political cohesion. These discrepancies also indicate the comparative difficulty of extending access to interactive communications, which requires links to each customer and switching facilities, compared to broadcasting, which simply requires relaying and retransmitting the signal. (However, distribution of television is also difficult in isolated areas such as the interior of Vietnam and the Philippine archipelago, where satellite transmission may be the least cost solution.)

There are striking differences in access to television vs. telephones in the region (see Table 15.3). In both Hong Kong and Singapore, there are apparently more telephone lines than TV sets. A probable explanation is that the economies of both of these city states are highly information intensive, so that there is a higher proportion of business lines than would be found in other developed economies with greater land area. Also, the data on TV sets may underrepresent the actual number of sets per household. We can conclude that access to television and to telephone service is close to universal in both Singapore and Hong Kong.

However, television is more accessible than telephone service in the other countries. Malaysia has 1.6 times as many television sets as telephone lines; it is likely that the disparity is more significant in rural areas. The Philippines has more than 7 times as many television sets as telephone lines, whereas Vietnam has more than 18 times as many TV sets as telephone lines. There may be a partial historical explanation in the Philippines and Vietnam. The Marcos regime heavily controlled the privately owned mass media to portray a favorable image of the country and its government, and siphoned off PLDT funds that could have been invested in extending telephone networks. The Vietnamese government operated both broadcasting and telephone service, giving priority to the mass media as a means of disseminating government-approved information. The relative inaccessibility of telephone service may also reflect a contemporary view that broadcasting remains more important.

TABLE 15.3
Access to Television versus Telephone Service

	Teledensity	TV Density	Ration TV Sets/ Telephone Lines
Singapore	47.3	38.0	0.8
Hong Kong	54.0	35.9	0.7
Malaysia	14.7	23.1	1.6
Philippines	1.7	12.1	7.1
Vietnam	0.6	11.0	18.3

Derived from ITU, 1995.

In addition to reflecting policies that have ignored interactive telecommunications in the past, these data also indicate that a significant percentage of households in the Philippines and Vietnam has sufficient disposable income to purchase a television set, despite the very low average annual incomes. Of course, some people may place a higher priority on television as an investment for the family, as has been documented for some low-income U.S. households. However, it appears safe to assume that households with television sets have members who would use a telephone if it were available, for example, in a phone booth or at a kiosk, and that a significant percentage could afford to become individual subscribers.

INFORMATION TECHNOLOGIES: ACCESS OR CONTROL?

Conflicting Information Infrastructure Policies

Although committed to economic reform, Deng Xiaoping voiced his ambivalence about opening China's doors to the world: "When the door opens, some flies are bound to come in" (Schwankert, 1995, p. 112). Chinese government attempts to control access include banning satellite antennas, blocking access to Internet sites, and impeding access to the Internet itself and to other means of electronic communication. Other Asian countries appear to share these concerns that information from outside will contaminate their country, even as they encourage investment in information infrastructure.

In 1993, more than 3,000 international companies hubbed their telecommunications through Singapore. Singapore has also attracted several regional satellite uplink operators including HBO, ESPN, and ABN (Asian Business News), among others. Yet Singaporeans are not allowed to install satellite terminals to receive television from AsiaSat, Malaysia's new Measat, or other satellites with footprints covering Singapore, even though it has attracted uplink operators that transmit many of these channels from Singapore. Singapore is also anxious to promote Internet access as part of its IT 2000 strategy, but wants to know who is using the Internet and what information is being accessed. Customers must provide their identification number or passport number to get an Internet account. The government has recently decided to apply the standards of broadcasting content to the Internet, holding Internet access providers accountable for information on their networks, and blocking access to Internet sites deemed unsuitable for Singaporeans. Such policies would seem paradoxical for a country that has staked its economic future on becoming an information-based economy.

Malaysia has also identified telecommunications as central to its goal to becoming a developed country by the year 2020. A fiber backbone now runs the length of peninsular Malaysia along its motorway; Malaysia now has its own domestic satellite system, and an "intelligent town" is being constructed on the outskirts of Kuala Lumpur. Yet the Malaysian government remains concerned about content that its citizens may find on the Internet and other satellite systems. Satellite antennas have been officially banned, although the government has turned a blind eye in rural areas and in east Malaysia (Sarawak and Sabah), where thousands of mesh antennas are pointed at Indonesia's Palapa satellite. Now that Measat has been launched, Malaysia's solution is to limit sales to 60-cm antennas, with the assumption that the equipment will only be able to pick up signals from the high powered Measat system.

Such policies would not be surprising, perhaps, among countries with less developed economies or less entrepreneurial citizens. Of course, Singapore and Malaysia have found that it is virtually impossible to keep out information, whether by banning certain magazines and newspapers or by banning access to Internet sites and satellite channels. Yet their commitment to a policy of control seems directly at odds with industrial policies designed to upgrade their telecommunications infrastructure and attract more high-tech and information-based industries.

New Technologies: Limiting Access?

As cable entrepreneurs are well aware, the advantage of cable is that it allows the owner to control access to the television channels, which can be received only on payment of a fee. Of course, cable distribution can also be monitored, so that the government will know what content is being transmitted. Thus cable TV (whether delivered via coaxial cable or optical fiber) may be considered a much safer way to satisfy demand for entertainment than uncontrollable access to satellite channels. Singapore and China are spurring investment in cable systems as a means of satisfying demand for more channels stimulated by satellite television, but monitoring content. ST is also planning to invest in "fiber to the flat," so that it can deliver high-bandwidth Internet access and video on demand. Such a network has the advantage of supporting the "intelligent island" policy and at the same time offering the capability to monitor content and control access.

Satellite television is widely available throughout the region, most notably STAR TV on AsiaSat, but also other national and regional satellites such as Indonesia's Palapa, Thailand's Thaicom, Malaysia's new Measat, and PanAmSat. Countries larger than Singapore are already unable to enforce bans on satellite antennas; controlling access will be more difficult as signals on new high-powered satellites can be received with easy-to-hide wok-sized antennas.

An alternative solution is to precensor programs so that only those deemed suitable are delivered by satellite. Eager to reach Asian markets, satellite programmers are willing to comply. Ironically, video compression technology, which has been introduced for U.S. Direct Broadcast Satellites (DBS) services to deliver more channel choices than are typically available on cable, now provides a cost-effective means to limit choice by distributing precensored programming. Networks with regional uplinks in Singapore, for example, can digitize programming, add subtitles or soundtracks in different languages, edit out material deemed offensive by various national governments, compress the channels, and transmit several versions of the same program on a single transponder (Hudson, 1997). The decoders sold in each country will be programmed to allow reception of only those channels approved by its government.

Of course, many Westerners are also searching for technological means to counter the proliferation of content they consider offensive such as violence, pornography, ethnic or racial slurs, or instructions on making bombs. For example, the "v chip" mandated by the 1996 Telecommunications Act will enable parents to block access to offensive TV programs or channels. New Internet software contains "network nannies" that can be set to block access to Internet sites. Yet, unlike the policies adopted in several Asian countries, these approaches put the power of the technology in the hands of the individual, rather than the state.

PROMISES AND PITFALLS

The Southeast Asian countries reviewed in this chapter have introduced policies to increase access to information and telecommunications services throughout their countries and to strengthen their own telecommunications and information technology sectors. Their experiences to date point out both the promise of increasing access to telecommunications and the pitfalls of policies that may turn out to be inequitable to new competitors or to users, or may actually serve to limit access to information.

The lessons from these varying approaches to telecommunications policy reform may be relevant both to other developing countries and to many industrialized economies. Separating policy and regulation from operation must mean more than changing names and separating offices. Introducing competition requires setting and enforcing the rules of the game in interconnection, settlements, and standards. Extending service to rural areas will increasingly require incentives rather than subsidies.

Perhaps most significant is the need to acknowledge that the inevitable result of investing in information infrastructure is to increase access to information. Telecommunications planners and policymakers in both developing and industrialized countries must recognize that the sharing and utilization of information, and not the mere extension of networks, should be the ultimate purpose of telecommunications policy reform.

REFERENCES

Asia-Pacific Telecommunity. (1995). *APT yearbook.* Bangkok: Author.

Hudson, H. E. (1997). *Global connections: International Infrastructure and Policy.* New York: Van Nostrand Reinhold.

Hukill, M. A. (1993, June). *The Privatisation of Singapore Telecom: Planning, Policy and Regulatory Issues.* Paper presented at the 1993 Pan-Asian Summit, Hong Kong.

Hukill, M. A. & Jussawalla, M. (1991). *Trends in Policies for Telecommunications Infrastructure Development and Investment in the ASEAN Countries.* Honolulu: Institute for Communication and Culture, East-West Center.

International Telecommunication Union. (1994). *World Telecommunication Development Report.* Geneva: Author.

International Telecommunication Union. (1995). *World Telecommunication Development Report.* Geneva: Author.

Mueller, M. (1994). One Country, Two Systems: What Will 1997 Mean in Telecommunications? *Telecommunications Policy, 18*(3), 243-253.

Parker, E. B. & Hudson, H. E. (1995). *Electronic Byways: State Policies for Rural Development Through Telecommunications* (2nd ed.). Washington, DC: Aspen Institute.

Schwankert, S. (1995, November). Dragons at the Gates. *Internet World*, pp. 109–112.

Ure, J. (Ed.). (1995). *Telecommunications in Asia: Policy, Planning and Development.* Hong Kong: Hong Kong University Press.

Author Index

Numbers in parentheses indicate the footnote number; italic numbers indicate the page where the complete reference is given.

A

ACLU v. Reno, 239(54, 55, 56), *239*
Aguilar, R., 240(58), *240*
Allard, T., 146, *151*
Altschull, J. H., 263(100), *263*
American Civil Liberties Union v.
    Reno, 225(1), *225*, 227(1),
    239(54, 55, 56)
American Library Association, Inc., et
    al. v. United States, et al., 208,
    222
American Trucking Association v.
    United States, 261(90), *291*
Ameritech Corp. v. United States,
    246(3), *246*, 253(43)
Anderson, R. H., 79, *93*, 240(60), *240*
Andrews, W., 237(50), *237*
Anick, D., 57, *74*
Anspacher, J., 16(5), 24(26), *32*,
    34(4), *34*
Armstrong, M., 24(26), *32*
Asia-Pacific Telecommunity, 304,
    309, *314*
Aspden, P., 81, *94*
Atkinson, A. B., 27(35), *32*
ATM Forum Technical Committee,
    100, *111*
Auerbach, A., 27(35, 36), *32*

Australia New Zealand Banking
    Group Ltd., 150, *151*
Australian Bureau of Statistics,
    135(1), 141, *151*
Australian Payments System Council,
    139, 140, *151*

B

Bank for International Settlements,
    140, *151*
Barrett, A. C., 246(7), *246*, 253(34),
    253, 266(129), *266*, 271(152)
Barron, J. A., 213, *222*, 261(94), *261*,
    264(104), 268(138), *268*,
    269(144), 273(155)
Batteau, A., 80, *93*
Baumol, W. J., 45(17), *45*, 261(92,
    96), *261*, 276, *299*
Bayes, J. R., 246(6), *246*
Beall, R., 205, *222*
Beard, T., 278, *299*
Becker, G. S., 277, 281(4), *299*
Becker, L. E., 205, *222*
Belinfante, A., 28(38), *32*
BellSouth Corp. v. United States,
    246(3), *246*, 254(42, 43)
Berlin, E., 229(17), *229*
Berman, J., 205, 211, *222*, 225(4), *225*

Subject Index

A

Academic competitiveness, Internet usage, 80, *see also* Competition

Access, *see also* Internet access providers
geographic rate averaging and cost recovery of incumbent local exchange carriers, 25
obstacle for getting started on Internet, 88, 92
providers of Internet telephony and FCC regulation, 190–191
real-time and needs/difficulties for mobile wireless carriers, 50

Accuracy, rating of Internet, 229–236

ACTA, *see* America's Carriers Telecommunications Association

Address
scheme and Internet telephony, 187
shared payment scheme for Internet access, 108

Adults, blocking software effects, 239–241, *see also* Blocking software

Advice, sources and Internet user survey, 86–87, 92

African-Americans, reasons for Internet usage, 84–85

Age
electronic money and online purchases, 145
Internet awareness, 90, 91

Agency theory, regulatory economics, 276, 277

All-or-nothing payment scheme, drawback and Internet pricing, 100

American Library Association, unrestricted access to Internet sites and blocking software, 240

America's Carriers Telecommunications Association (ACTA), issues of Internet telephony regulation, 183–186, 188, 194–195, 196, 203–204

Ancillary jurisdiction, FCC over Internet telephony, 195

Anticompetitive behavior, profit-seeking license holders, 51, *see also* Competition

Appeals Court, *see* Court system

Arbitration
development of local competition, 2–3
Telecommunications Act of 1996 issues, 17–18

Asians, reasons for Internet usage, 84–85

Asymmetric information, *see* Information

Atomic transactions, digital payment mechanisms in digital libraries, 125

AT&T
divestiture
empirical results of model of regulatory economics, 290–291, 294, 295

LEC, *see* Local exchange carriers
Licensing
  first amendment rights, 264, 271
  real-time access to spectrum, 49,
    51
  Southeast Asia, 306, 309
Limitations, implications of FCC ap-
    proach and local competition, 8
Listen before talk (LBT) approach,
    see also Etiquette; Spectrum
    sharing, unlicensed
  distributed control of real-time
    sharing, 53
  greedy behavior as causation of
    tragedy of the commons, 55, 56
Local access charges, high cost and In-
    ternet telephony, 198
Local area networks (LANs), spec-
    trum access, 49, 55
Local competition policy
  development, 1–7
  implications
  Appeals court decision, 10–12
  FCC approach, 7–10
Local exchange carriers (LEC)
  bargaining power effect on market
    entry, 29–30
  bargaining theory and
    Telecommunications Act of
    1996 issues, 17–18
  cost savings and competitive local
    exchange carriers, 38–42
  development of competition, 3–4
  implications
      ATT divestiture, 1
      FCC approach and local
        competition, 9
      opposition to FCC
        regulations, 5
  local interconnection, 42–45
  Southeast Asia, 307
Local exchange lines, cellular mobile
    telephone service in Southeast
    Asia, 309
Local interconnection, *see also* Inter-
    connection
  efficient competition, 37–42

pricing guidelines for local
    exchange carriers, 42–45
Local loops
  total element long-run
    incremental cost comparison
    with total service long-run
    incremental cost, 31
  provisions of
    Telecommunications Act of
    1996, 16
Log-infrequency of purchase model,
    postal delivery service,
    164–169, 174, 175, 176, 177
Log-linear models, Internet usage,
    89–91, 92
Long distance services, problems in
    Southeast Asia, 308
Long-run incremental costs (LRIC),
    local interconnection pricing,
    44–45
Loop plant, cost savings for local ex-
    change carriers versus market
    entry of competitors, 38–42
LRIC, *see* Long-run incremental costs
Lump-sum charges, bargaining theory
    and Telecommunications Act of
    1996 issues, 18

M

Majority-voting approach, regulatory
    economics, 277, 280, 281
Malaysia
  conflicting mandates in
    telecommunications, 305
  economic structure, 302, 303
  information technology
    infrastructure, 312
  rules of competition, 306
  tariff reform, 307
  telecommunications restructuring,
    305
  telephone vs. television access,
    311
Management/control, electronic
    money, 147–150
Mandates, conflicting in Southeast
    Asia, 305–306

Printed in the United States
by Baker & Taylor Publisher Services